AF556688

PLANT GENETIC ENGINEERING

Volume 2
Improvement of Food Crops

The Editors

Dr Pawan K. Jaiwal is a Reader in the Department of Biosciences, M. D. University, Rohtak, India. He is teaching Molecular Biology, rDNA technology and Plant Biotechnology courses to graduate and postgraduate students since 1986. He has throughout first division academic career with Ph.D. in Botany (Plant Physiology) from Kurukshetra University, Kurukshetra, India. He has more than 70 research publications including original research papers, reviews and book chapters in reputed International research journals and books. He has edited three books published by International publishers. Dr Jaiwal has been awarded DST Young Scientist Project, INSA Visiting Fellowship, New Delhi and DBT Overseas Associateship (Long Term) for advance research at laboratory of Prof Ingo Potrykus, ETH, Switzerland. Dr Jaiwal is a member of several academic bodies and on the Editorial Board of four International research journals. His has guided many students for Ph.D., M. Phil. and M. Sc. degrees. He has completed several research projects on Biotechnology for improvement of important grain legumes. His group has developed gene transfer protocols in *Vigna radiata* and *V. mungo* for the first time. His current research interests are metabolic engineering for resistance to abiotic and biotic stresses, improvement of nutrient use efficiency and nutritional quality of grain legumes.

Dr Rana P. Singh is a Reader in the Department of Biosciences, M. D. University, Rohtak, India. He is teaching courses in Plant Biology and Biotechnology to graduate and postgraduate students since 1986. He earned his first class M. Sc. in Botany from DDU Gorakhpur Univ., Gorakhapur and Ph.D. in Life Sciences (Plant Biochemistry) from Devi Ahilya University, Indore, India. He has more than 70 research publications including original research papers, reviews and book chapters in reputed International research journals and books. He has edited 5 books published by International publishers. He is fellow and member of various academic societies. Dr Singh has been awarded R. D. Asana Medal from Indian Soc. Plant Physiol., New Delhi and JEB Young Scientist award of The Academy of Environmental Biology, India for significant research contributions. He is recipient of INSA, New Delhi and UNESCO, Paris Biotech. Fellowships and visited twice at University of Guelph, Canada for advance research. He has visited Yunnan Agri. Univ., Kunming, China and hosted three Senegalese Scientists for research collaborations. He has guided many students for Ph.D., M. Phil. and M. Sc. degrees and completed many projects in areas of his interest. His group has contributed significant work in nitrogen relation to plants. Dr Singh is Editor-in-chief of an International research journal, Physiology and Molecular Biology of Plants and on editorial board of three other journals. His current research interests are metabolic engineering for nutrient use efficiency, stress tolerance and nutritional improvement in plants.

Other Volumes in this series:

Volume 1 **Applications and Limitations :** *Eds.* Rana P. Singh & Pawan K. Jaiwal

Volume 3 **Improvement of Commercial Plants - I :** *Eds.* Rana P. Singh & Pawan K. Jaiwal

Volume 4 **Improvement of Commercial Plants - II :** *Eds.* Pawan K. Jaiwal & Rana P. Singh

Volume 5 **Improvement of Vegetables :** *Eds.* Rana P. Singh & Pawan K. Jaiwal

Volume 6 **Improvement of Fruits :** *Eds.* Pawan K. Jaiwal & Rana P. Singh

PLANT GENETIC ENGINEERING

Volume 2

Improvement of Food Crops

Editors

Pawan K. Jaiwal **Rana P. Singh**

Department of Biosciences
M. D. University,
Rohtak-124001 India

2003

SCI TECH PUBLISHING LLC, U.S.A.

ISBN : 1-930813-03-1
SERIES ISBN : 1-930813-17-1

First Published 2003

Published by
SCI TECH PUBLISHING LLC.
9207, Country Creek Drive
Houston, Texas - 77036 (U.S.A.)
Tele. : 713-541-94-- Fax : 713-541-9401
E-mail : sci@hotmail.com

Printed by
RAINBOW PROCESSORS & PRINTERS
New Delhi-110059 India

PREFACE

The world population is predicted to rise from about 6.5 billion at present to 10 billion by 2050. Much of this rise will be in the developing countries. The burgeoning population has put a tougher challenge before man of food and nutrition security in wake of the degradation of environment and depletion of the natural resources. Scientific advances in crop improvement through the introduction of dwarf varieties and the exploitation of other hybrid vigors based on Mendelian genetics and conventional breeding has played a significant role in the 20th Century. The yield of major cereal crops over the past half century has increased three fold which is primarily due to input (nutrient, water and agrochemical) responsive hybrid varieties. The fatique of green revolution with intensive cultivation of few selected genotypes has resulted in the depletion of soil nutrients, and biodiversity, increased salinity due to extensive crop irrigation and vulnerability to the diseases, insect pests and other abiotic stresses. One of the major limitations of the conventional breeding is the restricted gene pool and the sexual incompatibility of the cultigens with wild relatives which are reservoirs of the desirable genes. Recent developments in *in vitro* regeneration and new genetic technologies during the last two decades have shown that they can help to improve the crops in more precise and faster ways as compared to traditional Mendelian methods. The transgenic crops based on novel genetic combinations created by moving genes across sexual barriers are now becoming available and some of them are moving to market for the end users. The first generation transgenic plants with resistance/tolerance to biotic and abiotic stresses, improvement in nutrient utilization efficiency and nutritional qualities, enhancing durability of products during harvesting and storage and for cleaning up of the environment are particularly important achievements of this technology.

However, one of the major constraints to have significant impact of transgenic technology on plant biology has been to understand the key regulatory steps involved in the various metabolic processes related to plant productivity. The isolation, characterization and cloning of the regulatory genes from the desired sources is another limiting factor. The recent upsurge in structural and functional genomics will make available genes for a large number of qualitative and quantitative traits. Thus the engineering of complex traits (yield, drought tolerance, grain quality and regulation of metabolic pathways etc.) involving many genes is likely to gain momentum in future. Development of tissue specific and stress inducible promoters and vectors for transfer of fairly large fragment of DNA, introduction of genes in chloroplast for higher gene expression, marker free transgenic plants for reduced health and environmental risks, gene silencing problem etc. need focal attention.

This volume contains 14 chapters on current status, state of art, applications and limitations of gene technology to increase the productivity and quality of major food crops to meet the increasing demand for food in the future. Each chapter in this volume is prepared by distinguished scientists who have themselves made pioneering and significant contributions to their respective field of study. We are grateful to them for their diligence in preparing their comprehensive and insightful accounts.

Rohtak, India

Pawan K Jaiwal
Rana P Singh

CONTENTS

Contents of other volumes in the series

PLANT GENETIC ENGINEERING

SERIES ISBN : 1-930813-17-1

ISBN : 1-930813-02-2

Volume 1 : **Applications and Limitations**
Editors : Rana P. Singh & Pawan K. Jaiwal

ISBN : 1-930813-04-X

Volume 3 : **Improvement of Commercial Plants - I**
Editors : Rana P. Singh & Pawan K. Jaiwal

ISBN : 1-930813-04-X

Volume 4 : **Improvement of Commercial Plants-II**
Editors : Pawan K. Jaiwal & Rana P. Singh

ISBN: 1-930813-14-7

Volume 5 : **Improvement of Vegetables**
Editors : Rana P. Singh & Pawan K. Jaiwal

ISBN: 1-930813-15-5

Volume 6 : **Improvement of Fruits**
Editors : Pawan K. Jaiwal & Rana P. Singh

Chapter 1

AGROBACTERIUM-MEDIATED TRANSFORMATION OF CEREALS

Anna Nadolska-Orczyk* and Waclaw Orczyk

Plant Breeding and Acclimatization Institute, Radzików, 05-870 Blonie, Poland

Summary

First reports on Agrobacterium-mediated transformation of monocot cells were published in the late 1980s. These attempts undertaken by some researchers were not able to brake the general opinion about the lack of susceptibility of cereals to Agro-infection. At the beginning of 1990s the first transgenic fertile plants of wheat, maize and rice were obtained. Turning point was the article of Hiei (1994). This highly efficient method of Agrobacterium-mediated rice transformation was, for the first time, comparable to well developed methods for dicot plants. In the next years several reports on successful transformation of different japonica and indica rice cultivars and single works describing transformation of maize, wheat and barley were published. An alternative to Agrobacterium – a transformation by direct delivery of naked DNA was exploited intensely from the late 1980s. Comparison of results obtained after the direct and Agrobacterium-mediated transformation indicated better applicability of the latter. Most of the advantages such as relatively precise transfer and integration of T-DNA, frequently as one copy, good expression of transgene and Mendelian transmission to the next generation were related to the natural process of transmission. Additionally, as shown in the articles, the method of Agro-transformation was efficient and most or all transgenic plants were fertile. Efficiency of Agrobacterium-mediated transformation of cereals is influenced by some factors associated with plant or

*Corresponding author : E-mail : a.oczyk@ihar.edu.pl

Agrobacteria. The first group includes a type of tissue, plant regeneration system, cultivar susceptibility, selection system and promoters. The second involves A. tumefaciens strains, vectors and activators of virulence system. All of them are discussed basing on the results of published articles. Novel plant transformation technologies combining elements from Agro and other transformation systems (agrolistic, sonication-assisted Agrobacterium-mediated transformation; SAAT) are presented. Application of Agrobacterium-mediated transformation method of cereals for basic research and commercial purposes is reviewed. Two types of genes expressing resistance to insect pests (striped stem borer and yellow stem borer) and sheath blight were introduced to rice. Another example of agrobiotechnological application is an engineering of barley for production of industrial microbial feed enzymes. Additionally a great population of T-DNA insertional mutants to study functional genomics was obtained. Research of a gene function by protein suppression is presented in another report. Future prospects of Agrobacterium-mediated transformation method are discussed.

Keywords : Genetic transformation, cereals, *Agrobacterium tumefaciens*, rice, maize, wheat, barley.

1. INTRODUCTION

Genetic transformation of plants has become a basic tool for cultivar improvement and research. First report on successful gene transfer to plant using *Agrobacterium tumefaciens* as a natural vector was published seventeen years ago by Zambryski *et al.* (1983). In subsequent years the knowledge gained about the process of natural *Agrobacterium* mediated transformation resulted in the far reaching modifications of the system. Detailed analyses of structure and function of T-DNA border sequences and enhancers, *vir* genes, factors influencing their induction and expression resulted in designing and refining cointegration and binary systems. Vectors available now cover the broad range of possible applications: integration and expression of a transgene, T-DNA tagging (Koncz *et al.*, 1989), enhancer/promoter trap systems (Pasquali *et al.*, 1994), integration of very long pieces of DNA (Hamilton *et al.*, 1996). Broad collection of bacterial strains makes possible preselection of a proper one that matches a particular plant chosen for transformation. All this resulted in a set of detailed procedures for routine transformation of many plant species. This, however, concerned only dicot plants. Monocots were believed to be outside the host range of *Agrobacterium*. Even single attempts undertaken by some researchers (Boulton *et al.*, 1989, Graves and Goldman, 1986, Grimsey *et al.*, 1989) were not able to brake this notion.

Strong requirement of getting transgenic cereals and grasses resulted in development and application of direct transformation methods to these plants. They were the methods of choice in late '80 and '90 and resulted in obtaining the first transgenic rye plants (DNA injected into young floral tillers by De la Pena *et al.*, 1987) followed by rice (Zhang and Wu, 1988, Shimamoto *et al.*, 1989) and maize plants (Rhodes *et al.*, 1988) both *via* protoplast transformation. Development of biolistic approach resulted, already a decade ago, in getting transgenic maize plants by Gordon-Kamm *et al.* (1990) and Fromm *et al.* (1990).

Although first reports of successful cereal transformation *via Agrobacterium* were published in early '90 (Chan *et al.*, 1993, Gould *et al.*, 1991, Hess *et al.*, 1990), most researchers at that time were still focused on direct transformation methods. Turning point was the article of Hiei *et al.* (1994) describing highly efficient system of *Agrobacterium*-mediated transformation of rice. Drawbacks of direct transformation methods and clear advantages of *Agrobacterium*-mediated techniques applied for some cereals (review by Nadolska-Orczyk *et al.*, 2000) and particularly for rice attracted more attention to this system. Nowadays, rice is a model cereal plant not only because of its relatively small genome, availability of physical and molecular maps, advanced genome sequencing, large-scale analysis of expressed sequence tags (ESTs) but also very efficient system for *Agrobacterium*-mediated transformation.

2. ADVANTAGES OF A GENETIC TRANSFORMATION BY *AGROBACTERIUM* VS. OTHER METHODS

Agrobacterium-mediated transformation is based on a natural process evolved in bacterium in order to transfer tumor inducing and opine synthesis genes. Transformed plant cells form tumor galls while opines supply bacteria with their main nutrient. To achieve this sophisticated goal, *Agrobacterium* developed a Ti (tumor inducing) plasmid system containing T-DNA (transferred DNA) and the *vir* genes. The T-DNA is delimited by 25 base pairs direct repeats. Any DNA inserted between these borders could be transferred to the plant cell. *Vir* gene products facilitate T-DNA processing, its transfer to plant nucleus and integration with plant DNA. Wounded plant cells release phenolic compounds, which activate *virA*. Phosphorylated products of *virA* and *virG* genes control expression of other *vir* genes involved in T-DNA transfer. T-DNA borders are recognized by VirD2 protein, which together with VirE2 mediate the transfer of the T-DNA from the cytoplasm to the nucleus of the plant cell. Additionally VirE2 bound to T-DNA forms T-complex resistant to nuclease degradation (for review see Zupan and Zambryski, 1995).

This short description exemplifies that each step of the process is tightly controlled by *Agrobacterium*. The main differences between direct and *Agrobacterium* based methods are the result of this feature. This particularly applies to transfer of T-DNA to nucleus and its integration with plant DNA. Analyses of rearrangements of transgene-chromosome junctions indicated higher precision of T-DNA integration *versus* integration of directly delivered DNA. Hiei *et al.* (1994) and Ishida *et al.* (1996) examined sequenecs of DNA at T-DNA junctions in transgenic rice plants. They reported lack of notable rearrangements (Ishida *et al.*, 1996) or minor rearrangements in two out of eight tested plants (Hiei *et al.*, 1994). Multiple complex patterns of long T-DNA (T-DNA and non-T-DNA) integration into the rice genome were also evidenced (Yin and Wang, 2000). In this experiment 33% of transgenic plants contained non-T-DNA sequences. There was no major difference in frequency of non-T-DNA transfer among three *A. tumefaciens* strains. One of the possible reasons for the transfer of long T-DNA may be inefficient nicking of T-DNA borders or insufficient amount of Vir proteins for the binary system (Wenk *et al.*, 1997). However, almost half of the transgenics contained proper T-DNA insertions.

Protoplast transformation and particle bombardment often leads to integration of multiple copies, fragmentation and/or rearrangements of a transgene which result in frequent regeneration of sterile plants (Finnegan and McElroy, 1994; Flavell, 1994). Analysis of transgenic plants of maize, rice and oat obtained through one of direct transformation methods (electroporation or biolistic) showed excessive rearrangements of integrated DNA involving both the vector and the gene sequence (Gordon-Kamm *et al.*, 1990; Kumpatla and Hall, 1998; Pawlowski and Somers, 1998; Register *et al.*, 1994). Complex pattern of *bar* gene integration in wheat after microprojectile bombardment was also reported by Uze *et al.* (1999).

Integration of a single, intact copy of a transgene is a prerequisite of its stability and proper expression both in primary transgenic and its sexual progeny. In wheat, rice and maize plants transformed by

Agrobacterium, a single copy T-DNA occurred at a frequency between 30 and 60% (Cheng *et al.*, 1997; Hiei *et al.*, 1994; Ishida *et al.*, 1996; Jeon *et al.*, 2000). Similar analysis of wheat transgenics obtained after particle bombardment based on results from 14 independent reports revealed that less than 10% were single-copy (Srivastava *et al.*, 1999).

Efficiency of cereal transformation by *Agrobacterium* depends on various factors and ranges from 0.4 up to 30 plants or cell lines selected from 100 explants cocultivated with bacteria (see Tables 1a,b). Incorporated traits were stable in T_0 plants and were transmitted to the next generations in Mendelian fashion. Almost all transgenic Japonica rice (Hiei *et al.*, 1994) and maize plants (Ishida *et al.*, 1996) showed unchanged morphology and 70 % were fully fertile. All transgenic plants of wheat obtained by Cheng *et al.* (1997) were fertile. Problems with fertility and somaclonal variation are common especially when direct techniques are applied for protoplasts. Some of them might be also associated with biolistic methods. Detailed analysis of cytological aberrations of transgenic plants of oat produced by microprojectile bombardment and regenerated control plans was done by Choi *et al.* (2000). Chromosomal aberrations (aneuploidy, deletion of chromosome segments, but no change in ploidy level) were detected in 58% of transgenic oat plants. Level of these changes in control plants ranged from 0-14%. Authors discuss the possible influence of transformation associated stresses on cytological variation.

3. EXAMPLES OF SUCCESSFUL TRANSFORMATION OF CEREALS (RICE, MAIZE, WHEAT AND BARLEY)

Well documented results of obtaining transgenic, fertile plants of cereals showing Mendelian transmission to the next generations after *Agrobacterium*-mediated transformation are presented in Tables 1a,b. First transgenic plants were obtained in 1990 after spikelets transformation of wheat (Hess *et al.*, 1990). One year later several transgenic plants of maize were regenerated from infected shoot apices (Gould *et al.*, 1991). It took next several years to publish well-developed and efficient techniques of *Agro*-transformation for wheat (Cheng *et al.*, 1997), barley (Tingay *et al.*, 1997) and maize (Ishida *et al.*, 1996). For rice the development was more fluent. The first report on regeneration of transgenic japonica varieties was published in 1993 by Chan *et al.* (1993) and one year later very efficient system was presented by Hiei *et al.* (1994). Methods of *Agro*-transformation for more recalcitrant indica varieties were developed by Rashid *et al.* (1996), Park *et al.* (1996) and

Mohanty *et al.* (1999). This progress, spectacular in the case of rice, allowed for basic research and commercial applications (Table 2).

4. FACTORS AFFECTING *AGROBACTERIUM*-MEDIATED TRANSFORMATION

4.1. Type of tissue, plant regeneration system and cultivar variability

In rice, the big differences in transformation ability of various explants like shoot apices, segments or calli from young roots, calli from scutella, immature embryos and suspension culture cells were reported by Hiei *et al.* (1994). Blue spots indicating transgene GUS expression were detected in all types of tissue except root segments. However, the best for bacteria inoculation and plant regeneration were scutellum-derived calli; up to 28.6% of them gave transgenics (Hiei *et al.*, 1994). High usability of calli derived from scutella of japonica rice varieties (Hiei *et al.*, 1994) was confirmed for the same tissue of indica varieties (Rashid *et al.*, 1996; Mohanty *et al.*, 1999). Despite that the later varieties were reported to be much more difficult in tissue culture, regeneration efficiency of transgenics was on a similar 22 to 25% level. Freshly isolated immature embryos, precultured immature embryos and embryogenic calli of wheat were used as starting material for *Agro* inoculation by Cheng *et al.* (1997). Transformation efficiency for freshly isolated explants (1.12%) was slightly lower than for cultured ones (1.55 and 1.56%). In other works, only one type of highly regenerable, meristematic tissue was chosen assuming that the *in vitro* response was the most important criterion. Beside scutellum, shoot apices (Gould *et al.*, 1991; Park *et al.*, 1997) and spikelets (Hess *et al.*, 1990) were also successfully used for transformation.

Mitotic activity of plant cells is probably one of the basic requirements for successful transformation with *Agrobacterium* (Binns and Thomashow, 1988). This activity in monocot tissue culture is observed only when the explants originate from meristematic, immature cells. Examples of successful transformation (shown in tables) confirm this hypothesis. According to Citovsky *et al.* (1994) a lack of susceptibility of mature cells to *Agrobacterium* could depend on nuclear import of T-DNA - protein complex. Molecular pathway of this process was studied in maize leaf cells, tobacco protoplasts and mature as well as immature root cells of both species. The complex was not accumulated in nuclei of mature root cells indicating the repression of protein transport system.

Earlier studies demonstrated that plant nuclear uptake required specific nuclear localization signals (NLSs) (Citovsky *et al.*, 1992). These NLSs, active in developing tissue remained nonfunctional in mature tissue. The lack of T-DNA-protein complex transport might be explained by the absence of specific cellular VirD2 and VirE2 NLS-binding proteins in mature cells. Noteworthy, mature cells of dicot plants are able to dedifferentiate and regain their mitotic activity in tissue culture. Wounding additionally induce this process. Good examples are leaf-discs of tobacco or potato - highly competent to transformation *via Agrobacterium*. Opposite to this, monocotyledonous mature and wounded cells tend to become lignified or sclerified without entering mitotic division.

Efficient plant regeneration system is another important requirement for plant transformation. In cereals it depends directly on the type and source of the explant. Immature cereal embryos are the most often used because they are highly regenerable and, what is important, this trait is only partly genotype dependent. Shoot apices (in maize reported by Gould *et al.*, 1991; in rice by Park *et al.*, 1997) might be an important alternative particularly for genotypes recalcitrant in tissue culture.

Any improvement of plant regeneration may positively affect transformation efficiency. Water stress applied by the use of highly concentrated agarose in rice regeneration medium increased the frequency of plant regeneration (Jain *et al.*, 1996) as well as the efficiency of transformation (Lee *et al.*, 1999).

Interdependence of transformation efficiency and the genotype varies in different cereals. The strongest one was reported for maize. According to Armstrong (1999) successful *Agrobacterium*-mediated transformation was limited to only A188 line and a germplasm which included this inbred as one parent. In other cereals, especially in rice, transformation susceptibility is not restricted to certain genotypes. It is rather correlated with regeneration efficiency (see Tables 1a,b; 2). Lee *et al.* (1999) presented stable integration of marker and reporter genes in thirteen rice cultivars. Two cultivars, which were highly susceptible to *A. tumefaciens,* were also highly responsive to plant regeneration. High regeneration ability is certainly an advantage in transformation procedures but, as it has been shown by Hiei *et al.* (1994), it can not be used as a final criterion. They found low regenerating genotype, which was efficiently transformed by *Agrobacterium*.

Table 1a. Evidence of *Agrobacterium*-mediated rice transformation methods that resulted in obtaining fertile transgenic plants showing Mendelian transmission to R_2 generation.

Variety	Strain (binary vector)	Promoter/ marker, reporter	Tissue	Selection/ activation of *vir*	Trans-formation efficiency	Reference
Japonica Tainung 62	A281 (pTiBo542 + pAG8)	*nos/nptII* α - *Amy* 8/*uidA*	Cultured immature embryos (wounded)	G418 (40 mg l^{-1}) PSC	R_2 generation tested based on single plant	Chan *et al.*, 1993
Japonica Tsukinohikari Asanohikari Koshihikari	LBA4404 (pIG121Hm) EHA101 (pIG121Hm) LBA4404 (pTOK233) EHA101 (pTOK233) - super-binary vector	CaMV35S/*hpt* CaMV35S/*int-gus* *nos/nptII*	Calli derived from scutella, immature embryos, suspension cells (not wounded)	Hygromycin (50 - 100 mg l^{-1}) Acetosyringone (100 μM)	up to 28.6%	Hiei *et al.*, 1994
Indica Basmati 370, 385 and 6129	EHA101 (pIG121Hm)	CaMV35S/*hpt* CaMV35S/*int-gus*	Calli derived from scutella (not wounded)	Hygromycin (50 mg l^{-1}) Acetosyringone (50 μM)	up to 22%	Rashid *et al.*, 1996

Table 1a. Continued

Variety	Strain (binary vector)	Promoter/ marker, reporter	Tissue	Selection/ activation of *vir*	Trans- formation efficiency	Reference
Japonica Taipei 309 **Javanica** Gulfmont, Jefferson	LBA4404 (pTOK233) EHA101 (pIG121)	CaMV35S/*hpt* CaMV35S/*int-gus*	Calli derived from scutella (not wounded)	Hygromycin (50 mg l^{-1}) Acetosyringone (150 μM)	up to ~ 30%	Dong *et al.*, 1996
Japonica Radon **Indica** TCS10, IR72	LBA4404 (pTOK233) A.t. 656 (pCNL56)	CaMV35S/*hpt* CaMV35S/*int-gus nos/nptII*	Immature embryos (not wounded)	G418 (30 mg l^{-1}) Hygromycin (30 - 50 mg l^{-1}) Acetosyringone (100 - 200 μM)	Japonica variety - 27% Indica varieties - 1 to 5%	Aldemita and Hodges, 1996
Tropical Japonica Maybelle	EHA101 (pTiBo542)	*actin1/int-bar nos/nptII* CaMV35S/*bar*	Isolated shoot apices (extra wounding)	PPT (0.5 mg l^{-1}) Acetosyringone (50 μM) PSC	0.7%	Park *et al.*, 1996
Indica Pusa Basmati 1	LBA4404 (pTOK233)	CaMV35S/*hpt* CaMV35S/*int-gus nos/nptII*	Calli from scutella	Hygromycin (50 mg l^{-1}) Acetosyringone (100 μM)	ab. 25%	Mohanty *et al.*, 1999

4.2. *Agrobacterium tumefaciens* strains and vectors

Genes placed on both virulence plasmid (*vir* genes) and chromosome of bacteria control the process of *Agro*-infection. Some of them are responsible for bacteria - host plant recognition. This determines the ability to infect and transform particular plant species. Earlier research on Gus expression in transformed maize cells indicated that nopaline C58C1 strain was more efficient then LBA4404 (Shen *et al.*, 1993). Authors suggested that factors responsible for this co-segregated with opine-utilization profile and were localized on Ti plasmid. Heath *et al.* (1997) studied bacterial factors important in *vir* gene induction and its role in maize agroinfection. Phosphorylated product of *virA* implicated other *vir* gene induction and was the major factor in this process. *Agrobacterium* strain specificity to infect maize was strongly determined by the source of this gene. Monosaccharide binding protein ChvE coded by chromosomal gene was another factor required for maize agroinfection (Heath *et al.*, 1997). The protein was found to interact with the periplasmic domain of VirA (Shimoda *et al.*, 1993) facilitating *vir* induction by phenolic compounds. The strains of *Agrobacterium* deficient in *chvE* were not able to transform maize.

Unsuccessful attampts to transform plants were the result of a limited host range of *Agrobacterium*. To solve this problem strains exhibiting high virulence as well as wider host range were selected. One of them was L,L succinamopine A281 strain hypervirulent on several solanaceus plants (Hood *et al.*, 1986). Experiments with binary system revealed that the hypervirulence was determined by the sequences placed on Ti (pTiBo542) plasmid. Based on the hypervirulence system two others helper plasmids: pEHA101 and pEHA105 were constructed (Hood *et al.*, 1993) and were successfully used for monocot and dicot transformation. Komari (1990) exploited the hypervirulence of pTiBo542 to construct super-binary vector pTOK233. He achieved this by inserting extra virulence genes into a small, T-DNA carrying vector.

Four combinations of *Agrobacterium* strains and binary vectors were successfully used for cereal transformation. They were supervirulent strains (EHA101, EHA105, AGL1) carrying binary or super-binary vectors as well as commonly used "ordinary" strains with binary or super-binary vectors (see Table 1a,b). All four systems were based on two original *Agrobacterium* strains: LBA4404 and EHA101 and these two vectors were compared by Hiei *et al.* (1994). Successful transformation of rice was obtained with each of them. Unexpectedly supervirulent strain containing super-binary vector was not very effective. There were no differences between ordinary and improved vector/strain combinations. The most effective combination in this experiment was LBA4404 (pTOK233). Very efficient transformation obtained with ordinary strain/super-binary vector combination suggested that the vector was a key factor for cereal transformation. Other results, however, did not confirm higher virulence of pTOK233. Commonly used strain LBA4404 and binary vectors (derivatives of pBin19) were able to transform rice at yields comparable to that of supervirulent strain and/or super-binary vector (Cheng *et al.*, 1998). In our research the ability to transform immature embryos of barley, wheat and *Triticale* by hypervirulent strain/binary vector was higher than that by ordinary strain/super-binary vector (Anna Nadolska-Orczyk, Anna Przetakiewicz, unpublished data).

4.3. Selection systems

A proper selection system was found to be important for transformation of pea where four binary vectors differing in selectable genes in the same *Agrobacterium* strain were tested (Nadolska-Orczyk and Orczyk 2000). Transgenic plants were obtained only after phosphinothricin (*bar*) or kanamycin *(npt*) selection; not after hygromycin (*hpt*) or methotrexate (*dhfr*) ones. Investigation of T-DNA transfer and integration in *Arabidopsis thaliana* root explants revealed that stable transformation without selection was below 0.5% and it rose to approximately 25% after the selection was applied (De Buck *et al.*, 2000).

There was no systematic study on how selection could influence regeneration of transgenic cereals. Li *et al.* (1997) compared two selectable genes *hpt* and *bar* under *ubi*1 promoter in stable transformed rice calli. They found better growth on hygromycin as a selection agent. Transgenic rice plants were obtained after selection on G418, hygromycin or phosphinothricin (PPT) (Table 1a), wheat plants were selected on kanamycin and G418, barley on bialophos and maize on kanamycin or hygromycin (Table 1b). The same as above were used for selection after direct transformation (reviewed by McElroy and Brettel, 1994; Vasil 1994). The most frequently reported were PPT and hygromycin. According to Armstrong (1999), transient expression of a reporter gene after co-cultivation of cereal cells with *Agrobacterium* was relatively easy to observe, but usually it did not help (sometimes it was even misleading) in finding the most suitable conditions for recovery of stable transformed plants.

Table 1b. Evidence of *Agrobacterium*-mediated wheat, barley and maize transformation methods that resulted in obtaining fertile transgenic plants showing Mendelian transmission to R_2 generation.

Variety	Strain (binary vector)	Promoter/ marker, reporter	Tissue	Selection/ activation of *vir*	Trans-formation efficiency	Reference
Wheat : Turbo	C58C1 (pGV3850 :: 1103*neo*)	*nos/nptII*	Spikelets	Kanamycin (10 – 500 mg l^{-1})	up to 2.6%.	Hess *et al.*, 1990
Wheat : Bobwhite	C58 (pMON18365)	E35S/*int-gus* (enhanced 35S) E35S/*nptII*	Freshly isolated immature embryos, precultured immature embryos, embryogenic calli	2 – 5 days with-out selection, G418 (25 mg l^{-1}) Acetosyringone (200 μM)	1.12, 1.56 and 1.55.	Cheng *et al.*, 1997
Barley : Golden Promise	AGL1 (pDM805)	*Act1/gus* *Ubi1/bar*	Immature embryos injured	Bialophos (3 mg l^{-1})	4.2%	Tingay *et al.*, 1997
Maize : Funk's G90	EHA1 (pTiBo542)	CaMV35S/*gus* *nos/nptII*	Shoot apices	Kanamycin (7.5 mg l^{-1}) Acetosyringone (30 μM)	Several transgenic plants	Gould *et al.*, 1991

Table 1b. Continued

Variety	Strain (binary vector)	Promoter/ marker, reporter	Tissue	Selection/ activation of *vir*	Trans- formation efficiency	Reference
Maize : *Inbred lines :* A188, W117, W59E A554, W153R, H99 BMS *Hybrids :* W117×A188, W59E×A188 A554×A188 W153R×A188 H99×A188	LBA4404 (pTOK233) LBA4404 (pSB131) LBA4404 (pIG121Hm)	CaMV35S/*int-gus* CaMV35S/*hpt* CaMV35S/*bar* *nos/nptII*	Immature embryos	Hygromycin (10 – 30 mg l^{-1}) PPT (5 – 10 mg l^{-1}) Acetosyringone (50 µM)	hybrids - 0.4% - 5.3% A188 line – 5% – 30%	Ishida *et al.*, 1996

4.4. Promoters

Cereal specific promoters became widely available from the beginning of 1990. McElroy *et al.* (1990) reported isolation of rice actin promoter – *Act1*, Last *et al.* (1991) - maize alcohol dehydrogenase 1 promoter – *Emu*, Christiansen *et al.* (1992) - maize ubiquitin promoter -*Ubi1* and Yu *et al.* (1992) - rice amylase promoter – *amy*. Relative activity of these promoters in directly transformed cereal cells was reported as high for *Ubi1-Ubi1 intron1*, moderate for *Emu* or *Act-Act1 intron1* and low for CaMV35S (McElroy and Brettell, 1994).

Park *et al.* (1996) compared PPT selection of rice shoot apices transformed with *bar* under actin or 35S promoters. They found higher expression of *bar* under *Act1* promoter which facilitated better selection. In experiments of Cheng *et al.* (1998) three different promoters: maize ubiquitin, CaMV35S, and *Brassica Bp10* gene were used to achieve high and tissue-specific expression of lepidopteran-specific δ-endotoxins (Bt) in rice. The highest level of Bt protein (up to 3% of total soluble proteins) was obtained when the gene was under maize *Ubi1* promoter. Much lower expression (0.024% total soluble proteins) was obtained with the same promoter after protoplast transformation (Fujimoto *et al.* 1993) or particle bombardment (Nayak *et al.*, 1997). Cheng *et al.* (1998) attributed those big differences to the method of gene transfer, gene copy number, host genotype, number of transformants screened and plant growth conditions.

To study the expression of feed enzyme – a fungal xylanase gene in the endosperm of developing grains of transgenic barley, the promoters of two cereal storage protein genes, rice *GluB-1*and barley *Hor2-4* were used (Patel *et al.*, 2000). The rice glutein B-1 (*GluB-1*) provided more than twice-higher expression level of recombinant protein in barley grains. Authors suggested that higher expression of the gene in barley under rice promoter could be explained by avoiding homologous sequence-dependent gene silencing.

To develop a high-level expression system in rice a synthetic gene (*sgfp*) encoding a green fluorescent protein (GFP) was inserted into two expression vectors – *Act1-sgfp* for an untargeted and *rbcS-Tp-sgfp* for a chloroplast targeted expression (Jang *et al.*, 1999). *RbcS* transit peptide specifically transferred GFP to chloroplasts and non green plastids of transgenic rice. Targeted expression significantly increased expression of *sgfp* in mature leaf tissues yielding about 10% of the total soluble protein in *rbcS-Tp-sgfp* lines *vs.* 0.5% in *Act1-sgfp* lines.

4.5. Activation of virulence system

In the natural process of *Agro*-infection, wounded plant cells release phenolic compounds (such as acetosyringone and α-hydroxyacetosyringone) which serve as signal molecules activating expression of *vir* genes (Stachel *et al.*, 1985). Usami *et al.* (1987) found that these chemicals were not produced or produced in insufficient quantities in grasses. Their later results (Usami *et al.*, 1988) indicated that undifferentiated, meristematic cells such as cereal immature embryos could produce phenolic compounds by themselves and wounding possibly stimulates this process further. Information that many monocotyledons might lack enough phenolics to stimulate *vir* genes created the need to test whether exogenously applied could induce or enhance the process. In some experiments potato suspension culture (PSC) was used as a natural and rich source of the compounds. The results were, however, ambiguous. Chan *et al.* (1993) reported that transformation frequency of rice calli was almost three times enhanced by PSC (transgenic plants were obtained only after PSC treatment) while Park *et al.*, (1996) did not find any significant influence on transformation of shoot apices of rice. Similar results concerning transformation of rice embryo-derived cultures were reported by Raineri *et al.* (1990).

More consistent results were obtained with acetosyringone added directly to transformation media. It stimulated transformation of rice scutella derived calli (Hiei *et al.*, 1994 and Rashid *et al.*, 1996) or rice immature embryos (Aldemita and Hodges, 1996). Similar, enhancing effect on maize immature embryo transformation was found by Ishida *et al.* (1996). In many cases, however, acetosyringone was supplied as a basic component of transformation without further testing (see Tables 1a,b). It is worthwhile to notice that in some procedures (barley by Tingay *et al.*, 1997 and wheat by Hess *et al.*, 1990) successful transformation was achieved without using this compound.

Tissue wounding as the alternative way to stimulate the release of phenolic compounds was tested in some experiments. Improved efficiency of transformation after this pretreatment *vs.* non treated tissue was reported by Park *et al.* (1996) and Tingay *et al.* (1997). In the first report, isolated shoot apices of rice were wounded with the hypodermic needle. This enhanced the frequency of transgenic plant regeneration and the effect was not further influenced by application of acetosyringone or PSC. In the second, immature embryos of barley were injured with a shooting of gold particles or axis removal and there was no extra

Table 2. Examples of practical application of developed *Agrobacterium*-mediated transformation methods

Variety	Method (strain/ promoter-marker)	Promoter/ transgene	target/trait	Results	Reference
Rice : Nipponbare, Zhong 8215 93VA, ZAU16 91RM, T8340 Pin92-528, T90502 Kaybonnet	Cheng *et al.*, 1997, modified from Hiei *et al.*, 1994 (LBA4404, EHA105/ CaMV35S/*hph* CaMV35S/*gus*)	*Ubi/cryIA(b)* *Ubi/cryIA(c)* *CaMV35S/cryIA(b)* *CaMV35S/cryIA(c)* *Bp/cryIA(b)* *Bp/cryIA(c)*	CryIA(b) and CryIA(c) proteins toxic to rice insect pests – striped stem borer and yellow stem borer	1. 2600 transgenic plants from 9 rice varieties. 2. Accumulation of up to 3% soluble proteins in R_0 plants. 3. R_1 transgenic plants highly toxic to insects.	Cheng *et al.*, 1998
Rice : Nipponbare, Shiokari, Kinmaze, Taichung, Daikoku	Toki, 1997, (EHA101/ *hpt*)	antisense RGA1	Antisense DNA for the α subunit of rice heterotri-meric G protein	1. Suppresion of heterotri-meric G protein leading to abnormal morphology - indication of G protein functions in the formation of normal internodes and seeds.	Fujisawa *et al.*, 1999
Rice : Basmati 122, Tulsi, Vaidehi	as described, (LBA4404, A281/ CaMV35S/*hph*)	CaMV35S/*Chi11*	The rice class-I *chitinase* gene	1. Constitutive expression of the gene resulted in enhanced resistance to sheath blight.	Datta *et al.*, 2000

Table 2. Continued

Variety	Method (strain/ promoter-marker)	Promoter/ transgene	target/trait	Results	Reference
Rice : Dongjin	Jeon *et al.*, 1999 and Lee *et al.*, 1999	promoterless *gus/hph* *int-gus/OsTub A1int/hph*	T-DNA insertional mutagenesis for functional genomics	1. 22090 primary transgenic plants. 2. 65% contained more than one copy of the inserted T-DNA. 3. 1.6 – 2.1% of tested organs were GUS positive.	Jeon *et al.*, 2000
Barley : Golden Promise	Tingay *et al.*, 1997, (AGL1/ *ubi1/bar*)	*GluB-1/xynA* *Hor2-4/xynA*	Modified xylanase gene from rumen fungus under endosperm-specific promoter for production of industrial microbial feed enzymes	1. 24 transgenic lines 2. The fungal xylanase was produced in the developing endosperm. 3. Enzyme was stably maintained in the grains.	Patel *et al.*, 2000

application of acetosyringone. Transformation efficiency was relatively high (4.2%) which might be explained by the presence of *vir* inducing compounds. Such compounds were identified in homogenized tissue of wheat and oats by Usami *et al.* (1988). Explants used in almost all other *Agrobacterium*-mediated transformations (calli derived from scutella, immature embryos) were not wounded. However, they were subjected to the influence of 30-200 μM of exogenous acetosyringone.

5. METHODS COMBINING ELEMENTS FROM *AGRO* AND OTHER TRANSFORMATION SYSTEMS

Hansen and Chilton (1996) combined advantages of precise T-DNA recognition, excision and integration mediated by VirD1 and VirD2 proteins with biolistic DNA delivery. In their system, called "agrolistic", plasmids containing *vir*D1 and *vir*D2 genes under the control of CaMV35S promoter were co-delivered with another plasmid carrying a transgene flanked by T-DNA borders. Analysis of T-DNA - plant DNA junctions in transgenic tobacco and maize cells showed that in 20% of foreign DNA was integrated in low copy number with intact T-DNA borders.

Exploring possible combinations of known transformation techniques some authors combined *Agrobacterium* infection with micro wounding mediated by particle gun (Bidney *et al.*, 1992; Tingay *et al.*, 1997) or sonication (Trick and Finer, 1997). The last approach called by the authors sonication-assisted *Agrobacterium*-mediated transformation (SAAT) was proved to increase transient transformation efficiency in several different plant tissues of dicot, monocot and gymnosperms. Stable transformation was obtained only in soybean (Ohio buckeye) embryogenic suspension culture.

6. RESULTS OF PRACTICAL APPLICATION OF DEVELOPED METHODS

Research and commercial application of developed *Agrobacterium*-mediated transformation methods are presented in table 2. T-DNA tagging by promoterless constructs is one of the known strategies to explore gene expression profiles *in planta*. Development of very efficient *Agrobacterium*-based transformation systems made this approach also applicable to cereals. Toward this end a population of 22,090 independent primary transgenic rice plants with promoterless *gus* gene was obtained (Jeon *et al.*, 2000). An average number of T-DNA inserts were estimated as 1.4 per plant. About two percent of tested organs were GUS positive, showing either specific (organ- or tissue-) or ubiquitous expression patterns. The large population of these plants would be useful for identifying insertional mutants of various genes and for discovering new genes in rice.

Fujisawa *et al.* (1999) suppressed expression of α-subunit of rice heterotrimeric G protein by transformation with a construct containing antisense cDNA of this gene. This protein has been proposed to function in various systems of signal transduction in diverse tissues or cells. The transformants exhibited morphological abnormalities including dwarfism and setting small seeds. They suggested that G protein was involved in morphogenic signal transduction.

Successful engineering of pest resistant rice was presented by Cheng *et al.* (1998). They obtained over 2600 plants transformed with two synthetic Bt toxin genes - *cryIA(b)* and *cryIA(c)*. Accumulation of toxic proteins (ranging from 2.3 to 3.1% of total soluble proteins) caused 100% mortality of the striped stem borer and yellow stem borer. Transgenic plants with directly introduced Bt toxin genes fused to the same *Ubi-1* promoter exhibited ten times lower accumulation of the proteins (Fujimoto *et al.*, 1993; Nayak *et al.*, 1997).

Incorporation of plants' own disease resistance genes with an altered mode of expression is another strategy to enhance their defense response. This approach was used by Lin *et al.* (1995) and repeated in the same lab by Datta *et al.* (2000). They reported increased resistance to sheath blight in indica rice constitutively expressing rice class-I *chitinase* gene. Similar results were obtained by both direct (Lin *et al.*, 1995) and *Agrobacterium*-based methods (Datta *et al.*, 2000). The later authors underlined, however, that their method was simpler as well as more cost and labor effective.

Production of industrial microbial enzymes in grains was approached by Patel *et al.* (2000). They obtained transgenic barley expressing modified xylanase gene (*xynA*) driven by an endosperm-specific promoter. Analysis of transgenic plants from twenty-four independent lines showed that the xylanase was stably maintained in the grain during maturation, desiccation and post-harvest storage. The results indicated that cereal grains could be used as a bioreactor for the production of feed enzymes.

7. CONCLUSIONS AND FUTURE PROSPECTS

Agrobacterium-mediated transformation technique quickly became a routine for a number of cereal species. Applicability of this method in crop improvement programs is correlated with the use of desirable, agronomically important, regenerable in tissue culture crop. Other requirements of transformation technology are related to integration and expression pattern of a transgene (for review see Gelvin 1998).The transgenes should be introduced in a single copy, without vector backbone sequences. Their expression must be predictable. Concerns of releasing selectable marker genes to the ecosystem (by outcrossing) initiated investigations on how to avoid or remove them from the systems. One of the solutions presented by Komari *et al.* (1996) was a production of marker-free transgenic plants by co-transformation. Vectors carrying two separate T-DNAs in *Agrobacterium*-mediated transformation system were used. Transgenic progeny free of selection markers were selected among GUS-positive and drug-sensitive segregants. Other solutions, like application of site-specific recombination are discussed in the review article (Puchta, 2000).

One of the great challenges of transformation technology is to develop methods of targeted gene integration or replacement. In the pioneering report, Ow and Medberry (1995) documented that site-specific recombination allowed direct integration of foreign DNA into pre-engineered sites in the genome. Some initial success in non-regenerable maize was obtained by Lyznik *et al.* (1996). The system was later applied to genetically engineered male sterility (Lyznik *et al.*, 2000). Another possibility of integrating the transgene into the targeted chromosomal site, additionally allowing gene replacement, is a homologous recombination. According to recent reports, the probability of homologous recombination event to occur is very rare, up to one per 750 illegitimate recombination events (Kempin *et al.*, 1997).

In summary, the results of *Agrobacterium*-mediated transformation of cereals indicate that the system is very attractive. It offers a relatively easy and effective way of obtaining transgenic plants without the necessity of using special equipment. Precise recognition and transmission of T-DNA, frequently integrated as a single copy facilitates efficient selection of plants with the expected mode of transgene expression. This also influences the Mendelian transmission of a transgene, which occurs in the majority of modified plants. Relatively high efficiency of obtaining transgenic, fertile and phenotypically normal plants enables application of the method for basic and commercial research purposes.

REFERENCES

Aldemita RR and Hodges TK (1996) *Agrobacterium tumefaciens*-mediated transformation of *japonica* and *indica* rice varieties. *Planta,* **199** : 612-617.

Armstrong CL (1999) The first decade of maize transformation: a review and future perspective. *Maydica,* **44** : 101-109.

Bidney D, Scelonge C, Martich J, Burrus M, Sims L and Huffman G (1992) Microprojectile bombardment of plant tissues increases transformation frequency by *Agrobacterium tumefaciens. Plant Mol. Biol.,* **18** : 301-313.

Binns AN and Thomashow MF (1988) Cell biology of *Agrobacterium* infection and transformation of plants. *Annu. Rev. Microbiol.,* **42** : 575-606.

Boulton MI, Buchholz WG, Marks MS, Markham PG and Davies J (1989) Specifity of *Agrobacterium*-mediated delivery of maize streak virus DNA to members of the Graminae. *Plant Mol. Biol.,* **12** : 31-40.

Chan MT, Chang HH, Ho SL, Tong WF and Yu SM (1993) *Agrobacterium*-mediated production of transgenic rice plants expressing a chimeric α-amylase promoter/ β-glucuronidase gene. *Plant Mol. Biol.,* **22** : 491-506.

Cheng M, Fry JE, Pang S, Zhou H, Hironaka CM, Duncan DR, Conner TW and Wan Y (1997) Genetic transformation of wheat mediated by *Agrobacterium tumefaciens. Plant Physiol.,* **115** : 971-980.

Cheng V, Sardana R, Kaplan H and Altosaar I (1998) *Agrobacterium*-transformed rice plants expressing synthetic *cryIA(b)* and *cryIA(c)* genes are highly toxic to striped stem borer and yellow stem borer. *Proc. Natl. Acad. Sci. USA.,* **95** : 2767-2772.

Choi HW, Lemaux PG and Cho MJ (2000) High frequency of cytogenetic aberration in transgenic oat (*Avena sativa* L.) plants. *Plant Sci.,* **156** : 85-94.

Christiansen AH, Sharrock RA and Quail PH (1992) Maize polyubiquitin genes: structure, thermal perturbation of expression and transcript splicing, and promoter activity following transfer to protoplasts by electroporation. *Plant Mol. Biol.,* **18** : 675-689.

Citovsky V, Warnick D and Zambryski P (1994) Nuclear import of *Agrobacterium* VirD2 and VirE2 proteins in maize and tobacco. *Proc. Natl. Acad. Sci. USA,* **91** : 3210-3214.

Citovsky V, Zupan J, Warnick D and Zambryski P (1992) Nuclear localization of *Agrobacterium* VirE2 protein in plant cells. *Science,* **256** : 1803-1805.

Datta K, Koukolikova-Nicola Z, Baisakh N, Oliva N and Datta SK (2000) *Agrobacterium*-mediated engineering for sheath blight resistance of *indica* rice cultivars from different ecosystems. *Theor. Appl. Genet.,* **100** : 832-839.

De Buck S, De Wilde Ch, Van Montagu M and Depicker A (2000) Determination of the T-DNA transfer and the T-DNA integration frequencies upon cocultivation of *Arabidopsis thaliana* root explants. *Mol. Plant-Microbe Interact.,* **13** : 658-665.

De la Pena A, Lorz H and Schell J (1987) Transgenic rye plants obtained by injecting DNA into young floral tillers. *Nature,* **325** : 274-276.

Dong J, Teng W, Buchholz WG and Hall TC (1996) *Agrobacterium*-mediated transformation of Javanica rice. *Mol. Breed.,* **2** : 267-276.

Finnegan J and McElroy D (1994) Transgene inactivation: Plants fight back. *Bio/ Technology,* **12** : 883-888.

Flavell RB (1994) Inactivation of gene expression in plants as a consequence of specific sequence duplication. *Proc. Natl. Acad. Sci. USA,* **91** : 3490-3496.

Fromm ME, Morrish F, Armstrong C, Williams R, Thomas J and Klein TM (1990) Inheritance and expression of chimeric genes in the progeny of transgenic maize plants. *Bio/Technology,* **8** : 833-839.

Fujimoto H, Itoh K, Yamamoto M, Kyozuka J and Shimamoto K (1993) Insect resistant rice generated by introduction of a modified δ-endotoxin gene of *Bacillus thuringiensis. Bio/Technology,* **11** : 1151-1155.

Fujisawa Y, Kato T, Ohki S, Ishikawa A, Kitano H, Sasaki T, Asahi T and Iwasaki Y (1999) Suppresion of the heterotrimeric G protein causes abnormal morphology, including dwarfism, in rice. *Proc. Natl. Acad. Sci. USA,* **96** : 7575-7580.

Gelvin SB (1998) The introduction and expression of transgenes in plants. *Curr. Opin. Biotech.,* **9** : 227-232.

Gordon-Kamm WJ, Spencer TM, Mangano ML, Adams TR, Daines RJ, Start WG, O'Brien JV, Chambers SA, Adams WR Jr, Willetts NG, Rice TB, Mackey CJ, Krueger RW, Kausch AP and Lemaux PG (1990) Transformation of maize cells and regeneration of fertile transgenic plants. *Plant Cell,* **2** : 603-618.

Gould J, Davey M, Hasegawa O, Ulian EC, Peterson G and Smidh RH (1991) Transformation of *Zea mays* L. using *Agrobacterium tumefaciens* and the shoot apex. *Plant Physiol.,* **95** : 426-434.

Graves ACF and Goldman SL (1986). The transformation of *Zea mays* seedlings with *Agrobacterium tumefaciens. Plant Mol. Biol.,* **7** : 43-50.

Grimsley N, Hohn B, Ramos C, Kado C and Rogovsky P (1989) DNA transfer from *Agrobacterium* to *Zea mays* or *Brassica* by agroinfection is dependent on bacterial virulence functions. *Mol. Gen Genet.,* **217** : 309-316.

Hamilton CM, Frary A, Lewis C and Tanksley SD (1996) Stable transfer of intact high molecular weight DNA into plant chromosomes. *Proc. Natl. Acad. Sci. USA,* **93** : 9975-9979.

Hansen G and Chilton MD (1996) "Agrolistic" transformation of plant cells: integration of T-strands generated *in planta. Proc. Natl. Acad. Sci.,* **93** : 14978-14983.

Hansen G, Shillito RD and Chilton MD (1997) T-strand integration in maize protoplasts after codelivery of a T-DNA substrate and virulence genes. *Proc. Natl. Acad. Sci. USA,* **94** : 11726-11730.

Heath JD, Boulton MI, Raineri DM, Doty SL, Mushegian AR, Charles TC, Davies JW and Nester EW (1997) Discrete regions of the sensor protein Ira determine the strain-specific ability of *Agrobacterium* to infect maize. *Mol. Plant-Microbe Interact.,* **10** : 221-227.

Hess D, Dressler K and Nimmrichter R (1990) Transformation experiments by pipetting *Agrobacterium* into the spikelets of wheat (*Triticum aestivum* L.). *Plant Sci.,* **72** : 233-244.

Hiei Y, Ohta S, Komari T and Kumashiro T (1994) Efficient transformation of rice (*Oryza sativa* L.) mediated by *Agrobacterium* and sequence analysis of the boundaries of the T-DNA. *Plant J.,* **6** : 271-282.

Hood EE, Gelvin ST, Melchers LS and Hoekema A (1993) New *Agrobacterium* helper plasmids for gene transfer to plants. *Transgenic Res.,* **2** : 208-218.

Hood EE, Helmer GL, Fraley RT and Chilton MD (1986) The hypervirulence of *Agrobacterium tumefaciens* A281 is encoded in a region of pTiBo542 outside of T-DNA. *J. Bacteriol.,* **168** : 1291-1301.

Ishida Y, Saito H, Ohta S, Hiei Y, Komari T and Kumashiro T (1996) High efficiency transformation of maize (*Zea mays* L.) mediated by *Agrobacterium tumefaciens. Nature Biotechnol.,* **14** : 745-750.

Jain RK, Jain S and Wu R (1996) Stimulatory effect of water stress on plant regeneration in aromatic *Indica* rice varieties. *Plant Cell Rep.,* **15** : 449-454.

Jang I Ch, Nahm BH and Kim JK (1999) Subcellular targeting of green fluorescent protein to plastids in transgenic rice plants provides a high-level expression system. *Mol. Breed.,* **5** : 453-461.

Jeon JS, Lee S, Jung KH, Jun SH, Jeong DH, Lee J, Kim Ch, Jang S, Lee S, Yang K, Nam J, An K, Han MJ, Sung RJ, Choi HS, Yu JH, Choi JH, Cho SY, Cha SS, Kim SI and An G (2000) T-DNA insertional mutagenesis for functional genomics in rice. *Plant J.,* **22** : 561-570.

Kempin SA, Liljegren SJ, Block LM, Rounsley SD, Yanofsky MF and Lam E (1997) Targeted disruption in *Arabidopsis. Nature,* **389** : 802-803.

Komari T (1990) Transformation of cultured cells of *Chenopodium quinoa* by binary vectors that carry a fragment of DNA from virulence region of pTiBo542. *Plant Cell Rep.,* **9** : 303-306.

Komari T, Hiei Y, Saito Y, Murai N and Kumashiro T (1996) Vectors carrying two separate T-DNAs for co-transformation of higher plants mediated by *Agrobacterium tumefaciens* and segregation of transformants free from selection markers. *Plant J.,* **10** : 165-174.

Koncz C, Martini N, Mayerhofer R, Koncz-Kalman Z, Korber H, Redei G and Schell J (1989) High-frequency T-DNA-mediated gene tagging in plants. *Proc. Natl. Acad. Sci. USA,* **86** : 8467-8471.

Kumpatla SP and Hall TC (1998) Longevity of 5-azacytidine-mediated gene expression and re-establishment of silencing in transgenic rice. *Plant Mol. Biol.,* **38** : 1113-1122.

Last DI, Bretell RIS, Chamberlain DA, Chaundhury AM, Larkin PJ, Marsh EL, Peacock WJ and Dennis ES (1991) pEmu: an improved promoter for gene expression in cereal cells. *Theor. Appl. Genet.,* **81** : 581-588.

Lee SH, Shon YG, Lee SI, Kim Ch Y, Koo J Ch, Lim Ch O, Choi Y J, Han Ch D, Chung Ch H, Choe ZR and Cho MJ (1999) Cultivar variability in the *Agrobacterium*-rice cell interaction and plant regeneration. *Physiol. Plant.,* **107** : 338-345.

Lyznik LA, Rao KV and Hodges TK (1996) FLP-mediated recombination of *FRT* sites in the maize genome. *Nucleic Acid Res.,* **24** : 3784-3789.

Li Z, Upadhyaya NM, Meena S, Gibbs AJ and Waterhouse PM (1997) Comparison of promoters and selectable marker genes for use in *Indica* rice transformation. *Mol. Breed.,* **3** : 1-14.

Lin W, Anuratha CS, Datta K, Potrykus I, Muthukrishnan S and Datta SK (1995) Genetic engineering of rice for resistance to sheath blight. *Bio/Technology,* **13** : 686-691.

Luo H, Lyznik LA, Gidoni D, Hodges TK (2000) FLP-mediated recombination for use in hybrid plant production. *Plant J.,* **23** : 423-430.

McElroy D and Brettell S (1994) Foreign gene expression in transgenic cereals. *TIBTECH,* **12** : 62-68.

McElroy D, Zhang W, Cao J and Wu R (1990) Isolation of an efficient actin promoter for use in rice transformation. *Plant Cell,* **2** : 163-171.

Mohanty A, Sarma NP and Tyagi AK (1999) *Agrobacterium*-mediated high frequency transformation of an elite indica rice variety Pusa Basmati 1 and transmission of the transgenes to R2 progeny. *Plant Sci.,* **147** : 127-137.

Nadolska-Orczyk A, Orczyk W and Przetakiewicz A (2000) *Agrobacterium*–mediated transformation of cereals – from technique development to its application. Acta Physiol. Plant., **22** : 77-88.

Nadolska-Orczyk A and Orczyk W (2000) Study of the factors influencing *Agrobacterium*-mediated transformation of pea (*Pisum sativum* L.). *Mol. Breed.,* **6** : 185-194.

Nayak P, Basu D, Das S, Basu A, Ghosh D, Ramakrishnan NA, Ghosh M and Sen SK (1997) Transgenic elite *indica* rice plants expressing CryIAc δ-endotoxin of *Bacillus thuringiensis* are resistant against yellow stem borer (*Scirpophaga incertulas*). *Proc. Natl. Acad. Sci. USA,* **94** : 2111-2116.

Ow DW and Medberry SL (1995) Genome manipulation through site-specific recombination. *Crit. Rev. Plant Sci.,* **14** : 239-261.

Park SH, Pinson RM and Smith RH (1996) T-DNA integration into genomic DNA of rice following *Agrobacterium* inoculation of isolated shoot apices. *Plant Mol. Biol.,* **32** : 1135-1148.

Pasquali G, Ouwerkerk PBF and Memelink J (1994) Versatile transformation vectors to assay the promoter activity of DNA elements in plants. *Gene,* **149** : 373-374.

Patel M, Johnson JS, Brettell RIS, Jacobsen J and Xue GP (2000) Transgenic barley expressing a fungal xylanase gene in the endosperm of the developing grains. *Mol. Breed.,* **6** : 113-123.

Pawlowski WP and Somers DA (1998) Transgenic DNA integrated into the oat genome is frequently interspersed by host DNA. *Proc. Natl. Acad. Sci. USA,* **95** : 12106-12110.

Puchta H (2000) Removing selectable marker genes: taking the shortcut. *Trends Plant Sci.,* **5** : 273-274.

Raineri DM, Bottino P, Gordon MP and Nester EW (1990) *Agrobacterium*-mediated transformation of rice (*Oryza sativa* L.). *Bio/Technology.,* **8** : 33-38.

Rashid H, Yokoi S, Toriyama K and Hinata K (1996) Transgenic plant production mediated by *Agrobacterium* in *Indica* rice. *Plant Cell Rep.,* **15** : 727-730.

Register JC III, Peterson DJ, Bell PJ, Bullock WP, Evans IJ, Frame B, Greenland AJ, Higgs NS, Jepson I, Jiao S, Lewnall CJ, Sillick JM and Wilson HM (1994) Structure and function of selectable and non-selectable transgenes in maize after introduction by particle bombardment. *Plant. Mol. Biol.,* **25** : 951-961.

Rhodes CA, Pierce DA, Mettler IJ, Mascarenhas D and Detmer JJ (1988) Genetically transformed maize plants from protoplasts. *Science,* **240** : 204-207.

Shen WH, Escudero J, Schlappi M, Ramos C, Hohn B and Koukolikova-Nicola Z (1993) T-DNA transfer to maize cells: histochemical investigation of β-glucuronidase activity in maize tissues. *Proc. Natl. Acad. Sci. USA,* **90** : 1488-1492.

Shimamoto K, Terada R, Izawa T and Fujimoto H (1989) Fertile transgenic rice plants regenerated from transformed protoplasts. *Nature,* **338** : 274-276.

Shimoda N, Toyoda-Yamamoto A, Aoki S and Machida Y (1993) Genetic evidence for an interaction between the VirA sensor protein and the ChveE sugar-binding protein of *Agrobacterium. J. Biol. Chem.,* **268** : 26552-26558.

Smith RH and Hood EE (1995) *Agrobacterium tumefaciens* transformation of monocotyledons. *Crop Sci.,* **35** : 301-309.

Srivastava V, Anderson OD and Ow DW (1999) Single-copy transgenic wheat generated through the resolution of complex integration patterns. *Proc. Natl. Acad. Sci. USA,* **96** : 11117-11121.

Stachel SE, Messens E, Van Montagu M and Zambryski P (1985) Identification of the signal molecules produced by wounded plant cells that activate T-DNA transfer in *Agrobacterium tumefaciens. Nature,* **318** : 624-629.

Tingay S, McElroy D, Kalla R, Fieg S, Wang M, Thornton S and Brettell R (1997) *Agrobacterium tumefaciens*-mediated barley transformation. *Plant J.,* **11** : 1369-1376.

Tinland B (1996) The integration of T-DNA into plant genomes. *Trends Plant Sci.,* **1** : 178-184.

Trick HN and Finer J (1997) SAAT: sonication-assisted *Agrobacterium*-mediated transformation. *Transgenic Res.,* **6** : 329-336.

Usami S, Morikawa S, Takebe I and Machida Y (1987) Absence in monocotyledonous plants of the diffusible plant factors inducing T-DNA circularization and *vir* gene expression in *Agrobacterium. Mol. Gen. Genet.,* **209** : 221-226.

Usami S, Okamoto S, Takebe I and Machida Y (1988) Factor inducing *Agrobacterium tumefaciens vir* gene expression is present in monocotyledonous plants. *Proc. Natl. Acad. Sci. USA,* **85** : 3748-3752.

Uze M, Potrykus I and Sautter C (1999) Single-stranded DNA in the genetic transformation of wheat (*Triticum aestivum* L.): transformation frequency and integration pattern. *Theor. Appl. Genet.,* **99** : 487-495.

Vasil IK (1994) Molecular improvement of cereals. *Plant Mol. Biol.,* **25** : 925-937.

Wenck A, Czako M, Kanevski I and Marton L (1997) Frequent collinear long transfer of DNA inclusive of the whole binary vector during *Agrobacterium*-mediated transformation. *Plant Mol. Biol.,* **34** : 913-922.

Yin Z and Wang GL (2000) Evidence of multiple complex patterns of T-DNA integration into the rice genome. *Theor. Appl. Genet.,* **100** : 461-470.

Yu SM, Tzou WS, Lo WS, Kuo YH, Lee HT and Wu R (1992) Regulation of α-amylase-encoding gene expression in germinating seeds and cultured cells of rice. *Gene,* **122** : 247-253.

Zambryski P, Joos H, Genetello C, Leemans J, Van Montagu M and Schell J (1983) Ti plasmid vector for the introduction of DNA into plant cells without alteration of their normal regeneration capacity. *EMBO J.,* **2** : 2143-50.

Zhang W and Wu R (1988) Efficient regeneration of transgenic plants from rice protoplasts and correctly regulated expression of foreign gene in the plants. *Theor. Appl. Genet.,* **76** : 835-840.

Zupan JR and Zambryski P (1995) Transfer of T-DNA from *Agrobacterium* to the plant cell. *Plant Physiol.,* **107** : 1041-1047.

Chapter 2

TRANSGENIC RICE

Rajinder Kumar Jain

Department of Biotechnology and Molecular Biology, CCS Haryana Agricultural University, Hisar 125 004, India.

Summary

Rice feeds more than a third of world's population. World population that stands at ~6 billion now is likely to stabilize at ~10 billion by 2050 with major increases being in rice consuming areas. Rice production needs to be more than doubled on less land and with less resources. Transgenic technology can greatly complement and strengthen rice breeding programs enabling the scientists to introduce of one or more genes irrespective of their origin into a leading cultivar without affecting its genetic background. During the last fifteen years, a rapid progress has been made towards the development of transformation methods in rice. Several transformation methods including Agrobacterium, biolistic, and DNA uptake by protoplasts, have been employed for transgenic plant production in rice. Novel genes for several important traits are now available and many of these have already been transferred in rice to develop resistance against insect pests, fungal, bacterial and viral diseases, tolerance against abiotic stresses such as drought, salinity and low temperature stresses, and to manipulate source-sink relationships, and improve the nutritional quality. The studies on rice transformation suggests that numerous factors such as genotype of plants, morphogenetic potential of the target tissues, method of transformation, constitutive versus targeted or tissue specific expression of transgenes, kind of reporter and selection marker genes, gene silencing, etc are of critical importance for the recovery

*Corresponding author : E-mail : rkjain@hau.nic.in

of fertile, genetically stable transgenic plants. The future role of transgenic research to engineer complex agronomic traits (e.g. yield, drought tolerance, plant type, grain quality) and metabolic or regulatory pathways involving many genes or gene complexes and in genomic analysis/functional genomics in rice has been discussed. Impact of issues such as biosafety and environmental concerns, public acceptance and patent rights, etc on rice transformation research has been highlighted.

Keywords : Rice transformation, *Agrobacterium*, particle bombardment, agronomic traits

1. INTRODUCTION

Rice is a primary food for more than a third of the world's population. Rice production and consumption is mainly concentrated in Asia where >90% of rice is produced and consumed; China and India account for about 50% of the world's rice area and 56% of production (David, 1991; Hossain, 1996; Hossain and Pingali, 1998). World population that stands at ~6.0 billion now is likely to exceed 8 billion by 2020 and may stabilize at ~10 billion by 2050 (Evans, 1998). The population in rice consuming areas continues to increase at an exceptionally fast rate. Evans (1998) correctly warns that in 2050, 4 billion more people need to be fed; 800 million currently remain undernourished and there are increasing environmental problems associated with high input agriculture. Massive famines expected due to rapid growth in population from 3 billion in 1960 to 5 billion in 1986, were averted primarily because of the 'green revolution'. Scientists used plant breeding techniques to develop broadly adapted semi-dwarf varieties of wheat and rice that respond to fertilizer by investing a greater proportion of their photosynthates in grain; these varieties produced yields that were double or higher than their previous levels. Worldwide, rice production doubled in a 25-year period, from 257 million tons in 1965 to 520 million tons in 1990. But the spectre of food shortages looms once again with ever increasing population, loss of productive land to rising sea levels, salinization, erosion, urbanization, etc. To meet future needs, yield increases of 3% per year-which have never been achieved with any major food crop- will be needed. Any shortfall in rice-growing countries could be a disaster for food security. Toenniessen (1999) expressed his concerns about the future population growth and food shortages in Asia and Africa especially when the system is under constraint by the lack of funds for research and scientists access to newer technologies and genetic resources marred by various regulations and intellectual property rights. Many scientists feel that in rice a yield plateau has been achieved and to break this barrier using conventional breeding alone would be difficult.

Biotechnology and molecular biology research can greatly strengthen the rice breeding programs and help produce varieties with higher yield potential, excellent grain quality, resistance to biotic and abiotic stresses, with a new value-added trait and lowering of production costs. Application of biotechnology shall enable rice breeders to achieve results more quickly and efficiently and to attain breeding goals not feasible using conventional techniques. In the last fifteen years, a rapid progress has been made towards the basic and applied research in biotechnology and molecular biology in rice (Khush, 1995b, 1996; Khush and Toenniessen, 1991; IRGS4A, 2000). A useful knowledge base on rice molecular genetics, tissue culture and cellular biology has been established. In fact, progress towards development of molecular techniques has been more rapid with rice than any other cereal. Advantage has been taken of the facts that rice is a diploid (2n = 24) and has a relatively small genome (4.2×10^8 bp per haploid genome). In rice, saturated molecular maps using RFLP and PCR based markers have been prepared (Causse *et al.*, 1994; Kurata *et al.*, 1994; Temnykh *et al.*, 2000). Recently, two biotech companies, Monsanto (USA) and Syngenta (Switzerland) have announced that they have sequenced the rice genome. Transgenic rice plants containing useful alien genes has been produced and in some cases transgenic plants have been field-tested. A number of research groups are actively working on map-based cloning, transformation of important rice varieties using genes that can potentially improve the stress tolerance, and on basic research concerning gene expression, genome analysis/sequencing, etc.

Genetic transformation approaches provide novel ways to transfer useful genes in commercially important rice varieties without disrupting their otherwise desirable genetic make-up. Improvement in the transformation technology for rice including *indicas,* has been remarkable in the past few years (for reviews see Ayres and Park, 1994; Jain and Jain, 2000; Datta, 1999; Roy *et al.*, 2000; Toenniessen, 1991; Tyagi *et al.*, 1999). In addition to the direct gene-delivery methods (protoplast transformation, biolistic, electroporation), rice can also be transformed efficiently using the *Agrobacterium* method. An array of useful genes has been transferred in different rice varieties to improve its resistance/

tolerance against insect pests, fungal diseases, drought and salinity and to improve its nutritional quality.

In this chapter, the priorities and problems of rice breeding and the progress made towards the development, as well as application, of transgenic technology for the genetic improvement of different types of rices, have been briefly reviewed.

2. TAXONOMY OF RICE

Rice belongs to Gramineae or grass family. The genus *Oryza* to which cultivated rice belongs, has twenty-one wild and two cultivated species; nine of the wild species are tetraploid and the remaining diploid (Khush, 2000). The two cultivated rice species are: *O. sativa* (common rice) and *O. glaberrima* (African red rice). *O. sativa* is a tremendously variable species, more popular due to higher yield and better grain quality and has worldwide distribution; it originated in the humid tropics of South and Southeast Asia. Chinese have recognized two rice varietal groups, Hsein and Keng, which correspond to *indica* and *japonica* classification introduced by Kato *et al.* (1928). *Indica* and *japonica* varieties differ in many characters when typical varieties are compared but show overlapping variations. Morinaga (1954) proposed a third group to include bulu and gundil varieties of Indonesia under the name *javanica*; later *javanica* varieties were kept within the *japonica* group and referred to as tropical *japonicas* and the so-called typical *japonicas* are referred to as temperate *japonicas* (Glaszmann, 1987). *O. glaberrima* originated in Niger basin and still confined mainly to the middle Niger delta.

Glaszmann (1987) examined 1,688 rice cultivars from different countries for allelic variation at 15 isozyme loci and classified them into 6 groups. Group I corresponded to the *indica* and Group VI to the *japonica*. Group VI also included the *javanicas*. Groups II, III, IV, and V were atypical but were classified as *indicas* in the conventional classification. Groups II corresponds to early maturing, drought tolerant upland rices called as 'Aus' varieties (grown in Bangladesh and West Bengal state of India). Floating rices of Bangladesh and India called 'Ashinas' and 'Rayadas' belongs to groups III and IV, respectively. Group V includes aromatic rices of Indian sub-continent including the world famous high quality Basmati rices. Various levels of sterility are observed in the F_1 hybrids of inter-group crosses but not in the intra-group crosses. In general, Group VI *japonicas* have been reported to be more tissue culture responsive and Group I *indicas* were found recalcitrant. As a result, most successes of gene transfer have been reported in *japonica* rice varieties.

3. RICE BREEDING : RESEARCH PRIORITIES AND PROBLEMS

Rice breeders face a major challenge of improving the rice varieties being grown in widely diverse environments (Pingali and Hossain 1998). Aside from the irrigated areas, there are four other types of environments, based mainly on the water regime – rainfed lowland including tidal wetland, deepwater, and upland or dryland areas. David (1991) and Fischer and Cordova (1998) briefly reviewed the progress made in rice breeding and challenges ahead for sustainable rice production and food security. It was after 1966 that dramatic changes occurred in the nature of rice production in South and Southeast Asia. The key contributor was, IR8, the first of the modern indica varieties with an improved plant type - short, stiff strawed, fertilizer-responsive, and photoperiod-insensitive – bred at the International Rice Research Institute. Development and adoption of such modern rice varieties coupled with the higher application of fertilizers, an increase in rice harvested area, expansion of irrigated area and double- or triple-crop rice systems led to the increased rice productivity in South and Southeast Asia. Improvements in the initially released improved rice varieties, such as multiple resistance to insects and diseases, better grain quality, and shorter growth duration, were largely accomplished prior to the 1980s. Little progress has been made in incorporating tolerance to abiotic stresses such as drought and salinity. Much of the work in the later years can be characterized as maintaining the yield potential, which has not been surpassed substantially since IR48 was released, the only exception being the development of hybrid rice with 15-20% yield advantage. Let us look at rice-harvested area and production figures for Asia; rice production grew from 240 million t in 1966 to 513 million t in 1996, an increase of 114% over 30 years (Hossain and Pingali, 1998). The average rice yield in Asia increased from 2.1 t ha^{-1} in 1966 to 3.8 t ha^{-1} in 1996, an increase of 2.0% per year. In Asia, the rice-harvested area increased from 96 million ha in 1951 to 128 million ha in 1975 and has expanded very little since than, fluctuating around 132 million ha during the last decade.

The major rice insect pests, which cause economic losses, are: the stemborer, rice bugs, leaf folder, leaf and plant hoppers, gall midge and rice hispa. Several studies have reported yield looses ranging between 10 to 44% in rice due to insects in Asian countries (Ramaswamy and

Jatileksono, 1996). Herdt and Riley (1987) assessed the relative intensity of problems of insects and other diseases in rice production. The stemborer, brown planthopper, gall midge and leafhopper were among the most important insects in Southeast Asia and China, and in South Asia the gall midge, brown planthopper and yellow stemborer were most important. Tungro and other virus diseases appear to be much less important problems in Asian rice production and the other diseases such as bacterial leaf blight, sheath blight, and blast seem far more serious. Bacterial leaf blight has been a serious problem in eastern India, Bangladesh, China, and Indonesia; sheath blight has been much more damaging than any other disease in China; rice blast has caused severe yield losses in eastern and southern India, Bangladesh and China.

The data show that, among the technical constraints, abiotic constraints are more prominent and yield limiting than biotic constraints; and among abiotic constraints, drought, salinity and floods are the most devastating to the yield in rice (Hossain, 1996; Dey and Upadhyaya, 1996). In Asia including India, a significantly large land area has been degraded due to one or more of the above stresses. Such marginal lands are either less productive or remains un-utilized for agriculture. Drought, cold and submergence al-together cause a loss of about 8% of total rice production in Asia, amounting to 36 million tons. About 43% of rice area in Asia is completely dependent on rainfall and is largely prone to drought (Dey and Upadhyaya, 1996). In the remaining 53% irrigated area, rice crop is occasionally subjected to drought due to poor quality of water and inadequate water supply at critical stages. Almost every alternate year, drought appears as a major yield-limiting constraint in many parts of the South Asia region. According to Dey and Upadhyaya (1996), the yield loss due to drought (in relation to the yield loss due to all technical constraints) is 4% in Southern India (largely irrigated) and highest (22%) in Eastern India (largely un-irrigated). No such data are available for Northern India. However, rice cultivation in Northern India is restricted to irrigated areas only. According to an FAO study (Massoud 1974), salt affected soils occupy nearly 7% of the world's land area with great differences between continents, countries and climatic regions (Africa 3.5%, America 0.7-7.6%, Asia 21.0%, Australia 42.3%, Europe 4.6%).

Yield of rice varieties can be increased in two major ways :

(i) raise yield potential of irrigated as well as rain-fed rices, and

(ii) incorporate durable resistance to insect pests and diseases and tolerance to salinity, drought, low-temperature, floods, etc to minimize the losses.

Substantial yield gains in tropical rice have already been reaped. In fact growth in rice production, particularly in the irrigated system, has declined in the last decade. This decline has occurred mostly as a result of a reduction in yield growth from 2.3% per annum during 1967-85 to 1.4% during 1985-96 (Hossain and Pingali, 1998). It is becoming difficult to sustain earlier rates of growth in rice production because most farmers are now using the best modern varieties. Despite of higher inputs, further yield increases are becoming more difficult to achieve and also more costly in terms of input, particularly the labor. Growth in rice yields has slowed where irrigated ecosystems predominate, as in China, Indonesia, the Philippines, and Sri Lanka. Yield growth has turned negative in Japan and South Korea. In India, Bangladesh and Thailand with large rainfed areas, rice yield growth accelerated between1986-96 compared to that during the first two decades (1966-86) of green revolution. There is much less basic and applied research on rainfed than on irrigated rice. The progress in developing improved varieties for un-favourable environments, has been extremely slow and very little improvements in the productivity of these environments have occurred (Khush, 1995a). The national rice research capacity has grown stronger in many countries, and some measure of progress is apparent, dramatic changes in rice productivity at this stage are not expected with the conventional research approaches.

4. TRANSGENIC TECHNOLOGY

A rapid progress has been made towards the development of genetic transformation technology in rice. Initially, rice transformation was carried out by the direct gene-delivery methods such as DNA uptake into protoplasts by polyethylene treatment or electroporation and microprojectile bombardment. Hiei *et al.* (1994) reported that rice could also be efficiently transformed using *Agrobacterium*-mediated gene transfer method. Most success regarding rice transformation has been achieved in case of *japonica* rice cultivars where a number of potentially useful genes have been transferred and in some cases transgenic plants have been field tested for a number of years. Subsequently, a number of *indica, javanica* and Basmati rice varieties have also been transformed. Transformation technology can greatly strengthen rice breeding programs by transferring several of the available novel genes to improve its resistance against insect/pests, fungal diseases, salinity, drought, etc. (Figure 1). The other advantage of transformation is that a useful trait for example resistance to yellow stem borer, can be engineered in to an elite rice cultivar without altering its genetic background/ quality traits.

4.1. Plant regeneration from cells, tissues and protoplasts

Before targeting a particular rice variety to *in vitro* genetic transformation, there is a need to develop efficient procedure(s) for regeneration of green, fertile plants from explant tissues, cells or protoplasts with minimal somaclonal variation. Indeed, it is the totipotency of plant cells that underlies the efficiency of most plant transformation systems. *Japonica* rice varieties have shown to be more responsive to *in vitro* culture and plant regeneration than *indica* rice varieties. As a result, it has been easier to transform *japonica* rice varieties. However, a lot of research has been done to develop tissue culture protocols in *indica* rice varieties to achieve a workable efficiency of gene transfer, selection and regeneration of transformants. A number of different explants/ tissues/ cells have been used in rice depending upon the transformation method and genotype; these include protoplasts derived from embryogenic cell suspensions (Datta *et al.*, 1990; Jain *et al.*, 1995, 1997a; Shimamoto *et al.*, 1989), embryogenic cell suspension cultures (Cao *et al.*, 1992; Jain *et al.*, 1996a, b, c, 1997b; Xu *et al.*, 1996a, b), immature embryos as such and calli initiated from them (Christou *et al.*, 1991; Hiei *et al.*, 1994; Jain *et al.* unpublished data) and embryogenic calli initiated from mature seed scutella (Hiei *et al.*, 1994), microspores (Datta *et al.*, 1990; Chair *et al.*, 1996), shoot apices (Park *et al.*, 1996), etc. Most success has been achieved in this regard by using the explants such as immature embryos with many regenerable cells, freshly initiated scutella-derived embryogenic calli or embryogenic cell suspension cultures. The choices were made based on the easy and round the year availability of the explant tissues with the option to keep tissue culture regime to the minimum. There have been remarkable improvements in media required for efficient plant regeneration in *indica* rice varieties (Jain, 1997; Jain *et al.*, 1997a; Ogawa, 2000). Recently, efficient protocols for microspore culture of some *indica* rice varieties have been reported (Bishnoi *et al.*, 2000; Raina and Zapata, 1997). Microspores directly or haploid calli/ suspensions derived from may have an added advantage as the starting material for the production of stable transgenic plants. Detailed consideration of optimization of tissue culture systems useful for rice transformation is beyond the scope of this review, but the progress made and protocols developed in this regard have been addressed in several

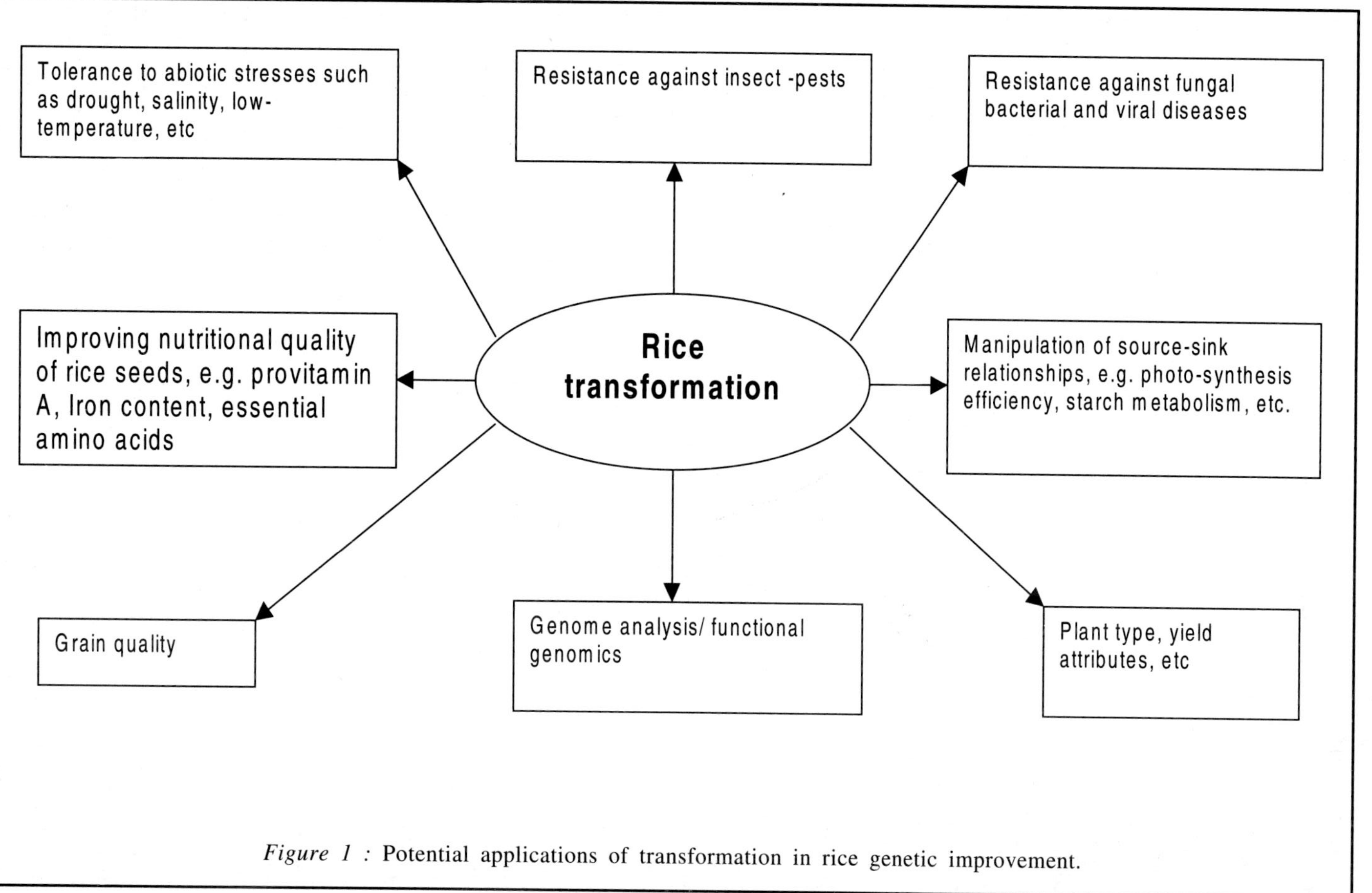

Figure 1 : Potential applications of transformation in rice genetic improvement.

reviews and technical manuals (Ayres and Park, 1994; Christou, 1994; Jain, 1997; Potrykus and Spangenberg, 1995).

4.2. Gene transfer methods

Protoplast-based direct gene transfer, biolistic transformation and *Agrobacterium*-mediated gene transfer are the major techniques that are being routinely used for transgenic plant production in rice. Other methods such as tissue electroporation (Chaudhury *et al.*, 1995; Arencibia *et al.*, 1998), laser mediation (Guo *et al.*, 1995), pollen tube pathway (Luo and Wu, 1989), DNA uptake by imbibition of dry embryos (Yoo and Jung, 1995), silicon carbide fiber-mediated transformation (Matsushita *et al.*, 1999), etc. have been reported to produce the transgenic rice plants but have not found widespread use. Several excellent reviews (Ayres and Park, 1994; Christou, 1997; Hiei *et al.*, 1997; Songstad *et al.*, 1995; Tyagi *et al.*, 1999) and laboratory manuals (Gamborg and Philips, 1995; Potrykus and Spangenberg, 1995) on transformation methods have been published. There are still many problems that have to be solved in terms of reproducibility. Most of the above mentioned methods could deliver DNA into the cell, but the subsequent events in cell, nuclei and organelles are not controlled and the integration of foreign DNA is random. Gene silencing and interactions between multiple copies of the same transgene or different transgenes result in unexpected expression pattern of foreign genes (Chareonpornwattana *et al.*, 1999; Depicker and Van Montagu, 1997; Kumpatla *et al.*, 1998; Matzke and Matzke, 1995; Morino *et al.*, 1999). Several independent transformants with a specific gene cassette are necessary, to find one transgenic plant with the desired transgene expression pattern and level.

4.2.1. Protoplast-mediated transformation

The first approach to transfer foreign DNA into rice cells was the direct gene transfer into protoplasts. Both polyethylene glycol (PEG) and electroporation methods have been used for rice transformation (for review see Ayres and Park, 1994). Protoplasts are the ideal material for gene transfer. While the delivery of DNA into protoplasts has been quite easy, major problems relate to the establishment of embryogenic cell suspension cultures (source of protoplasts) and regeneration of green, fertile plants from protoplasts. Protoplast culture and regeneration procedure(s) are quite labor intensive and time consuming; for example it may take up to six months to establish embryogenic cell suspensions that are ideal for protoplast isolation. This method is strongly genotype dependent and has not been worked out for many elite *indica* rice varieties. In many cases, transgenic rice plants produced through protoplast transformation were reported to sterile, carry extensive somaclonal variation, or multiple copies of the transgene. In spite of these limitations, protoplasts have been successfully used for the production of transgenic plants containing agronomically useful genes in some responsive *japonica* and *indica* rice varieties (Table 1; Lin *et al.*, 1995; Irie *et al.*, 1996). This method of transformation is no longer a method of choice for rice transformation. However, protoplasts may have an advantage to accomplish the transfer of large amounts of DNA and organelles into a single cell.

4.2.2. Biolistic transformation

Biolistic method employs high-velocity metal particles to deliver biologically active DNA into regenerable plant cells, tissues and organs. This method depends far less on sophisticated tissue culture procedures and provides a fairly genotype-independent transformation system bypassing *Agrobacterium* host specificity and protoplast-related regeneration difficulties. Chances of tissue culture raised somaclonal variation can also be minimized using the explants exhibiting direct shoot regeneration with a minimum of callusing phase. The transformation of several elite *indica* rice varieties became possible due to the extended range of cell tissues/ explants that can be targeted by microprojectiles. In rice, biolistic method has been successfully used for the transfer of many useful genes in over 40 *japonica* and *indica* rice varieties (Table 1, Christou, 1997; Datta, 1999). In most cases, embryogenic calli or suspension cells derived from immature and mature embryos, were used as the target tissue. A number of factors including osmotic preconditioning of the target tissues, nature and size of metallic particles used for DNA delivery, humidity and temperature, DNA loading on microprojectiles, target cell distance, accelerating force and depth of penetration, have been reported to influence the biolistic transformation frequency (Christou, 1994, 1997; Jain *et al.*, 1996a). Biolistic has also been used for organelle transformation especially the plastids which is an attractive target for crop engineering due to the high level transgene expression (McBride *et al.*, 1995; Sakamoto *et al.*, 1998). Besides plasmids, RNA, YAC (yeast artificial chromosome) DNA clones, or even *E. coli* or *Agrobacterium* cells have been successfully used for biolistic transformation (for review see Siemens and Schieder, 1996). Biolistic method has some drawbacks also and that includes the high cost of the equipment and consumables, integration of transgene in high copy number, transgene rearrangement, and gene silencing.

4.2.3. *Agrobacterium*-mediated transformation

Several monocots including rice, are now among the plants that can be transformed routinely by *A. tumefaciens*. *Agrobacterium*-mediated gene transfer system is becoming the method of choice in rice due to its low cost, convenience, high transformation efficiency, high probability of single copy integration with minimal rearrangements, transfer of relatively large segments of DNA and high fertility of transgenic plants. In 1994, Hiei *et al.* reported an efficient *Agrobacterium*-mediated transformation procedure in *japonica* rice. Since then a rapid progress has been made towards the development of *Agrobacterium*-mediated transformation protocols in various rice varieties including *indicas* and several useful genes have been transferred using this approach (Aldemita and Hodges, 1996; Azhakanandam *et al.*, 2000; Hiei *et al.*, 1997; Rashid *et al.*, 1997; Roy *et al.*, 2000; Zhang *et al.*, 1997). In general, the procedure involved the cocultivation of actively dividing, embryogenic cells such as immature embryos, isolated shoot apices and calli induced from scutella in the presence of acetosyringone, which is a potent inducer of the virulence genes. Other rice tissues such as leaf segments, long-term culture cells or liquid suspension cells, have been reported to be unsuitable for *Agrobacterium*-mediated transformation (Hiei *et al.*, 1994; Vijaychandra *et al.*, 1995). Monocots either do not produce phenolic compounds such as acetosyringone or, if they do, the levels are insufficient to serve as a signal for the induction of *vir* gene(s). The addition of acetosyringone is essential for transformation in rice, but the concentration of acetosyringone may vary between different genotypes of rice (Roy *et al.*, 2000).

The development and use of super-virulent binary vectors containing virulence genes (*virB*, *virC* and *virG*) from pTibo542, the Ti plasmid of the supervirulent *Agrobacterium* strain A281, have been the keys to the success of rice transformation (Hiei *et al.*, 1997; Komari, 1989). Further, small-size (11-15 kb) hybrid vectors were developed for cloning the gene of interest. Recombinant vectors were transferred into an *Agrobacterium* strain using a helper plasmid through triparental mating (Komari *et al.*, 1996) and used for rice transformation. Toki (1997) developed a new pSMABuba binary vector for rice transformation, which is a modified version of pSAMB70 (Yamashita *et al.*, 1995). Ku *et al.* (1999) used EHA101 strain containing the pSMABuba vector for high-efficiency rice transformation of an intact maize *phosphoenolpyruvate carboxylase* gene (8 kb). A Bin19-derived binary vector, pKGH4, was used by Cheng *et al.* (1998) to transfer the *cryIA(b)* and *cryIA(c)* genes in to rice through *Agrobacterium* (LBA4404 and EHA105) transformation. Sakamoto *et al.* (1998) used the superbinary vector 'pIGI121Hm' to transfer chimeric fusion choline oxidase (COD) gene cassette with or without a transit peptide sequence in rice. Goto *et al.* (1999) transferred a soybean ferritin gene into rice using a modified binary vector, pGPTV-35S-*bar*. Chin *et al.* (1999) constructed *Ac* and gene-trap *Ds* vectors and introduced them in to rice through *Agrobacterium* transformation. Ye *et al.* (2000) developed pB19hpc, pZPsC and pZLcyH vectors carrying the genes of provitamin A (ß-carotene) biosynthetic pathway enzymes; these vectors were electroporated separately into *Agrobacterium* strain LBA4404. Park *et al.* (2000) reported that *Agrobacterium*-mediated transformation efficiency could be significantly increased by using shorter T-DNA and/or additional virulence genes (*virG*, *virGN54D*, *virE*, or *virE/virG*) in tobacco, cotton and rice. Of the various *vir* genes, *virE/virG* gene was more and consistently effective.

In conjunction with an improved system for *Agrobacterium*-mediated plant transformation, a new binary bacterial artificial chromosome (BIBAC) vector has been developed that is capable of transferring at least 150 kb of foreign DNA into the tobacco nuclear genome (Hamilton *et al.*, 1996; Hamilton, 1997). The BIBAC vector has the minimal origin of replication of both *E. coli* and *A. rhizogenes* Ri plasmids, and it can replicate as a single-copy plasmid in both *E. coli* and *A. tumefaciens*. The large BIBAC T-DNAs, in conjunction with the helper that carries additional copies of virulence genes, *virG* and *virE*, has been used for high-frequency transformation in dicots. A modified BIBAC vector (BIBAC4) containing 60 kb or 90 kb rice BAC clone has been similarly used for rice transformation (Garg *et al.*, 1999). The ability to introduce high molecular weight DNA into plant chromosomes should accelerate the studies on gene identification, genome organization and genetic engineering of complex polygenic traits (Hamilton *et al.*, 1996).

Although T-DNA transfer has been assumed to only involve DNA sequences between the right border and the left border, however, evidences from some recent reports suggests that DNA sequences residing outside the T-DNA (non-T-DNA) may also be occasionally transferred to the plants using binary vectors (Ooms *et al.*, 1982; Ramanathan and Veluthambi, 1995; Wenck *et al.*, 1997). Yin and Wang (2000) reported the presence of non-T-DNA sequences in 33% of independent transgenic rice plants suggesting that the high frequency of non-T-DNA transfer should be considered when producing transgenic rice for commercial production. Hanson *et al.* (1999) reported a method to enrich an *Agrobacterium*-transformed population of plants containing only T-DNA sequences. The system involved the use of a binary vector having a lethal gene (CaMV 35S-*barnase* gene with an intron in the coding sequence) located outside the left border, which significantly reduced the frequency of transgenic plants having non-T-DNA sequences.

4.3. Choice of the promoter

A promoter drives the expression of a gene and may be a key factor in determining the success of a particular transformation experiment. Constitutive promoters direct the expression of genes in almost all tissues and are independent of any environmental or developmental conditions. The rice *Act1* (McElroy *et al.*, 1990, 1991, 1995), maize *ubiquitin* (Christensen *et al.*, 1992; Cornejo *et al.*, 1993) and chimeric maize *emu* (Chamberlain *et al.*, 1994) are strong promoters for gene expression in rice and have been used widely for the production of transgenic rice plants with higher-level, constitutive expression of transgene(s) (Toki, 1997; Cheng *et al.*, 1998; Cornejo *et al.*, 1993; Sakamoto *et al.*, 1998; Yokoi *et al.*, 1998; Nishizawa *et al.*, 1999). Li *et al.* (1997) compared the transient gene expression efficiency of several constitutive promoters; *Ubi1* was found to be the strongest followed by *Act1*, *Emu* and *CaMV35S*.

Besides constitutive promoters, tissue-specific promoters and inducible promoters have been used for rice transformation. Xu *et al.* (1993) and Duan *et al.* (1996) used potato protease inhibitor II (*PinII*) wound-inducible promoter to drive the *PinII* gene in transgenic rice. Yokoi *et al.* (1997) reported an anther-specific *Osg6B* promoter that directed tapetum-specific expression of the *gusA* gene in rice. Cheng *et al.* (1998) used a pollen-specific (*Bp10*) promoter to transform nine different *japonica* cultivars for high-level and tissue-specific expression of lepidopteran-specific δ-endotoxins. Okita *et al.* (1989) isolated the endosperm-specific promoters in rice. Wu *et al.* (1998) reported that all genes encoding the rice seed storage protein glutelin contain the GCN4 motif at similar sites in their 5′ flanking regions and demonstrated that the GCN4 motif functions as an essential cis-element for endosperm-specific gene expression and is activated by maize *Opaque-2* locus encoded protein (a transcriptor activator) in transgenic rice plants. The glutelin gene promoter, *GluB-1*, was used to drive the expression of the soybean ferritin transgene specifically in rice seeds (Goto *et al.*, 1999). Similarly, phytoene desaturase (*crtI*) gene was driven by the endosperm-specific glutelin promoter for provitamin A synthesis in rice grain (Ye *et al.*, 2000). Kyozuka *et al.* (1991) and Zhang and Wu (1988) reported anaerobic induction and tissue specific expression of maize *Adh1* (alcohol dehydrogense 1) gene promoter in transgenic rice plants. Su *et al.* (1998) developed an abscisic acid (ABA)-inducible promoter for rice transformation. Finally, the intron-mediated enhancement of gene expression in monocot cells, while increasing the activity of rice actin 1 (*Act1*), wound-inducible *pinII* or other stress-inducible promoters in transgenic rice, did not alter its pattern of regulation or expression (McElroy *et al.*, 1990, 1991; Su *et al.*, 1998; Wang *et al.*, 1997; Xu *et al.*, 1993).

4.4. Reporter and selectable marker genes

Several reporter and selection marker genes have been used to monitor transformation efficiency (for review see McElroy and Brettell, 1994). Reporter genes have also been used to analyze the expression potential of different promoters in transformed tissues. An ideal reporter gene should exhibit low background activity in plant tissues and should not have any detrimental effects on metabolism. The reporter gene product should have moderate stability *in vivo*, so that down-regulation of gene expression as well as gene activation can be detected. Finally, the reporter gene should be assayed by a non-destructive, quantitative and sensitive procedure. Reporter genes currently in use do not fulfill all these criteria. ß-glucuronidase (GUS) gene encoded by the *uidA* locus of *E. coli*, has been the most commonly used reporter gene in rice transformation (Jefferson *et al.*, 1987). The *gusA* gene with an intron in the N-terminal region of the coding sequence is more useful, because the chimeric *gusA* gene is expressed only in the plant cells and not in the cells of *A. tumefaciens* (Hiei *et al.*, 1994, 1997). The green fluorescent protein gene (*gfp*) has also been used as a reporter gene in rice transformation experiments (Chung *et al.*, 2000; Jang *et al.*, 1999; Harper *et al.*, 1999; Vain *et al.*, 1998). Baruah-Wolff (1999) used the firefly luciferase gene (*luc*) as a reporter gene for rice transformation. Use of *gfp* and *luc* genes provides the opportunity to recover the putative transformed tissue after its identification.

Identification and selection of transformation events generally rely on the differential growth of transformed versus non-transformed tissues in the presence of a selection agent. The selection process may take 6 to 10 weeks and require the use of different levels of the selection agent to eliminate escapes. A number of antibiotics and herbicides have been used as selective agents for rice transformation. Different rice varieties exhibit different sensitivities towards selection agents. The hygromycin phosphotransferase (*hpt*) gene has been widely used as a selectable marker gene in *Agrobacterium* transformation experiments (Hiei *et al.*, 1994; Jain *et al.*, 1996a; Ku *et al.*, 1999; Cheng *et al.*, 1998; Ye *et al.*, 2000). Herbicides are also potential selective agents. Genes have been isolated for their resistance to various commercially important herbicides. Among them, the *bar* gene, which encodes phosphinothricin acetyltransferase, is the most commonly used selectable marker gene in rice (Cao *et al.*, 1992; Christensen *et al.*, 1992; Goto *et al.*, 1999; Park *et al.*, 1996; Rathore *et al.*, 1993; Toki, 1997). The *bar* gene confers resistance to L-phosphinothricin (PPT), glufosinate (an ammonium salt of PPT) and bialaphos (a derivative of PPT).

New and benign selectable marker systems for plants using mannose and cytokinin glucuronides have been developed that may replace antibiotic resistance and other markers currently in use (Joersbo *et al.*, 1998). The selection strategy involves the use of mannose that cannot be metabolized by many plant species. After uptake, mannose is phosphorylated by a hexokinase to mannose-6-phosphate that accumulates resulting in severe growth inhibition. Introduction of a phosphomannose isomerase gene allows transgenic cells/ shoots to regenerate on mannose containing media enabling mannose to be used as selective agent. Another selection method in this category employs a ß-glucuronidase gene as selectable as well as screenable gene (Okkels *et al.*, 1997). Cytokinin glucuronides are used as selective agents, which are inactive as cytokinins, but, upon hydrolysis by β-glucuronidase, active cytokinin is released stimulating regeneration of the transformed cells.

4.5. Fertility, genetic stability and transgene expression in transgenic plants

Stable inheritance and expression of foreign genes are of critical importance for the successful application of genetically engineered rice in agriculture. A variety of molecular and biochemical tools including polymerase chain reaction (PCR), quantitative PCR, reverse transcription PCR, southern hybridization, northern hybridization, ELISA, and western immunoblot have been used to confirm the presence, copy number and expression of transgenes in rice (for review see Cao *et al.*, 1991; Register

III, 1997). The next step is to obtain a complete picture of transgene inheritance, segregation and expression and trait development that may require use of multiple complementary molecular, biochemical and breeding approaches.

Normally, insertion of one or a few foreign genes should not adversely affect the agronomic characteristics of the host cultivars except for the expression of the introduced genes. However, large-scale sterility problems have been observed in transgenic plants especially those obtained through protoplast transformation (Li *et al.*, 1990; Datta *et al.*, 1992). Some of the problems with fertility may be due to the use of suspension cultures that were too old (Jain, 1997). Most of the transgenic rice plants obtained using *Agrobacterium* and biolistic transformation methods have been reported to show normal morphology, and higher degree of fertility (for review see Christou, 1997; Hiei *et al.*, 1997). Although no reports are yet available on a systematic comparison of protoplast, biolistic and *Agrobacterium* transformation methods with regard to the number of copies of integrated genes, gene rearrangement, gene silencing, etc, but published literature do suggests *Agrobacterium* method having an edge over the other transformation procedures in producing the low-copy number, normal transgenic plants.

4.6. Production of marker-free transgenic plants

Selectable marker genes are always included in transformation systems, but they are not usually required once transgenic plants have been produced. In fact, commercialization of transgenic plants has generated a debate regarding the desirability and biosafety implications of the transgenic products containing selectable marker genes (Flavell *et al.*, 1992; Goldsbrough, 1992; Malik and Saroha, 1999; Sawahel, 1994; Yoder and Goldsbrough, 1994).

Several transformation techniques/strategies for eliminating marker genes have been developed; they include (1) co-transformation, (2) a site-specific recombination system, (3) intra-genomic relocation of transgenes via transposable elements and (4) use of non-hazardous selectable marker systems. The co-transformation system appears to be the simplest among them and is widely used in plant transformation systems. In co-transformation, a plant cell is transformed with two separate DNAs, one incorporating a gene of interest and the other containing the selectable marker gene. Consequently, the gene of interest segregates from the marker gene in successive generations. Novel super-binary vectors carrying two separate T-DNAs for co-transformation have been constructed (Komari *et al.*, 1996). One T-DNA contains a drug-resistance, selectable marker gene and the other contains the ß-glucuronidase (GUS) gene. More than 50% of the co-transformants were GUS-positive and drug sensitive. Both the efficiency of co-transformation and the frequency of unlinked integration of the two T-DNAs into plants were reasonably higher through the *Agrobacterium* transformation method than a direct gene delivery method. In site-specific recombination, the gene of interest is cloned between the asymmetric inverted repeat sequences, which are the sites for specific recombination. The recombination event is identified in the transformants by selecting the presence of cloned genes (Yoder and Goldsbrough, 1994). Zubko *et al.* (2000) reported intra-chromosomal recombination between *attP* regions to remove selectable markers genes. Maize transposable elements *Ac/Ds* maintain transposition competence when transferred to other plant cells and provide a useful alternative to carrying out multiple independent transformations in order to achieve optimal transgene expression. Selectable markers can be removed via transposition of the *Ds*; only the gene of interest flanked the T-DNA repeats remains in the plant (Chin *et al.*, 1999). Recently, new non-hazardous selectable marker systems have been developed which need not to be removed from the transgenic plants (for details see the section 4.4).

4.7. Gene silencing

Transgenic research in plants has revealed that many factors can prevent the gene expression such as, the transgene may be rearranged, it may not be linked in the right order to sequences that control its activity, or it may be present and intact but somehow 'silenced'. Transgene silencing has been a focus of research for several years and only now is mechanisms by which this occurs starting to be understood. While both the copy number and the position of the integrated gene(s) can influence gene expression in transgenic plants, other explanations have been invoked to account for the non-mendelian inheritance of introduced genes, and for the observations that some plants do not express all copies of the integrated foreign gene (Kumpatla *et al.*, 1998; Vaucheret *et al.*, 1998). Transgene-induced gene silencing can occur at the transcriptional or post-transcriptional level. Transcriptional gene silencing (TGS) occurs when multiple repeats of a transgene are inserted in the genome of transgenic plants. In rice transformation, especially by using direct gene transfer methods (protoplast transformation, biolistic), there is a tendency towards the integration of multiple copies of the introduced genes. Although methylation, especially of repeated sequences, is widely associated with gene inactivation, other attributes, including chromatin modification, may be involved. Kumpatla *et al.* (1997) reported that treatment of R_2 progeny of silenced plants with 5-azacytidine resulted in demethylation of the *Ubi1* promoter and reactivation of *bar* gene expression, demonstrating a functional relationship for methylation in gene silencing. The recognition and inactivation of introduced genes might be regarded as a defense mechanism that protects plants against the expression of potentially deleterious foreign DNA.

Post-transcriptional gene silencing mechanisms affect mRNA stability or turnover, and in many cases have been shown to cause degradation of homologous mRNAs of viral, endogenous and introduced genes (Vaucheret *et al.*, 1998). Such a mechanism can be conceived as a host-defense response to 'intrusive' RNA, and has been hypothesized to evolve as a response to viral RNA (Ratcliff *et al.*, 1997). Morino *et al.* (1999) reported post-transcriptional silencing of aleurone-specific *Ltp2-gus* transgene in transgenic rice that may have caused by interaction of partial antisense *gus* transcripts with normal sense transcripts.

Several strategies have been suggested to avoid the silencing of introduced DNA and these includes: (i) use of the gene constructs with as little DNA sequence similarity as possible to putative endogenous sequences, or to similar sequences in the same construct (Kumpatla *et al.*, 1998), (ii) by preferably using the *Agrobacterium* method of transformation (Hiei *et al.*, 1997), (iii) selecting the transgenic plants with only one copy of the transgene (Kumpatla *et al.*, 1998), or (iv) by using the MAR (Matrix Attachment Regions) sequences which has been shown to reduce the position effect (Spiker and Thompson, 1996; Holmes-Davis and Comai, 1998). Transgene silencing is a matter of great concern for the application of transgenic technology in developing genetically stable value-added transgenic plants (Matzke and Matzke, 1995).

4.8. Novel useful genes transferred in rice

Since the first report of fertile transgenic plant production in 1989, remarkable progress has been made towards the transfer of useful genes in *japonica* as well as *indica* rice varieties. A variety of useful genes conferring resistance against abiotic and biotic stresses have been transferred in rice (Table 1). The Table 1 does not contain the exhaustive list of the reports on rice transformation, but it serves to illustrate the type of genes that can be used for the genetic improvement of rice.

Table 1. A list of useful genes transferred in rice

Useful trait/Gene		Transformation method	Rice type	Reference
Insect and pest control				
Bt, δ-endotoxin	*cry1A(b)*	Protoplast electroporation	Japonica	Fujimoto *et al.*, 1993
	cry1A(b)	Biolistic	Indica	Wünn *et al.*, 1996; Alam *et al.*, 1998
	cry1A(b)	Biolistic	Aromatic Iranian	Ghareyazie *et al.*, 1997
	cry1A(b)	Biolistic	Indica, Japonica	Datta *et al.*, 1998
	cry1A(b)	Biolistic	Japonica	Wu *et al.*, 1997
	cry1A(b) & *cry1A(c)*	*Agrobacterium*	Japonica	Cheng *et al.*, 1998; Shu *et al.*, 2000
	cry1A(c)	Biolistic	Indica	Nayak *et al.*, 1997; Gahakawa *et al.*, 2000
	cry2A	Biolistic	Indica, Basmati	Maqbool *et al.*, 1998; Gahakawa *et al.*, 2000
Potato protease inhibitor II *(pinII)*		Biolistic	Japonica	Duan *et al.*, 1996
		Biolistic	Basmati	Jain *et al.*, 1996a
Cowpea trypsin inhibitor (*CpTi*)		PEG-Protoplast	Japonica	Xu *et al.*, 1996a
Corn cystatin (*cc*)		Protoplast electroporation	Japonica	Irie *et al.*, 1996
Oryzacystatin (*oc*)		Protoplast transformation	Japonica	Hosoyama *et al.*, 1995
Snowdrop lectin (*GNA*)		Protoplast-electroporation	Japonica	Rao *et al.*, 1998
Nematode resistance				
OC-IΔD86		Biolistic	African Indica	Vain *et al.*, 1998

Table 1. Continued

Useful trait/Gene	Transformation method	Rice type	Reference
Bacterial leaf blight resistance			
Xa21	Biolistic	Japonica	Wang *et al.*, 1996
	Biolistic	Indica	Tu *et al.*, 1998, 2000; Zhang *et al.*, 1998
CecB	*Agrobacterium*	Japonica	Sharma *et al.*, 2000a
Fungal diseases,			
1. Sheath blight			
Chitinase gene, *Chi11*	PEG-Protoplast	Indica	Lin *et al.*, 1995
Chitinase gees, *cht-2, cht-3*	*Agrobacterium*	Japonica	Nishizawa *et al.*, 1999
2. Rice blast			
Grapevine stilbene synthase gene	PEG-Protoplast	Japonica	Stark-Lorenzen *et al.*, 1997
Trichosanthin	*Agrobacterium*	Japonica	Ming *et al.*, 2000
Resistance against viral diseases			
RNA2, RNA3 & capsid genes from BMV	Protoplast electroporation	Japonica	Huntley and Hall, 1996
Outer coat protein (*S8*) gene of RDV	Biolistic	Japonica	Zheng *et al.*, 1997
RTSV coat protein genes	Biolistic	Indica, Japonica	Sivamani *et al.*, 1997

Table 1. Continued

Useful trait/Gene	Transformation method	Rice type	Reference
RNA-dep. RNA Polymerase of RYMV	Biolistic	African Indica	Pinto *et al.*, 1999
Salt, drought and cold tolerance			
P5CS	Biolistic	Japonica	Zhu *et al.*, 1997
mt1D, gutD	*Agrobacterium*	Japonica	Wang *et al.*, 2000
HVA1	Biolistic	Japonica	Xu *et al.*, 1996b
	Biolistic	Basmati	Jain *et al.*, 1996a
codA	*Agrobacterium*	Japonica	Sakamoto *et al.*, 1998
adc	Biolistic	Japonica	Capell *et al.*, 1998, 2000; Noury *et al.*, 2000
GPAT	*Agrobacterium*	Japonica	Yokoi *et al.*, 1998
OsCDPK7	*Agrobacterium*	Japonica	Saijo *et al.*, 2000
Nutritional quality			
1. Provitamin A			
Daffodil phytoene synthase, Bacterial phytoene desaturase, daffodil lycopene β cyclase gene cassettes	Biolistic	Japonica	Burkhardt *et al.*, 1997; Ye *et al.*, 2000
2. Iron content			
Soybean ferritin gene	*Agrobacterium*	Japonica	Goto *et al.*, 1999

Table 1. Continued

Useful trait/Gene	Transformation method	Rice type	Reference
Dwarfism			
Rgp1	Protoplast electroporation	Japonica	Kisaka *et al.*, 1998
Photosynthetic efficiency			
PEPC (maize)	*Agrobacterium*	Japonica	Ku *et al.*, 1999
PCK (*Urochloa panicoides*)	*Agrobacterium*	Japonica	Suzuki *et al.*, 2000
Starch metabolism			
E. coli glgC-TM gene	*Agrobacterium*	Japonica	Sakulsingharoj *et al.*, 2000; Zhang *et al.*, 1996

adc, arginine decarboxylase gene; *Bar,* phosphinothricin acetyltransferase gene; *CecP,* cecropin B (antibacterial peptide) gene from *Bombyx mori*; *CodA,* choline oxidase gene from a soil bacterium *Arthrobacter globiformis*; *E. coli glgC-TM,* a gene for mutant ADP glucose unregulated by allosteric factors; *gutD,* glucitol-6-phosphate dehydrogenase gene; *HVA1,* Barley late embryogenesis-abundant (LEA 3) protein gene; *mtlD,* mannitol-1-phosphate dehydrogenase gene; *OC-IΔD86,* an engineered cysteine proteinase inhibitor (Oryzacystatin-IDΔ86) gene; *OsCDPK7,* a riceCa^{2+}-dependent protein kinase gene; *PCK,* phosphoenolpyruvate carboxylase gene; *P5CS,* D^1-pyrroline-5-carboxylate synthetase gene; *Rgp1,* a small GTP –binding protein gene from rice; RTSV, rice tungro spherical virus; RYMV, rice yellow mottle virus.

4.8.1. Resistance against insect/pests

The δ-endotoxin crystal protein genes from *Bacillus thuringiensis* and plant defensive genes, protease inhibitors and lectins, are among the insect resistance genes (Estruch *et al.*, 1997; Peferoen, 1997; de Maagd *et al.*, 1999; Ryan, 1990; Schuler *et al.*, 1998; Sharma *et al.*, 2000b) that have been transferred in *japonica* and *indica* rice varieties (Table 1). The transgenic plants have been reported to be genetically stable and display the improved resistance against a variety of stem borers, leaf-hoppers or leaf folder insects.

B. thuringiensis δ-endotoxin crystal (Cry) proteins are part of a large and still growing family of homologous proteins. About 130 *cry* genes have been identified to date. Most *Bt* crystal proteins are synthesized in a protoxin form and are proteolytically converted into smaller toxic polypeptides in the insect midgut. These toxins kill insects by binding to midgut membranes causing lesions. These *cry* genes have been recently classified on the basis of amino acid homology into four major classes, *cry1*, *cry2*, *cry3* and *cry4*, which are, respectively, specific for Lepidotera (caterpillars), Lepidotera and Diptera (flies and mosquitos), Coleoptera (beetles) and Diptera. In rice, *cry* genes belonging to first two classes have been used for transgenic plant production. Fujimoto *et al.* (1993) produced the transgenic japonica rice plants containing truncated and codon-modified δ-endotoxin gene using the protoplast-electroporation method. The R_2 generation transgenic plants were more resistant to striped stem borer (SSB) and leaf-hopper than the non-transgenic plants. Wünn *et al.* (1996) reported increased insecticidal effect on several lepidoteran insect pests (yellow stem borer, striped stem borer and leaf folder) in the R_0, R_1 and R_2 generation transgenic IR58 plants expressing the synthetic *cry1A(b)* gene. Datta *et al.* (1998) produced transgenic plants of several *indica* and *japonica* rice varieties containing *cryIA(b)* gene driven by either constitutive (*CaMV35S, Actin 1*) or tissue-*specific (PEP carboxylase, pith specific*) promoters using the biolistic and protoplast transformation systems. The transgene, *cry1A*, driven by different promoters showed a wide range of expression (low to high) but conferred enhanced resistance to yellow stem borer (YSB). Out of 800 Southern-positive plants that were bio-assayed, 81 transgenic plants showed 100% mortality of insect larvae of the YSB. Cheng *et al.* (1998) reported the production of over 2600 transgenic japonica rice plants containing the *cry1A(b)* or *cry1A(c)* synthetic gene driven by maize *ubiquitin*, *CaMV35S* and *Brassica Bp 10* gene promoters using *Agrobacterium*-mediated transformation method. The transgenic R_0

plants accumulated the CryIA(b)/CryIA(c) protein at higher levels (up to 3% of soluble protein). The R_1 generation transgenic plants were highly toxic to SSB and YSB with mortalities of 97-100%. Shu *et al.* (2000) reported that transgenic plants of rice variety KMD1 with a synthetic *cry1Ab* gene under the control of a maize ubiquitin promoter were highly resistant to eight lepidopteran rice pest species. The eight lepidopteran pest species were: four *Pyralidae* species: *Chilo suppressalis* (striped stem borer, SSB), *Scirpophaga incertulas* (yellow stem borer, YSB), *Cnaphalocrocis medinalis* (leaf folder), *Herpitogramma licarisalis*; two Noctuidae: *Sesamia inferens* (pink stem borer, PSB) and *Naranga anescens*; one Stayridae: *Mycalesis gotama*; and one Hesperiidae, *Parnara guttata*. Ghareyazie *et al.* (1997) reported the transfer of the *cryIA(b)* gene driven by the tissue-specific maize C_4 PEP carboxylase gene promoter in an Iranian aromatic rice variety Tarom Molaii belonging to isozyme group V. The transgene was expressed in the leaf blades but was not expressed to a detectable level in dehulled mature grain. Transgenic plants showed enhanced resistance against to first-instar larvae of SSB and YSB. Maqbool *et al.* (1998) reported the production of Basmati 370 and M7 expressing the *cry2A* gene via biolistic method. The gene product was expressed up to 5% of total leaf protein and the transgenic plants had higher insecticidal activity against YSB and rice leaf folder. Gahakwa *et al.* (2000) reported stable inheritance and expression of *cry1Ac* and *cry2A* genes over four generations of transgenic rice plants; the primary transformants showed moderate or low-level expression.

The other classes of genes that have been used for improving insect/pest resistance are the protease inhibitor or lectin genes of plant origin, whose expression have shown anti-metabolic effects against certain insects. Expression of protease inhibitor genes have been useful for control of insects which feed by chewing plant tissues, such as insects belonging to Lepidotera and Coleoptera, while that of lectin genes have been toxic to the sap-sucking insects belonging to the order Homoptera (brown plant hopper, green plant hopper). Duan *et al.* (1996) and Xu *et al.* (1996b) produced transgenic *japonica* rice plants containing potato proteinase inhibitor II (*pinII*) and cowpea trypsin inhibitor (*CpTi*) genes, respectively. Transgenic plants had high level accumulation of the PinII/CpTi proteins and their progenies showed increased resistance against pink stem borer and/or SSB. Hosoyama *et al.* (1995) and Irie *et al.* (1996) produced transgenic rice plants oryzacystatin (*oc*) and corncystatin (*cc*), respectively, which showed potent inhibitory activity against the cysteine proteinases that occur in the gut of insect pests. Vain *et al.* (1998) reported increased nematode resistance in transgenic plants containing an engineered cysteine proteinase inhibitor (*Oryzacystatin – IΔD86*) gene. These reports clearly indicate that proteinase inhibitor genes can be used to engineer the resistance against insect pests and nematodes in rice.

Among the various plant lectins, snowdrop lectin (*Galanthus nivalis* agglutinin: GNA) has been reported to be most toxic to the brown plant hopper (BPH) and non-toxic to mammals (Boulter *et al.*, 1990). GNA is a tetrameric protein consisting of identical sub-units of 12 kD. Rao *et al.* (1998) reported the production of transgenic rice plants containing snowdrop lectin gene driven by either phloem-specific rice sucrose synthase-1 gene promoter (*RSs1*; Shi *et al.*, 1994) or the constitutive maize ubiquitin-1 promoter (*Ubi*). Phloem-specific promoters have been used to specifically express this gene in phloem tissue of transgenic rice plants conferring resistance against sap-sucking insects. Some of the transgenic plants had GNA at levels of up to 2.0% of total protein. Insect bioassays and feeding studies showed that GNA expressed in the transgenic plants decreased the survival, retarded the insect development, and had a deterrent effect on BPH feeding.

4.8.2. Resistance against fungal diseases

In response to fungal attack, plants synthesize an assortment of new proteins commonly known as pathogenesis-related (PR) proteins including chitinases and β-1,3-glucanases (Linthorst, 1991). Chitinase, which hydrolyzes the polymer chitin, have been reported to inhibit fungal growth *in vitro* (Schlumbaum *et al.*, 1986). Transgenic approach based on constitutive expression of a chitinase gene in transgenic tobacco have been shown to result in an increased ability to survive in soil infested with fungal pathogens and delayed development of disease symptoms (Broglie *et al.*, 1991). Lin *et al.* (1995) developed transgenic *indica* rice plants that expressed chitinase gene constitutively. The plants with high levels of chitinase exhibited increased resistance to infection by the sheath blight pathogen, *R. solani*. Nishizawa *et al.* (1999) reported the transfer of class I chitinase genes, *cht-2* and *cht-3*, through *Agrobacterium* transformation in *japonica* rice cultivars Nipponbare and Koshihikari. R_0 transgenic rice plants that constitutively expressed either chitinase gene under the control of the 35S CaMV promoter, showed significantly higher resistance to the rice blast pathogen *Magnaporthe grisea*. Both high-level expression of chitinase and the blast resistance characteristic were stably inherited in the next generation. Datta *et al.* (2000) reported *Agrobacterium*-mediated transfer of rice class-I chitinase gene in *indica* rice varieties to enhance sheath blight resistance.

Stark-Lorenzen *et al.* (1997) reported active transcription of grapevine stilbene-synthase- gene in transgenic rice plants after incubation with the fungus of the rice blast *P. oryzae*. Preliminary results indicated an enhanced resistance of the transgenic rice to *P. oryzae*. Stilbene synthase in some plant species synthesize a phytoalexin trans-resveratrol that seems to have a role in the early protection of plants against fungal pathogens. The effectiveness of stilbene synthase genes in enhancing resistance to another fungal pathogen, *Phytophthora infestans*, has been demonstrated in transgenic tomatoes (Thomzik, 1997). Ming *et al.* (2000) reported a delay in the infection of rice blast pathogen in transgenic rice plants expressing the trichosanthin gene. Tada *et al.* (1996) reported decreased symptoms of rice blast disease caused by *Magnporthe grisea* following bialaphos treatment in transgenic rice plants expressing *bar*-gene.

4.8.3. Resistance against bacterial leaf blight

A significant progress has been made towards the genetic engineering for BLB resistance in rice. A dominant gene for resistance to BLB was transferred from a wild species, *O. longistaminata*, to the cultivated variety "IR24" (Khush *et al.*, 1990). This gene designated as *Xa21*, confers resistance to all known races of *Xanthomonas oryzae* pv. *Oryzae* (Xoo). The *Xa21* encodes a receptor kinase-like protein, which may have a role for cellular signaling in plant disease resistance. Wang *et al.* (1996) reported that transgenic rice plants expressing *Xa21* gene conferred multi-isolate resistance to 29 diverse isolates from eight countries indicating that a single cloned gene is sufficient to confer multi-isolate resistance. Tu *et al.* (1998) reported higher resistance against two prevalent races (4 and 6) of Xoo in transgenic IR72 plants containing *Xa21* gene. Resistance against race 4 was higher due to the pyramiding of transgene *Xa21* and *Xa4* (already present in IR72). Zhang *et al.* (1998) produced transgenic *indica* (IR64, IR72), hybrid restorer line Minghui 63 and BG90-2 rice lines containing *Xa21* gene; about 10% of the transgenic R_0 plant lines displayed high levels of resistance to the pathogen. Tu *et al.* (2000) reported the results of field-testing of a promising IR72 homozygous line for *Xa21* gene to evaluate its levels of resistance to the BB pathogen using the *Xoo* isolates from the Philippines,

Japan and China. The results demonstrated that the transgenic homozygous line expressed the same resistance spectrum, but with a shorter lesion length to each inoculated isolates as the lesion length of the *Xa21* donor line IRBB21. Sharma *et al.* (2000) reported enhanced resistant to bacterial blight in transgenic rice producing cecropin B, an antibacterial peptide from *Bombyx mori*.

4.8.4. Viral resistance

Research efforts have been made to transfer genes for resistance against viruses (RTSV, rice tungro spherical virus; RTBV, rice tungro bacilliform virus; RDV, rice dwarf virus) in rice. A highly successful strategy, termed coat protein (CP)-mediated, has been employed for resistance against viral diseases. The chimeric gene consisting of a strong promoter with a virus gene for capsid protein, when expressed in the transgenic plant results in the accumulation of capsid protein in plant cells. Such plants are resistant to infection by the virus from which the gene stems (Beachy, 1990). Zheng *et al.* (1997) developed the transgenic plants expressing the rice dwarf virus coat protein gene (*S8*), which is an important step towards studying the function of RDV genes and obtaining RDV-resistant rice plants. Sivamani *et al.* (1999) produced transgenic *indica* and *japonica* rice plants containing RTSV coat protein genes. Most of transgenic plants, as well as their R_1, R_2 and/or R_3 progeny that contained the target gene, showed the moderate levels of protection to RTSV infection and a significant delay of virus replication. Similarly, Huntley and Hall (1996) reported interference with Brome Mosaic Virus (BMV) replication in some of the transgenic rice plants expressing BMV RNAs.

Pinto *et al.* (1999) reported that resistance against the disease caused by rice yellow mottle virus (RYMV) can be engineered in sensitive African rice varieties by transferring a transgene encoding the RNA-dependent RNA polymerase of RYMV. In the most resistant transgenic line, the transcription analysis indicated that the resistance derives from an RNA-based mechanism associated with post-transcriptional gene silencing.

4.8.5. Abiotic stress tolerance

Plants respond to environmental stresses such as salinity, drought and freezing at the molecular and cellular levels as well as at the physiological level. These stresses induce the expression of a variety of genes whose products are thought to function not only in stress tolerance but also in the regulation of gene expression and in the signal transduction (Yamaguchi-Shinozaki and Shinozaki, 1999). These gene products have been classified into two groups :

(i) The proteins that have a function in protecting cells from stress condition, *e.g.* protection factors of macromolecules such as LEA proteins, key enzymes for osmolyte biosynthesis, detoxification enzymes, membrane proteins, etc. and

(ii) Protein factors that are further involved in regulation of gene expression and signal transduction, *e.g.* transcription factors and protein kinases.

Detailed discussion on the role of these genes in stress tolerance is beyond the scope of this review. Many of these genes have already been transferred in to model plant species '*Arabidopsis*, tobacco and rice' to improve the tolerance to different abiotic stresses (for review see Bartels and Nelson, 1994; Bohnert and Jensen, 1996a,b; Hasegawa *et al.*, 2000; Holmberg and Bülow, 1998; Ito *et al.*, 1999; Zhu *et al.*, 1997). A few examples of genes transferred in rice are as given below:

4.8.5.1. Late-embryogenesis abundant protein (LEA) genes **:** LEA genes are normally expressed in the seed during maturation stage that involves desiccation and also in vegetative tissues during water deficit. *LEA* genes are induced by ABA and by osmotic stress resulting from drought, salinity stress, or cold temperatures. LEA proteins have long been suggested as important in water retention or protection, ion sequestration, and as molecular chaperones, although their precise functioning is not clear. Xu *et al.* (1996a) produced the transgenic *japonica* rice plants containing barley *LEA3* (*HVA1*) gene driven by a constitutive rice actin 1 promoter. The transgenic plants showed the constitutive accumulation of the HVA1 protein both in leaves and roots. The second generation transgenic plants showed increased tolerance to both water and salt stresses. Artus *et al.* (1996) reported that constitutive expression of a cold-regulated, *Arabidopsis thaliana Cor15a* (a *LEA*-related gene) in transgenic plants enhanced the freezing tolerance of both chloroplasts and protoplasts. *Cor15a* expression has been suggested to affect the cryostability of the plasma membrane possibly through the interaction of cor15a polypeptide with lipid bilayers.

4.8.5.2. Osmolyte biosynthesis genes **:** During osmotic stress, plant cells accumulate low molecular weight osmolytes to prevent water loss and maintain turgor. Transgenic approach has been used to engineer biosynthesis pathways of some of these osmolytes. Transgenic plants containing key genes encoding enzymes involved in the production of mannitol, proline, fructans, trehalose, glycine-betaine, D-ononitol and polyamines have been produced in model plant species like *Arabidopsis*, tobacco and *japonica* rice (Bartels and Nelson, 1994; Bohnert and Jensen, 1996b; Holmberg and Bülow, 1998). The transgenic plants have been reported to accumulate the corresponding osmolyte at higher levels and consequently showed a marginal to significant increase in dehydration, salinity and/or cold tolerance. To give a few examples in rice; Zhu *et al.* (1998) reported that overexpression of ABA/stress-inducible promoter driven mothbean pyrroline-5-carboxylate synthetase (*P5CS*) gene resulted in up to 2.5-fold higher proline accumulation under stress conditions. The second generation transgenic plants compared to the non-transgenic plants had higher shoot and root biomass under salt- and water-stress conditions. Sakamoto *et al.* (1998) reported that transgenic rice plants over-expressing the choline oxidase (*codA*; isolated from a soil bacterium *Arthrobacter globiformis*) gene accumulated glycine betaine at higher levels and were more tolerant to salt and low temperature stresses. The study also showed that such a gene if targeted to chloroplast was more effective in improving the stress tolerance. Wang *et al.* (2000) reported production and accumulation of mannitol and sorbitol in transgenic rice plants expressing *mtlD* (mannitol-1-phosphate dehydrogenase) and *gutD* (glucitol-6-phosphate dehydrogenase) genes; the transgenic plants showed enhanced tolerance to salinity.

4.8.5.3. Polyamine accumulation : Polyamines have been implicated in a variety of stress responses in plants . Polyamines accumulate under different stress situations and are known to provide protection to membrane structure (Lee *et al.*, 1999; Dobrovinskaya *et al.*, 1999). Capell *et al.* (1998) produced the transgenic rice plants over-expressing the oat arginine decarboxylase (*ADC*) gene. The transgenic plants showed improved drought tolerance in terms of chlorophyll loss, but constitutive expression of this gene severely affected the development patterns *in vitro*. Noury *et al.* (2000) demonstrated that over-expression of the oat *adc* cDNA in transgenic rice results in increased accumulation of polyamines at different stages of development. Roy and Wu (1999) have transferred two major polyamine biosynthesis genes, arginine decarboxylase (*adc*) and S-adenosylmethionine decarboxylase (*samdc*), through *Agrobacterium*-mediated transformation into *japonica* rice using the stress/ABA-inducible promoter. Large numbers of R_O and R_1 transgenic lines have been obtained, and R_2 plants are currently being analyzed for stress tolerance.

4.8.5.4. Oxidative stress-related genes **:** Much of the injury to plants caused by abiotic stresses, is associated with oxidative damage at the cellular level (Allen, 1995). Protection of sensitive metabolic reactions through maintaining the structures of protein complexes or membranes by an increased capacity for detoxification and scavenging reactive oxygen intermediates (ROIs) could be another important strategy to engineer stress tolerance, may be an important strategy for engineering stress tolerance. Several groups have developed transgenic alfalfa and tobacco plants expressing a oxidative stress-related gene such as superoxide dismutase (Gupta *et al.*, 1993; Mckersie *et al.*, 1996, 1999), glutathione-S- transferase (Roxas *et al.*, 1997), and iron-binding protein ferritin (Deak *et al.*, 1999). The transgenic plants were reported to display greater tolerance to oxidative damage, salt stress, water deficit, and/or freezing. Ferritin provides iron for the synthesis of iron proteins, such as ferredoxin and cytochromes, and prevents damage from the free radicals produced by iron/oxygen (Theil, 1990). Transgenic rice plants producing the iron-storage protein ferritin were found to be more tolerant to oxidative damages than control plants (Goto *et al.*, 1999).

4.8.5.5. Transcriptional activator genes **:** Jaglo-Ottosen *et al.* (1998) reported that over-expression of a regulator gene, *CBF1*, whose product is a transcriptional activator, in transgenic *Arabidopsis* plants induced *COR* gene expression and increased the freezing tolerance. Kasuga *et al.* (1999) transformed *Arabidopsis* with a gene encoding DREBIA, a homologue of CBFI, driven by a stress-inducible *rd29A* promoter. The overexpression of this gene in transgenic plants activated the expression of many stress-tolerance genes, *rd17* and *rd29A* (LEA like proteins), *Kin 1*, *Cor 6.6*, *Cor 15a*, and *P5CS*; the transgenic plants showed greater tolerance to drought, salt and freezing stresses. The transformation of plants using such regulatory genes could be more rewarding to develop tolerance against complex traits like drought and salt tolerance. Saijo *et al.* (2000) reported that over expression of a single rice gene encoding Ca^{2+}-dependent protein kinase (CDPK), *OsCDPK7*, confers both cold and salt/drought tolerance in transgenic rice plants. Over-expression of *OsCDPK7* led to enhanced induction of some stress-responsive genes in response to salinity/drought, but not to cold.

4.8.6. Manipulation of source-sink relationships

Plant productivity is dictated by source-sink relationships (Egli, 1998; Sonnewald *et al.*, 1994). This can be achieved either by manipulating the 'source' by increasing the net photosynthesis in leaf tissues or the 'sink strength' by increasing the plant's ability to convert photosynthates into storage reserves.

4.8.6.1. Photosynthetic efficiency : Genetic manipulation of photosynthetic processes has been a key target for crop improvement (Horton, 2000; Mann, 1999). Whilst current knowledge of the component reactions of photosynthesis are well understood, but the molecular basis of underlying processes is still not clear. As a result genetic manipulation to improve the photosynthetic efficiency is still in the infancy stage.

Phosphoenolpyruvate decarboxylase (*PEPC*) is the key enzyme of C_4 plants. It catalyzes the initial fixation of atmospheric CO_2, even in the presence of high concentrations of O_2. Atmospheric oxygen is known to reduce photosynthetic efficiency by as much as 40% through a process called photorespiration. Photorespiration increases under drought and high-temperature conditions, because the concentration of CO_2 decreases in the leaf due to closure of stomata. C_4 plants have evolved a biochemical mechanism to overcome the O_2 inhibition of photosynthesis. The C_4 photosynthetic pathway serves as a pump to concentrate the atmospheric CO_2 at the site of ribulose 1,5 bisphosphate (Rubisco), thus suppressing its oxygenase activity and the associated photophosphorylation (Hatch, 1987). The maize *PEPC* gene of the C_4 photosynthetic pathway, including all the exons, introns, promoter and terminator sequences, was transferred into C_3 rice through *Agrobacterium*-mediated transformation (Ku *et al.*, 1999). The transgenic rice plants showed high levels of expression of the maize *PEPC* gene and accumulated high levels of PEPC protein (12% of the total leaf soluble protein). High-level expression of the *PEPC* gene was also observed in R_2 and R_3 plants. Physiologically, the transgenic plants exhibited reduced oxygen inhibition of photosynthesis compared to non-transformed plants.

Suzuki *et al.* (2000) reported that expression of a cDNA encoding phosphoenolpyruvate carboxykinase (PCK) from *Urochloa panicoides* (a PCK-type C_4 plant) in rice plants under the control of the promoter of a maize gene for phosphoenolpyruvate carboxylase or pyruvate, orthophosphate dikinase with the transit peptide of the small subunit of Rubisco, resulted in high PCK activity and the newly expressed PCK was localized in chloroplasts. These results indicate that the ectopic expression of PCK in rice chloroplasts was able partially to change the carbon flow in mesophyll cells into a C_4-like photosynthetic pathway.

Such strategies may provide methods for enhancing the photosynthetic capacity of C_3 plants.

4.8.6.2. Starch metabolism during seed development **:** Rice has been reported to be sink-limited during the seed development (Chen and Sung, 1994). Starch is the major component of rice grain and constitutes about 89% of rice grain. Increasing the conversion of photosynthate into starch during seed development should increase the sink strength of this storage organ and it may translate into higher yields and productivity.

Starch synthesis is controlled by the activation and expression of several genes that code for AGPase (ADP glucose pyrophosphorylase), starch synthase, and the other branching enzymes but the key reaction is catalyzed by AGPase (Preiss *et al.*, 1991). Enzymatic analysis of developing rice seed extracts indicates that AGPase is allosterically activated by 3-phosphoglycerate (3-PGA) and inhibited by Pi. Since 3-PGA/Pi ratios are likely to be much lower in seed amyloplasts than in leaf chloroplasts, the in situ activity of AGP may be much lower than its maximum potential and, thereby, restricts the flow of carbon into starch in developing rice seeds (Zhang *et al.*, 1996). Sakulsingharoj *et al.* (2000) reported the production of transgenic rice plants containing the *E. coli glgC-TM* gene driven by the rice endosperm-specific glutelin *Gt1* promoter by *Agrobacterium*-mediated transformation. The *glgC-TM* gene codes for a mutant AGPase that is highly active and unregulated by allosteric effectors. Transgenic rice expressing the *glgC-TM* gene showed up to 3-fold higher levels of AGPase activity under Pi-inhibited conditions in developing seeds compared to that in non-transformed plants. Stable transgenic lines those displayed segregation patterns of 3:1 ratios and showed high levels of AGPase activity were selected. Segregation analysis of these lines indicated the correlation between the gene segregation patterns following Mendelian fashion and the levels of AGPase activity. Further work is in progress to correlate the AGPase activity, seed weight and starch biosynthesis in the transgenic plants.

4.8.7. Improving nutritional quality of rice seeds

Nutritional deficiency-induced diseases are widely prevalent in Asia (WFC, 1992) and these include child blindness due to vitamin A deficiency, anemia caused by iron deficiency, endemic goiter caused by iodine deficiency, etc. Bouis (1994) has shown that low micronutrient intakes are a more important constraint to better health and nutrition than low calories intakes for the Asian population. As Asian people consume substantial amounts of rice, increase in the amounts of these micronutrients could effectively control the incidence of these diseases.

4.8.7.1. Provitamin A : Milled rice completely lacks provitamin A. Insufficient dietary provitamin A leads to severe clinical symptoms (eye disease xeropthalmia leading to blindness, increased susceptibility to other diseases such as diarrhoea, respiratory problems, measles, etc). *Agrobacterium* transformation method has been used to transfer genes encoding several ß-carotene biosynthetic pathway enzymes into rice endosperm (carotenoid-free) using three vectors (Burkhardt *et al.*, 1997; Ye *et al.*, 2000). The vector pB19hpc combines the sequences from a plant (daffodil) phytoene synthase (*psy*) with the sequence coding for a bacterial phytoene desaturase (*crtI*) placed under control of the endosperm-specific glutelin (*Gt1*) and the constitutive CaMV35S promoter, respectively; it also contained the *aphIV* expression cassette. The phytoene synthase cDNA contains a 5'-sequence coding for a functional transit peptide, and the *crtI* gene contains the Rubisco small subunit transit peptide (*tp*) sequence. Vector pZPsC carries *psy* and *crtI* , as in plasmid pB19hpc, but lacks the selectable marker *aphIV* expression cassette. Vector pZLcyH provides daffodil lycopene ß-cyclase controlled by rice glutelin promoter and the *aphIV* gene controlled by the CaMV35S promoter; Lycopene ß-cyclase carried a functional transit peptide allowing plastid import. The pB19hpc single transformants, engineered to synthesize only lycopene (red), were similar in color to the pZPsC/pZLcyH co-transformants engineered for ß-carotene (yellow synthesis). The combination of these transgenes enabled biosynthesis of provitamin A in the rice endosperm. Efforts are in progress to transfer these transgenes into other commercially important *indica* rice varieties.

4.8.7.2. Iron content : To improve the iron content of rice, Goto *et al.* (1999) reported the transfer of the entire coding sequence of the soybean ferritin gene driven by the rice seed-storage protein glutelin promoter (*GluB-1*) into rice by *Agrobacterium*-mediated transformation. The iron content of T_1 seeds was threefold higher than that of non-transformed seeds. Drakakaki *et al.* (2000) reported that expression of recombinant ferritin gene under the control of the maize ubiquitin promoter in transgenic wheat and rice significantly increases iron levels in vegetative tissues, but that the levels of recombinant ferritin are not sufficient to increase iron levels in seeds.

4.8.7.3. Essential amino acids : Milled rice is also deficient in essential amino acids (cysteine, methionine, lysine). Attempts are also being made to improve the nutritional quality of rice by modifying the coding sequences of a glutelin gene to increase the percentage of methionine, lysine and tryptophan residues (Moon and Wu, 1996). The procedure involved designing of the synthetic oligomers encoding met, lys and trp, insertion of oligomers into glutelin DNA, construction of a complete glutelin gene by adding the 2.3 kb glutelin promoter and the 0.4 kb nos transcription terminator, and introduction of the engineered glutelin genes by biolistic transformation into rice plants. Efforts are also in progress to engineer amino acid biosynthetic pathways using the rice endosperm specific promoters for increasing the content of lysine and sulphur-containing amino acids, cysteine and methionine (Potrykus *et al.*, 1995).

4.8.8. Grain quality

Demand of superior grain quality/ aromatic rices has increased with increase in prosperity among the Asians. Also farmers in low-income countries could gain by producing exportable surplus of superior quality grains for the markets in developed countries. It must be noted that preferences for grain quality varies from country to country. For instance, the middle east consumer prefers a long grain, well-milled rice with strong aroma, and European community generally prefers a long grain rice with no scent. A lot of basic research is required to improve the grain quality of indicas or to increase the yield potential of traditional, low-yielding, high-quality aromatic rices. Some progress has already been made towards the molecular tagging of genes/QTLs for quality traits including aroma, cooked kernel elongation, amylose content, length/ breadth ratio, gelatinization temperature, etc (Singh *et al.*, 2000a); research in this area is likely to gain momentum. Jain and Jain (2000) reviewed the various transgenic strategies that can be used for the genetic improvement of Basmati rices.

Shimada *et al.* (1993) used the antisense waxy gene construct and produced transgenic rice plants with a significant reduction in the amylose content of grain starch. Amylose content of rice is most important constitutional indices of rice cooking and processing behavior as it determines the hardness of cooked rice, gloss of the final product and rice water ratio (Singh *et al.*, 2000b).

4.8.9. Genome analysis/functional genomics

There has been much progress in the development of strategies to discover the function of plant genes (Martin, 1998; Sundaresan, 1996; Krysan *et al.*, 1999). Recently, techniques involving the gene inactivation by insertion of a transposon or T-DNA with a gene trap system in the model plants *Arabidopsis* and rice were shown to be very useful in functional genomics. This strategy involves production of a large population of transgenic plants with transposon/ T-DNA insertions, screening the insertions in individual genes, recovering sequences flanking the insertions and preparation of a database of flanking DNA sequences. A gene trap contains multiple acceptor sites fused to the coding region of the reporter gene. A fusion protein of the reporter gene with the N-terminal portion of a host gene will be produced when the element is inserted into either an exon or an intron of the host gene in the same transcriptional orientation. Chin *et al.* (1999) reported the construction of *Ac* and gene trap *Ds* vectors and their transfer into rice genome by *Agrobacterium*-mediated transformation. The transgenic plants that contained single and simple insertions of T-DNA were analyzed to evaluate the gene-tagging efficiency. Nearly 80% of *Ds* elements were excised from the original T-DNA sites, when Ac cDNA was expressed under a CaMV35S promoter. Repetitive ratoon culturing induced new transpositions of *Ds* in plants derived from cuttings. Eight per cent of transposed *Ds* elements expressed GUS in various tissues of rice panicles. The data show that A*c*/D*s*-mediated gene trap system could prove an excellent tool for the analysis of functions of genes in rice. Enoki *et al.* (1999) reported the *Ac* transposition in the rice population for three successive generations. Analysis of randomly cloned sequences flanking *Ac* showed a preferential transposition of *Ac* into protein-coding regions of rice. A three-dimensional PCR screening was used for gene knockouts in rice. Of 14 randomly selected genes, two gene knockouts were identified, and one encoding a rice cytochrome P450 gene was shown to be stably inherited to the progeny.

Jeon *et al.* (2000) reported the generation of 18,358 fertile transgenic rice plants that carry a T-DNA insertion. The binary vector used in the insertion contained the promoterless *gus* reporter gene with an intron and multiple splicing donors and acceptors immediately next to the right border. This gene trap vector was used to detect a gene fusion between *gus* and an endogenous gene, which is tagged by T-DNA. Histochemical GUS analysis of plant tissues revealed that 1.6-2.1% of them were positive and that their GUS expression patterns were organ- or tissue-specific or ubiquitous in all parts of the plant. These T-DNA tagged lines greatly facilitate to analyze the function of a number of valuable genes.

4.8.10. Dwarfism

One of the major breeding objectives in rice especially in aromatic rices (*e.g.* Basmati rice) is to develop short-stature plant type. Dwarf phenotype is important since a balance between vegetative and reproductive growth is a crucial factor that controls crop yield. Enhancement of harvest index in grain crops including rice, has been accomplished by the use of dwarfing genes. Kisaka *et al.* (1998) reported the production of transgenic japonica rice plants expressing the transgene, *rgp1* (encodes for a GTP binding protein). These transgenic plants had distinct characteristics; namely, dwarfism, early flowering and high grain yield, and these characteristics were stable and heritable. The exact physiological function of GTP-binding proteins remains to be worked out. However, *rgp1* gene has been shown to result in the reduction apical dominance and high-level synthesis of cytokinins in transgenic tobacco (Sano *et al.*, 1994). An (1994) reported that transgenic tobacco plants showing ectopic expression of rice MADS box genes (*e.g* OsMADS1) were shorter, early flowering and branched with no adverse effects on flower development. It will be worthwhile to study the impact of such dwarfing genes in Basmati rice.

5. CONCLUSIONS AND FUTURE PROSPECTS

Rice is already a model monocot plant species for research on genetic transformation and genomics. Both direct gene transfer (biolistic, DNA uptake by protoplasts) and *A. tumefaciens* methods are being routinely used for rice transformation. An array of useful genes has been transferred in rice to improve the resistance against insect pests, fungal and viral diseases, salinity and drought stresses and to improve its nutritional quality and to manipulate source to sink relationships. Transgenic research to engineer complex agronomic traits (*e.g.* yield, drought tolerance, grain quality) and metabolic or regulatory pathways involving many genes or gene complexes, is likely to gain momentum in future (Chen *et al.*, 1998). This is pretty evident from the progress made towards the genome mapping and gene tagging research in rice and other plant species, development of tissue specific or stress-inducible promoters, and development of the *Agrobacterium* vectors for the transfer of fairly large DNA fragments. Introduction of genes via chloroplast transformation or targeting of transgene products into chloroplasts, may be rewarding in terms of greater gene expression, reduction in the cytotoxicity of transgene products and diminished possibility of gene pollution to other crops (Daniell, 1999; Jang *et al.*, 1999). The problem of gene silencing can be tackled at least partially, by preferably using the *Agrobacterium* method of transformation (Hiei *et al.*, 1997), selecting the transgenic plants with only one copy of the transgene, or by using the MAR sequences. Further research may be necessary to transform elite, commercially important *indica* rice varieties including Basmati rice varieties (Jain and Jain, 2000). With the recent advances in rice molecular and transformation technologies, it should also be feasible to design various transgenic strategies to increase rice productivity and quality.

Transgenic technology combined with the genome sequencing, functional genomics and proteomics has the potential to advance crop breeding beyond imagination. The isolation of genes and regulatory components determining the complex agronomic traits is likely to gain momentum with the recent progress made in the area of rice genomics. The questions on regulatory issues, safety and environmental concerns, public acceptance of the perceived advantages and patent rights remained unsolved (Robinson, 1999). The bureaucratic difficulties and legal uncertainties encountered in meeting all requirements for the release of transgenic crops are not very encouraging for commercialization in most countries except Canada, China and USA. The problem is that currently ethical and social concerns tend to override the scientific concerns with respect to biosafety. IRRI's rice breeder Dr. Gurdev Khush worries that Asia will suffer if opposition to genetically modified crops continued and blocked transgenic research (Normile, 2000). "Asia needs to use every tool available to produce food for growing populations," Khush says. Scientists and policy makers should not forget that public acceptability is the most important component to ensure the continuous flow of funds from both private and public enterprises/ Institutions for transgenic research. More fundamental research/ scientific data on risks and benefits of transgenics, is required to remove the people's fear and to demonstrate the usefulness of genetic engineering research in agriculture. Intellectual property rights (IPR) shall help in attracting large financial investments with the promise of returns on investments by commercialization, but the current minefield of patents is seriously affecting the Scientist's freedom to undertake transgenic projects.

ACKNOWLEDGEMENTS

Research grants from the Rockefeller Foundation, New York, USA, for rice biotechnology research at CCS Haryana Agricultural University, Hisar (India), are gratefully acknowledged.

REFERENCES

Alam MF, Firoz M, Datta K, Abrigo E, Vasquez A, Senadhira D, Datta SK (1998) Production of transgenic deepwater indica rice plants expressing a synthetic *Bacillus thuringiensis cry1A(b)* gene with enhanced resistance to yellow stem borer. *Plant Sci.,* **135** : 25-30.

Aldemita RR, Hodges TK (1996) *Agrobacterium tumefaciens*-mediated transformation of indica and japonica rice varieties. *Planta,* **199** : 612-617.

Allen RD (1995) Dissection of oxidative stress tolerance using transgenic plants. *Plant Physiol.,* **107** : 1049-1054.

An G (1994) Regulatory genes controlling flowering time or floral organ development. *Plant Mol. Biol.,* **25** : 335-337.

Arencibia A, Gentinetta E, Cuzzoni E, Castiglione S, Kohli A, Vain P, Leech M, Christou P, Sala A (1998) Molecular analysis of the genome of transgenic rice (*Oryza sativa* L.) plants produced via particle bombardment or intact cell electroporation. *Mol. Breed.,* **4** : 99-109

Artus NN, Uemura M, Steponkus PL, Gilmour SJ, Lin C, Thomashow MF (1996) Constitutive expression of the cold-regulated *Arabidopsis thaliana Cor15a* gene affects both chloroplast and protoplast freezing tolerance. *Proc. Natl. Acad. Sci. USA,* **93** : 13404-13409.

Ayers NM, Park WD (1994) Genetic transformation of rice. *Crit. Rev. Plant Sci.,* **13** : 219-239.

Azhakanandam K, McCabe MS, Power JB, Lowe KC, Cocking EC, Davey MR (2000) T-DNA transfer, integration, expression and inheritance in rice: effects of plant genotype and *Agrobacterium* super-virulence. *J. Plant Physiol.,* **157** : 429-439.

Bartels D, Nelson D (1994) Approaches to improve stress tolerance using molecular genetics. *Plant Cell Environ.,* **17** : 659-667.

Baruah-Wolff J, Harwood WA, Lonsdale DA, Harvey A, Hull R, Snape JW (1999) Luciferase as a reporter gene for transformation studies in rice (*Oryza sativa* L.). *Plant Cell Rep.,* **18** : 715-720.

Beachy RN (1990) Plant transformation to confer resistance against virus infection. In: *Gene Manipulation in Plant Improvement II* (Ed. Gustafson JP), Plenum Press, New York, pp 305-313.

Bishnoi U, Jain RK, Rohilla JS, Chowdhury VK, Gupta KR, Chowdhury JB (2000) Anther culture of recalcitrant indica x Basmati rice hybrids. *Euphytica* **114** : 93-101.

Bohnert HJ, Jensen RG (1996a) Metabolic engineering for increased salt tolerance – the next step. *Aust. J. Plant Physiol.,* **23** : 661-667.

Bohnert HJ, Jensen RG (1996b) Strategies for engineering water-stress tolerance in plants. *Trends Biotech.,* **14** : 89-97.

Bouis HE (1994) Income, demand for food staples, and demand for diet quality: exploring the relationship between preferences for calories and micro-nutrient requirements. Washington, D.C.: International Food Policy Research Institute.

Boulter D, Edwards GA, Gatehouse AMR, Gatehouse JA, Hilder VA (1990) Additive protective effects of different plant-derived insect resistance genes in transgenic tobacco plants. *Crop Protection,* **9** : 351-354.

Broglie K, Chet I, Holliday M, Cressman R, Biddle P, Knowlton S, Mauvais J, Broglie RM (1991) Transgenic plants with enhanced resistance to the fungal pathogen *Rhizoctonia solani. Science,* **254** : 1194-1197.

Burkhardt PK, Beyer P, Wünn J, Klöti A, Armstrong GA, Schledz M, von Lintig J, Potrykus I (1997) Transgenic rice (*Oryza sativa*) endosperm expressing daffodil (*Narcissus pseudonarcissus*) phytoene synthase accumulates phytoene, a key intermediate of provitamin A biosynthesis. *Plant J.*, **11** : 1071-1078.

Cao J, Duan X, McElroy D, Wu R (1992) Regeneration of herbicide resistant transgenic rice plants following microprojectile-mediated transformation of suspension culture cells. *Plant Cell Rep.*, **11** : 586-591.

Cao J, Zhang W, McElroy D, Wu R (1991) Assessment of rice genetic transformation techniques. In: *Rice Biotechnology* (Eds. Khush GS, Toenniessen GH), CAB-International, IRRI, pp 175-198.

Capell T, Escobar C, Liu H, Burtin D, Lepri O, Christou P (1998) Overexpression of the oat arginine decaroxylase cDNA in transgenic rice (*Oryza sativa* L.) affects normal developmental patterns *in vitro* and results in putrescine accumulation in transgenic plants. *Theor. Appl. Genet.*, **97** : 246-254.

Causse M, Fulton TM, Cho YG, Ahn SN, Wu K, Xiao J, Chunwongse J, Yu Z, Ronald PC, Harrington SB, Second GA, McCouch SR, Tanksley SD (1994) Saturated molecular map of the rice genome based on an interspecific backcross population. *Genetics,* **138** : 1251-1274.

Cha¿r H, Legavre T, Guiderdoni E (1996) Transformation of haploid, microspore-derived cell suspension protoplasts of rice (*Oryza sativa* L.). *Plant Cell Rep.*, **15** : 766-770.

Chamberlain DA, Brettell RIS, Last DT, Witrzens B, McElroy D, Dolferus R, Dennis ES (1994) The use of *emu* promoter with antibiotic and herbicide resistance genes for the selection of transgenic wheat callus and rice plants. *Austr. J. Plant Physiol.*, **21** : 95-112.

Chareonpornwattana S, Thara KV, Wang L, Datta SK, Panbangred W, Muthukrishnan S (1999) Inheritance, expression, and silencing of a chitinase transgene in rice. *Theor. Appl. Genet.*, **98** : 371-378.

Chaudhury A, Maheshwari SC, Tyagi AK (1995) Transient expression of *gus* gene in intact seed embryos of indica rice after electroporation-mediated gene delivery. *Plant Cell Rep.*, **14** : 215-220.

Chen L, Marmey P, Taylor NJ, Brizard J-P, Espinoza C, D'Cruz P, Huet H, Zhang S, de Kochko A, Beachy RN, Fauquet CM (1998) Expression and inheritance of multiple transgenes in rice plants. *Nat. Biotechnol.*, **16** : 1060-1064.

Chen CL, Sung JM (1994) Carbohydrate metabolism enzymes in CO2-enriched developing rice grains varying in grain size. *Physiol. Plant.*, **90** : 79-85.

Cheng X, Sardana R, Kaplan H, Altosaar I (1998) *Agrobacterium*- transformed rice plants expressing synthetic *cry1A(b)* and *cry1A(c)* genes are highly toxic to striped stem borer and yellow stem borer. *Proc. Natl. Acad. Sci. USA,* **95** : 2767-2772.

Chin HG, Choe MS, Lee S-H, Park SH, Park SH, Koo JC, Kim NY, Lee JJ, Oh BG, Yi GH, Kim SH, Choi HC, Cho MJ, Han C-d (1999) Molecular analysis of rice plants harboring an Ac/Ds transposable element-mediated gene trapping system. *Plant J.*, **19** : 615-623.

Christensen AH, Sharrock RA, Quail PH (1992) Maize polyubiquitin genes: structure, thermal perturbation of expression and transcript splicing, and promoter activity following transfer to protoplasts by electroporation. *Plant Mol. Biol.*, **18** : 675-689.

Christou P (1994) *Rice Biotechnology and Genetic Engineering*. Technomic Pub. Co. Inc., Lancaster.

Christou P (1997) Rice transformation: bombardment. *Plant Mol. Biol.,* **35** : 197-203.

Christou P, Ford TL, Kofron M (1991) Production of transgenic rice (*Oryza sativa* L.) plants from agronomically important indica and japonica varieties via electric discharge particle acceleration of exogenous DNA into immature zygotic embryos. *Bio/Technology,* **9** : 957-962.

Chung, BC, Kim, JK, Nahm, BH, Lee, CH (2000) In planta visual monitoring of green fluorescent protein in transgenic rice plants. *Molecules Cells,* **10** : 411-414.

Cornejo M-J, Luth D, Blankenship KM, Anderson OD and Blechl HAVE (1993) Activity of a maize ubiquitin promoter in transgenic rice. *Plant Mol. Biol.,* **23** : 567-581.

Daniell H (1999) New tools for chloroplast genetic engineering. *Nat. Biotechnol.,* **17** : 855-856.

Datta SK (1999) Transgenic cereals: *Oryza sativa* (rice). In: *Molecular Improvement of Cereal Crops* (Ed. Vasil IK), Kluwer Academic Publishers, The Netherlands, p 149-187.

Datta SK, Datta K, Soltanifar N, Donn G, Potrykus I (1992) Herbicide-resistant indica rice plants from IRRI breeding line IR72 after PEG-mediated transformation of protoplasts. *Plant Mol. Biol.,* **20** : 619-629.

Datta K, Koukolíková-Nicla Z, Baisakh N, Oliva N, Datta SK (2000) *Agrobacterium*-mediated engineering for sheath blight resistance of indica rice cultivars from different ecosystems. *Theor. Appl. Genet.,* **100** : 832-839.

Datta SK, Peterhans A, Datta K, Potrykus I (1990) Genetically engineered fertile indica-rice recovered from protoplasts. *Bio/Technology,* **8** : 736-740.

Datta K, Vasquez A, Tu J, Torrizo L, Alam MF, Oliva N, Abrigo E, Khush GS, Datta SK (1998) Constitutive and tissue-specific differential expression of the *cryIA(b)* gene in transgenic rice plants conferring resistance to rice insect pest. *Theor. Appl. Genet.,* **97** : 20-30.

David CC (1991) The world rice economy: challenges ahead. In: *Rice Biotechnology* (Eds. Khush GS, Toenniessen GH), CAB International, Wallingford, Oxon, UK and International Rice Research Institute, Manila, Philippines, pp 1-18.

Deak M, Horvath GV, Davletova S, Torok K, Sass L, Vass I, Barna B, Kiraly Z, Dudits D (1999) Plants ectopically expressing the iron-binding protein ferritin, are tolerant to oxidative damages and pathogens. *Nat. Biotechnol.,* **17** : 192-196.

De Maagd RA, Bosch D, Stiekema W (1999) *Bacillus thuringiensis* toxin-mediated insect resistance in plants. *Trends Plant Sci.,* **4** : 9-13.

Depicker A, Van Montagu M (1997) Post-transcriptional gene silencing in plants. *Curr. Opin. Cell Biol.,* **9** : 373-382.

Dey MM, Upadhyaya HK (1996) Yield loss due to drought, cold and submergence in Asia. In: *Rice Research in Asia: Progress and Priorities* (Eds. Evenson RE, Herdt RW, Hossain M), CAB International and IRRI, Manila, Philippines, pp 291-303.

Dobrovinskaya OR, Muniz J, Pottosin II (1999) Inhibition of vacuolar ion channels by polyamines. *J. Mem. Biol.,* **167** : 127-140.

Drakakaki G, Christou P, Stoger E (2000) Constitutive expression of soybean ferritin cDNA intransgenic wheat and rice results in increased iron levels in vegetative tissues but not in seeds. *Transgenic Res.,* **9** : 445-452.

Duan X, Li X, Xue Q, Abo-El-Saad M, Xu D, Wu R (1996) Transgenic rice plants harbouring an introduced potato proteinase inhibitor II gene are insect resistant. *Nat. Biotechnol.,* **14** : 494-498.

Egli DB (1998) *Seed biology and the yield of grain.* CAB International Wallingford, Oxford.

Enoki H, Izawa T, Kawahara M, Komatsu M, Koh S, Kyozuka J, Shimamoto K (1999) Ac as a tool for the functional genomics of rice. *Plant J.,* **19** : 605-613.

Estruch JJ, Carozzi NB, Desai N, Duck NB, Warren GW, Koziel MG (1997) Transgenic plants: an emerging approach to pest control. *Nat. Biotechnol.,* **15** : 137-141.

Evans LT (1998) *Feeding the Ten Billion: Plants and Population Growth,* Cambridge Univ Press.

Fischer KS, Cordova VG (1998) Impact of IRRI on rice science and production. In: *Impact of Rice Research* (Eds. Pingali PL, Hossain M), International Rice Research Institute, Manila, Philippines, pp. 27-50.

Flavell RB. Dark E, Fuchs RI, Fraley RT (1992) Selectable marker genes; safe for plants? *Bio/Technol.,* **10** : 141-144.

Fujimoto H, Itoh K, Yamamoto M, Kyozuka J, Shimamoto K (1993) Insect resistant rice generated by introduction of a modified δ–endotoxin gene of *Bacillus thuringiensis. Bio/technology,* **11** : 1151-1155.

Gahakwa D, Maqbool SB, Fu X, Sudhakar D, Christou P, Kohli A (2000) Transgenic rice as a system to study the stability of transgene expression: multiple heterologus transgenes show similar behaviour in diverse genetic backgrounds. *Theor. Appl. Genet.,* **101** : 388-399.

Gamborg OL, Philips GC (1995) *Plant Cell, Tissue and Organ Culture: Fundamental Methods.* Springer-Verlag, Berlin.

Garg AK, Blair M, Ramanathan V, McCouch S, Wu R. 1999. Transfer of large rice genomic DNA fragments into *indica* and *japonica* rice plant chromosomes using a binary-BAC system for plant transformation. In: *General Meeting of the International Program of Rice Biotechnology,* September 20-24, 1999. Phuket, Thailand, p. 98.

Ghareyazie B, Ainia F, Menguito CA, Rubia LG, De Palma JM, Liwanag EA, Cohen MB, Khush GS, Bennett J (1997) Enhanced resistance to two stem borers in an aromatic rice containing a synthetic *cry1A(b)* gene. *Mol. Breed.,* **3** : 401-414.

Glaszmann JC (1987) Isozymes and classification of Asian rice varieties. *Theor. Appl. Genet.,* **74** : 21-30.

Goldsbrough A. 1992. Marker gene removal: a practical necessity? *TIBTECH.,* **10** : 417.

Goto F, Yoshihara T, Shigemoto N, Toki S, Takaiwa F (1999) Iron fortification of rice seeds by the soybean ferritin gene. *Nat. Biotechnol.,* **17** : 282-286.

Gupta AS, Heinen JL, Holaday AS, Burke JJ, Allen RD (1993) Increased resistance to oxidative stress in transgenic plants that overexpress chloroplastic Cu/Zn superoxide dismutase. *Proc. Natl. Acad. Sci. USA,* **90** : 1629-1633.

Guo Y, Liang H, Berns MW (1995) Laser-mediated gene transfer in rice. *Physiol. Plant.,* **93** : 19-24.

Hamilton CM (1997) A binary-BAC system for plant transformation with high-molecular-weight DNA. *Gene,* **200** : 107-116.

Hamilton CM, Frary A, Lewis C, Tanksley SD (1996) Stable transfer of intact high molecular weight DNA into plant chromosomes. *Proc. Natl. Acad. Sci. USA,* **93** : 9975-9979.

Hanson B, Engler D, Moy Y, Newman B, Ralston E, Gutterson N (1999) A simple method to enrich an *Agrobacterium*-transformed population for plants containing only T-DNA sequences. *Plant J.,* **19** : 727-734.

Harper BK, Mabon SA, Leffel SM, Halfhill MD, Richards HA, Moyer KA, Stewart Jr CN (1999) Green fluorescent protein as a marker for expression of a second gene in transgenic plants. *Nat. Biotechnol.,* **17** : 1125-1129.

Hasegawa PM, Bressan RA, Zhu J-K, Bohnert HJ (2000) Plant cellular and molecular responses to high salinity. *Annu. Rev. Plant Physiol. Plant Mol. Biol.,* **51** : 463-499.

Hatch MD (1987) C_4 photosynthesis: a unique blend of modified biochemistry, anatomy and ultrastructure. *Biochem. Biophys. Acta,* **895** : 81-106.

Herdt RW, Riley FZ (1987) International Rice Research Priorties: implications for Biotechnology initiatives. Rockefeller Foundation Workshop on Allocating Resources for Developing Country Agricultural Research. Bellazio, Italy, 6-10 July, 1987.

Hiei Y, Komari T, Kubo T (1997) Transformation of rice mediated *by Agrobacterium tumefaciens. Plant Mol. Biol.,* **35** : 205-218.

Hiei Y, Ohta S, Komari T, Kumashiro T (1994) Efficient transformation of rice (*Oryza sativa* L.) mediated by *Agrobacterium* and sequence analysis of the boundaries of the T-DNA. *Plant J.,* **6** : 271-282.

Holmberg N, Bülow L (1998) Improving stress tolerance in plants by gene transfer. *Trends Plant Sci.,* **3** : 61-66.

Holmes-Davis R, Comai L (1998) Nuclear matrix attachment regions and plant gene expression. *Trends Plant Sci.,* **3** : 91-97.

Horton P (2000) Prospects for crop improvement through the genetic manipulation of photosynthesis: morphological and biochemical aspects of light capture. *J. Exp. Bot.,* **51** : 475-485.

Hosoyama H, Irie K, Abe K, Arai S (1995) Introduction of a chimeric gene encoding an oryzacystatin-ß-glucuronidase fusion protein into rice protoplasts and regeneration of transformed plants. *Plant Cell Rep.,* **15** : 174-177.

Hossain M (1996) Economic prosperity in Asia: implications for rice research. In: *Rice Genetics III,* Proceedings of the Third International Rice Genetics Symposium (Ed. Khush GS), International Rice Research Institute, Los Baòos, Philippines, pp 3-29.

Hossain M, Pingali PL (1998) Rice research, technological progress, and impact on productivity and poverty: an overview. In: *Impact of Rice Research* (Eds. Pingali PL, Hossain M), International Rice Research Institute, Manila, Philippines, pp 1-25.

Huntley CC and Hall TC (1996) Interference with brome mosaic virus replication in transgenic rice. *Mol. Plant-Microbe Interact.,* **9** : 164-170.

IRGS4A (4th International Rice Genetics Symposium, Abstracts) (2000) 22-27 October, International Rice Research Institute, Los Baòos, Laguna, Philippines.

Irie K, Hosoyama H, Takeuchi T, Iwabuchi K, Watanabe H, Abe M, Abe K, Arai S (1996) Transgenic rice established to express corn cystatin exhibits strong inhibitory activity against insect gut proteinases. *Plant Mol. Biol.,* **30** : 149-157.

Ito O, O'Toole J, Hardy B (1999) *Genetic Improvement of Rice for Water-Limited Environments,* IRRI, Los Baòos, Laguna, Philippines.

Jaglo-Ottosen KR, Gilmour SJ, Zarka DG, Schabenberger O, Thomashow MF (1998) *Arabidopsis CBFI* overexpression induces *COR* genes and enhances freezing tolerance. *Science,* **280** : 104-106.

Jain RK (1997) Recent progress in plant regeneration from indica rice cells and protoplasts. *Indian J. Exp. Biol.,* **35** : 232-331.

Jain RK, Davey MR, Cocking EC, Wu R (1997a) Carbohydrate and osmotic requirements for high-frequency plant regeneration from protoplast-derived colonies of indica and japonica rice varieties. *J. Exp. Bot.,* **48** : 751-758.

Jain RK, Jain S (2000) Transgenic strategies for genetic improvement of Basmati rice. *Indian J. Exp. Biol.,* **38** : 6-17.

Jain RK, Jain S, Wang B, Wu R (1996a) Optimization of biolistic method for transient gene expression and production of agronomically useful transgenic Basmati rice plants. *Plant Cell Rep.,* **15** : 963-968.

Jain RK, Jain S, Wang BY, Wu R (1997b) An improved biolistic method for transformation and production of fertile transgenic Basmati rice plants. *Intl. Rice Res Notes,* **22** : 10-12.

Jain RK, Jain S, Wu R (1996b) Stimulatory effect of water stress on plant regeneration in aromatic Indica rice varieties. *Plant Cell Rep.,* **15** : 449-454.

Jain S, Jain RK, Wu R (1996c) A simple and efficient procedure for cryopreservation of embryogenic cells of aromatic indica rice varieties. *Plant Cell Rep.,* **15** : 712-717.

Jain RK, Khehra GS, Lee S-H, Blackhall NW, Marchant R, Davey MR, Power JB, Cocking EC, Gosal SS (1995) An improved procedure for plant regeneration from indica and japonica rice protoplasts. *Plant Cell Rep.,* **14** : 515-519.

Jang I-C, Nahm BH, Kim J-K (1999) Subcellular targeting of green fluorescent protein to plastids in transgenic rice plants provides a high-level expression system. *Mol. Breed.,* **5** : 453-461.

Jefferson RA, Kavanagh TA, Bevan MW (1987) Gus fusions: β-glucuronidase as a sensitive and versatile gene fusion marker in higher plants. *EMBO J.,* **86** : 3901-3907.

Jeon J-S, Lee S, Jung K-H, Jun S-H, Jeong D-H, Lee J, Kim C, Jang S, Lee S, Yang K, Nam J, An K, Han M-J, Sung R-J, Choi H-S, Yu J-H, Choi J-H, Cho S-Y, Cha S-S, Kim S-I, An G (2000) T-DNA insertional mutagenesis for functional genomics in rice. *Plant J.,* **22** : 561-570.

Joersbo M, Donaldson I, Kreiberg J, Guldager, Petersen SG, Brunstedt J and Okkels FT (1998) Analysis of mannose selection used for transformation of sugar beet. *Mol. Breed.,* **4** : 111-117.

Kasuga M, Liu Q, Miura S, Yamaguchi-Shinozaki K, Shinozaki K (1999) Improving plant drought, salt and freezing tolerance by gene transfer of a single stress-inducible transcription factor. *Nat. Biotechnol.,* **17** : 287-291.

Kato S, Kosaka H, Hara S (1928) On the affinity of rice varieties as shown by fertility of hybrid plants. *Bull. Sci. Fac. Agric. Kyushu Univ., Fukuoka, Japan,* **3** : 132-147.

Khan MS, Maliga P (1999) Fluorescent antibiotic resistance marker for tracking plastid transformation in higher plants. *Nat. Biotechnol.,* **17** : 910-919.

Khush GS (1995a) Modern varieties - Their real contribution to food supply and equity. *GeoJournal,* **35** : 275-284.

Khush GS (1995b) Biotechnology approaches to rice improvement. In: *Induced Mutations and Molecular Techniques for Crop Improvement,* Proc Intl Symp

"Use of Induced Mutations and Molecular Techniques for Crop Improvement," IAEA, FAO, Vienna, 19-23 June, p 455-475, IAEA-SM-340/28.

Khush GS (1996) *Rice Genetics III,* Proceedings of the Third International Rice Genetics Symposium, Manila, Philippines, 16-20 Oct 1995, IRRI, Manila, Philippines.

Khush GS (2000) Taxonomy and origin of rice. In: *Aromatic Rice* (Eds. Singh RK, Singh US, Khush GS), Oxford & IBH Publishing Co. Pvt. Ltd., New Delhi, pp 5-13.

Khush GS, Bacalangco E, Ogawa T (1990) A new gene for resistance to bacterial blight from *O. longistaminata. Rice Genet Newslett.,* **7** : 121-122.

Khush GS, Toenniessen GH (1991) *Rice Biotechnology,* CAB International, Wallingford, Oxon, UK and International Rice Research Institute, Manila, Philippines.

Kisaka H, Sano H, Kameya T (1998) Characterization of transgenic rice plants that express *rgp1*, the gene for a small GTP-binding protein from rice. *Theor. Appl. Genet.,* **97** : 810-815.

Komari T (1989) Transformation of callus cultures of nine plant species mediated by *Agrobacterium. Plant Sci.,* **60** : 223-229.

Komari T, Hiei Y, Saito Y, Murai N, Kumashiro T (1996) Vectors carrying two separate T-DNA for co-transformation of higher plants mediated by *Agrobacterium tumefaciens* and segregation of transformants free from selection markers. *Plant J.,* **10** : 165-174.

Krysan PJ, Young JC, Sussman MR (1999) T-DNA as an insertional mutagen in *Arabidopsis. Plant Cell,* **11** : 2283-2290.

Ku SBM, Agarie S, Nomura M, Fukayama H, Tsuchida H, Kazuko O, Hirose S, Toki S, Miyao M and Matsuoka M. 1999. High level expression of maize phosphoenolpyruvate carboxylase in transgenic rice plants. *Nat. Biotechnol.,* **17** : 76-80.

Kumpatla SP, Chandrasekharan MB, Iyer LM, Li G, Hall TC (1998) Genomic intruder scanning and modulation systems and transgene silencing. *Trends Plant Sci.,* **3** : 97-104

Kumpatla SP, Teng W, Bucholz WG, Hall TC (1997) Epigenetic transcriptional silencing and 5-azacytidine-mediated reactivation of a complex transgene in rice. *Plant Physiol.,* **115** : 361-373.

Kurata N, Nagamura Y, Yamamoto K, Harushima Y, Sue N, Wu J, Antonio BA, Shomura A, Shimizu T, Lin SY, Inoue T, Fukuda A, Shimano T, Kuboki Y, Toyama T, Miyamoto Y, Kirihara T, Hayasaka K, Miyao A, Monna L, Zhong HS, Tamura Y, Wang ZX, Momma Y, Umehara Y, Yano M, Sasaki T, Minobe Y (1994) A 300 kilobase interval genetic map of rice including 883 expressed sequences. *Nature Genet.,* **8** : 365-375.

Kyozuka J, Fujimoto H, Izawa T, Shimamoto K (1991) Anaerobic induction and tissue-specific expression of maize *Adh1* promter in transgenic rice plants and its progeny. *Mol. Gen. Genet.,* **228** : 40-48.

Lee JK, John SA, Weiss JN (1999) Novel gating mechanism of polyamine block in the strong inward rectifier K channel Kir 2.1. *J. Gen. Physiol.,* **113** : 555-563.

Li ZJ, Burrow MD, Murai N (1990) High frequency generation of fertile transgenic rice plants after PEG-mediated protoplast transformation. *Plant. Mol. Biol. Rep.,* **8** : 276-291.

Li Z, Upadhyaya NM, Meena S, Gibbs AJ, Waterhouse PM (1997) Comparison of promoters and selectable marker genes for the use in indica rice transformation. *Mol. Breed.,* **3** : 1-14.

Lin W, Anuratha CS, Datta K, Potrykus I, Muthukrisnan S, Datta SK (1995) Genetic engineering of rice for resistance to sheath blight. *Bio/technology,* **13** : 686-691.

Linthorst HJM (1991) Pathogenesis-related proteins of plants. *Crit. Rev. Plant Sci.,* **10** : 123-150.

Luo Z, Wu R (1989) A simple method for the transformation of rice via the pollen tube pathway. *Plant Mol. Biol. Rep.,* **7** : 69-77.

Malik VS, Saroha MK (1999) Marker gene controversy in transgenic plants. *J. Plant Biochem. Biotechnol.,* **8** : 1-13.

Mann GC (1999) Genetic engineers aim to soup up crop photosynthesis. *Science,* **283** : 314-316.

Martin GB (1998) Gene discovery for crop improvement. *Curr. Opin. Biotech.,* **9** : 220-226.

Massoud FI (1974) Salinity and alkalinity as soil degradation hazards. FAO/ UNDP-Expert Consultation on Soil Degradation. June 10-14, 1974, FAO, Rome, p. 21.

Matsushita J, Otani M, Wakita Y, Tanaka O, Shimada T (1999) Transgenic plant regeneration through silicon carbide whisker-mediated transformation of rice (*Oryza sativa* L.). *Breed Sci.,* **49** : 21-26.

Matzke MA, Matzke AJM (1995) How and why do plants inactivate homologous (trans) genes? *Plant Physiol.,* **107** : 679-685.

Maqbool SB, Husnain T, Riazuddin, Masson L, Christou P (1998) Effective control of yellow stem borer and rice leaf folder in transgenic rice indica varieties Basmati 370 and M7 using the novel δ–endotoxin *cry2A Bacillus thuringiensis* gene. *Mol. Breed.,* **4** : 501-507.

McBride KE, Svab Z, Schaaf DJ, Hogan PS, Stalker DM, Maliga P (1995) Amplification of a chimeric *Bacillus* gene in chloroplasts leads to an extraordinary level of an insecticidal protein in tobacco. *Bio/Technology,* **13** : 362-365.

McElroy D, Blowers AD, Jenes B, Wu R (1991) Construction of expression vectors based on the rice actin 1 (*Act1*) 5' region for use in monocot transformation. *Mol. Gen. Genet.,* **231** : 150-160.

McElroy D, Brettell RIS (1994) Foreign gene expression in transgenic cereals. *Trends Biotech.,* **12** : 62-68.

McElroy D, Chamberlain DA, Moon E, Wilson KJ (1995) Development of a *gusA* reporter gene construct for cereal transformation, availability of plant transformation vectors from the CAMBIA molecular genetic resource service. *Mol. Breed.,* **1** : 27-37.

McElroy D, Zhang W, Cao C, Wu R (1990) Isolation of an efficient actin promoter for use in rice transformation. *Plant Cell,* **2** : 163-171.

Mckersie BD, Bowley SR, Harjanto E, Leprince O (1996) Water-deficit tolerance and field performance of transgenic alfalfa overexpressing superoxide dismutase. *Plant Physiol.,* **111** : 1177-1181.

Mckersie BD, Bowley SR, Jones KS (1999) Winter survival of transgenic alfalfa overexpressing superoxide dismutase. *Plant Physiol.,* **119** : 839-847.

Ming XT, Wang LJ, An CC, Yuan HY, Chen ZL (2000) Resistance to rice blast (*Pyricularia oryzae*) caused by the expression of trichosanthin gene in transgenic rice plants transferred through *Agrobacterium* method. *Chin. Sci. Bull.,* **45** : 1774-1778

Moon E, Wu R (1996) Genetic modification of a rice glutelin cDNA and expression of the engineered glutelin gene in transgenic rice plants. In : *Rice Genetics III,* Proceedings of the third International Rice Genetics Symposium (Ed. Khush GS), IRRI, Manila, Philippines, pp 814-817.

Morinaga T (1954) Classification of rice varieties on the basis of affinity. In: *Reports for 5th Meeting of International Rice Commission's Working Party on Rice Breeding.* Ministry of Agric and Forestry, Tokyo, pp 1-4.

Morino K, Olsen O-A, Shimamoto K (1999) Silencing of an aleurone-specific gene in transgenic rice is caused by a rearranged transgene. *Plant J.,* **17** : 275-285.

Nayak P, Basu D, Das S, Basu A, Ghosh M, Sen KS (1997) Transgenic elite indica rice plants expressing CryA(c) δ–endotoxin of *Bacillus thuringiensis* are resistant against yellow stem borer (*Scirpophaga incertulas*). *Proc. Natl. Acad. Sci. USA,* **94** : 2111-2116.

Nishizawa Y, Nishio Z, Nakazono K, Soma M, Nakajima E, Ugaki M, Hibi T (1999) Enhanced resistance to blast (*Magnaporthe grisea*) in transgenic japonica rice by constitutive expression of rice chitinase. *Theor. Appl. Genet.,* **99** : 383-390.

Normile D (2000) Asia gets a taste of genetic food fights. *Science,* **289** : 1279-1280.

Noury M, Bassie L, Lepri O, Kurek I, Christou P, Capell T (2000) A transgenic rice cell lineage expressing the oat arginine decarboxylase (*adc*) cDNA constitutively accumulates putrescine in callus and seeds but not in vegetative tissues. *Plant Mol. Biol.,* **43** : 537-544.

Ogawa T (2000) Improvement of cell culture conditions for rice. *Jpn. Agric. Res. Q,* **34** : 215-223.

Okita TW, Hwang YS, Hnilo J, Kim WT, Aryan AP, larson R, Krishnan HB (1989) Structure and expression of the rice glutelin multigene family. *J. Biol. Chem.,* **264** : 12573-12581.

Okkels FT, Ward J, Joersbo M (1997) Synthesis of cytokinin glucuronides for the selection of transgenic plant cells. *Phytochem.,* **46** : 801-804.

Ooms G, Bakker A, Molendijk L, Wullems GJ, Gorden MP, Nester EW, Schilperoort RA (1982) T-DNA organization in homogeneous and heterogeneous octopine-type crown gall tissues of *Nicotiana tabacum. Cell,* **30** : 589-597.

Park SH, Lee B-M, Salas MG, Srivatanakul M (2000) Shorter T-DNA or additional virulence genes improve *Agrobacterium*-mediated transformation. *Theor. Appl. Genet.,* **101** : 1015-1020.

Park SH, Pinson SRM, Smith RH (1996) T-DNA integration into genomic DNA of rice following *Agrobacterium* inoculation of isolated rice shoot apices. *Plant Mol. Biol.,* **32** : 1135-1148.

Peferoen M (1997) Progress and prospects for field use of BT genes in crops. *Trends Biotech.,* **15** : 173-177.

Pingali PL, Hossain M (1998) *Impact of Rice Research,* International Rice Research Institute, Los Baòos, Philippines.

Pinto YM, Kok RA, Baulcombe DC (1999) Resistance to rice yellow mottle virus (RYMV) in cultivated African rice varieties containing RYMV transgenes. *Nat. Biotechnol.,* **17** : 702-707.

Potrykus I, Burkhardt PK, Datta SK, Fütterer Ghosh-Biswas GC, Klöti A, Spangenberg G, Wünn J (1995) Genetic engineering of Indica rice in support of sustained production of affordable and high quality food in developing countries. *Euphytica,* **85** : 441-449.

Potrykus I, Spangenberg G (1995) *Gene transfer to plants.* Springer-Verlag, Berlin.

Preiss J, Ball K, Smithwhite B, Iglesias A, Kakefuda G, Li L (1991) Starch biosynthesis and its regulation. *Biochem. Soc. Trans.,* **19** : 539-547.

Raina SK, Zapata FJ (1997) Enhanced anther culture efficiency of indica rice (*Oryza sativa* L.) through modification of the culture media. *Plant Breed.,* **116** : 305-315.

Ramanathan V, Veluthambi K (1995) Transfer of non-T-DNA portion of the *Agrobacterium tumefaciens* Ti plasmid pTiA6 from the left terminus of T_L-DNA. *Plant. Mol. Biol.,* **28** : 1149-1154.

Ramaswamy C, Jatileksono T (1996) Intercountry comparison of insect and disease losses. In: *Rice Research in Asia: Progress and Priorities* (Eds. Evenson RE, Herdt RW, Hossain M), CAB International and IRRI, Manila, Philippines, pp 305-316.

Rao KV, Rathore KS, Hodges TK, Fu X, Stoger E, Sudhakar D, Williams S, Christou P, Bharathi M, Bown DP, Powell KS, Spence J, Gatehouse AMR, Gatehouse JA (1998) Expression of snowdrop lectin (GNA) in transgenic rice plants confers resistance to rice brown planthopper. *Plant J.,* **15** : 469-477.

Rashid H, Yokoi S, Toriyama K, Hinata K (1996) Transgenic plant production mediated by *Agrobacterium* in indica rice. *Plant Cell Rep.,* **15** : 727-730.

Ratcliff F, Harrison BD, Baulcombe DC (1997) A similarity between viral defense and genome silencing in plants. *Science,* **276** : 1558-1560.

Rathore KS, Chowdhury VK, Hodges TK (1993) Use of *bar* as a selectable marker gene and for the production of herbicide-resistant rice plants from protoplasts. *Plant Mol. Biol.,* **21** : 871-884.

Register III JC (1997) Approaches to evaluating the transgenic status of transformed plants. *TIBTECH,* **15** : 141-146.

Robinson J (1999) Ethics and transgenic crops: a review. EJB Electronic J Biotechnol **2** : 71-81. http://www.ejb.org/content/vol2/issue2/full/3/.

Roxas VP, Smith RK Jr, Allen ER, Allen RD (1997) Overexpression of glutathione-S-transferase/ glutathione peroxidase enhances the growth of transgenic tobacco seedlings durng stress. *Nat. Biotechnol.,* **15** : 988-991.

Roy M, Jain RK, Rohila JS, Wu R (2000) Production of agronomically superior transgenic rice plants using *Agrobacterium* transformation methods: present status and future perspectives. *Curr. Sci.,* **79** : 954-960.

Roy M, Wu R (1999) Engineering rice with genes encoding arginine decarboxylase and s-adenosyl methionine decarboxylase to study stress tolerance. Abstracts, *General meeting of the International Program on Rice Biotechnology,* 20-24 September, Phuket, Thailand, p. 369.

Ryan CA (1990) Protease inhibitors in plants: genes for improving defense against insects and pathogens. *Annu. Rev. Phytopath.,* **28** : 25-49.

Saijo Y, Hata S, Kyozukz J, Shimamoto K, Izui K (2000) Over-expression of a single Ca^{2+}-dependent protein kinase confers both cold and salt/drought tolerance on rice plants. *Plant J.,* **23** : 319-327.

Sakamoto A, Alia, Murata N (1998) Metabolic engineering of rice leading to biosynthesis of glycinebetaine and tolerance to salt and cold. *Plant Mol. Biol.,* **38** : 1011-1019.

Sakulsingharoj C, Choi S-B, Okita TW (2000) Manipulation of ADP-glucose pyrophosphrylase in starch biosynthesis during rice seed development. *4th International Rice Genetics Symposium,* October 22-27, International Rice Research Institute, Los Baòos, Laguna, Philippines, Abstract, p 159.

Sano H, Seo S, Orudgev E, Yossefian S, Ishizuka K, Ohashi V (1994) Expression of the gene for a small GTP-binding protein in transgenic tobacco elevates endogenous cytokinin levels, abnormally induces salicylic acid in response to wounding, and increases resistance to tobacco mosaic virus infection. *Proc. Natl. Acad. Sci. USA* **91** : 10556-10560.

Sawahel WA (1994) Transgenic plants: performance, release and containment. *World J. Microbiol. Biotech.,* **10** : 139-144.

Schlumbaum A, Mauch F, Vögeli U, Boller T (1986) Plant chitinases are potent inhibitors of fungal growth. *Nature,* **324** : 365-367.

Schuler TH, Poppy GM, Kerry BR, Denholm I (1998) Insect-resistant transgenic plants. *TIBTECH,* **16** : 168-175.

Sharma A, Sharma R, Imamura M, Yamakawa M, Machii H (2000a) Transgenic expression of cecropin B, an antibacterial peptide from *Bombyx mori*, confers enhanced resistance to bacterial leaf blight in rice. *FEBS Lett.,* **484** : 7-11.

Sharma HC, Sharma KK, Seetharama N, Ortiz R (2000b) Prospects for using transgenic resistance to insects in crop improvement. EJB Electronic J Biotechnol **3** : 1-20. http://www.ejb.org/content/vol3/issue2/full/3/.

Shi Y, Wang M-B, Powell KS, Van Damme E, Hilder VA, Gatehouse AMR, Boulter D, Gatehouse JA (1994) Use of rice sucrose synthase-1 promoter to direct phloem-specific expression of ß-glucuronidase and snowdrop lectin genes in transgenic tobacco plants. *J. Exp. Bot.,* **45** : 623-631.

Shimada H, Tada Y, Kawasaki T, Fujimura T (1993) Antisense regulation of the rice waxy gene expression using a PCR-amplified fragment of the rice genome reduces the amylose content in grain starch. *Theor. Appl. Genet.,* **86** : 665-672.

Shimamoto K, Terada R, Izawa T, Fujimoto H (1989) Fertile transgenic rice plants regenerated from transformed protoplasts. *Nature,* **338** : 274-276.

Shu QU, Ye GY, Cui HR, Cheng XY, Xiang YB, Wu DX, Gao MW, Xia YW, Hu C, Sardana R, Altosaar I (2000) Transgenic rice plants with a synthetic *cry1Ab* gene from *Bacillus thuringiensis* were highly resistant to eight lepidopteran rice pest species. *Mol. Breed.,* **6** : 433-439.

Siemens J, Schieder O (1996) Transgenic plants: genetic transformation - re rice cent developments and state of the art. *Plant Tissue Cult. Biotechnol.* (IAPTC) **2** : 66-75.

Singh RK, Singh US, Khush GS (2000a) *Aromatic Rices,* Oxford & IBH Publishing Co. Pvt. Ltd., New Delhi.

Singh RK, Singh US, Khush GS, Rohilla R (2000b) Genetics and biotechnology of quality traits in aromatic rices. In: *Aromatic Rices* (Eds Singh RK, Singh US, Khush GS), Oxford & IBH Publishing Co. Pvt. Ltd., New Delhi, pp 47-69.

Sivamani E and Huet H with: Shen P, Ong CA, de Kocko A, Fauquet C, Beachy RN (1999) Rice plant (*Oryza sativa* L.) containing rice tungro spherical virus (RTSV) coat protein transgenes are resistant to virus infection. *Mol. Breed.,* **5** : 177-185.

Songstad DD, Somers DA, Griesbach RJ (1995) Advances in alternative DNA delivery techniques. *Plant Cell Tiss. Org. Cult.,* **40** : 1-15.

Sonnewald U, Lerchl J, Frommer W (1994) Manipulation of sink-source relations in transgenic plants. *Plant Cell Environ.,* **17** : 649-658.

Spiker S, Thompson WF (1996) Nuclear matrix attachment regions and transgene expression in plants. *Plant Physiol.,* **110** : 15-21.

Stark-Lorenzen P, Nelke B, Hänßler G, Mühlbach HP, Thomzik JE (1997) transfer of a grapevine stilbene synthase gene to rice (*Oryza sativa* L.). *Plant Cell Rep.,* **16** : 668-673.

Su J, Shen Q, Ho T-H D, Wu R (1998) Dehydration-stress-regulated transgene expression in stably transformed rice plants. *Plant Physiol.,* **117** : 913-922.

Sundaresan V, Springer P, Volpe T, Haward S, Jones JDG, Dean C, Ma H, Martienssen R (1995) Patterns of gene action in plant development revealed by enhancer trap and gene trap transposable elements. *Genes Dev.,* **9** : 1797-1810.

Suzuki, S, Murai, N, Burnell, JN, Arai, M (2000) Changes in photosynthetic carbon flow in transgenic rice plants that express C4-type phosphoenolpyruvate car boxykinase from *Urochloa panicoides. Plant Physiol.,* **124** : 163-172.

Tada T, Kanazaki H, Norita E, Uchimiya H, Nakamura I (1996) Decreased symptoms of rice blast disease on leaves of *bar*-expressing transgenic rice plants following treatment with bialaphos. *Mol. Plant-Microbe Interact.,* **9** : 758-759.

Temnykh S, Park WD, Ayres N, Cartinhour S, Hauck N, Lipovich L, Cho YG, Ishii T, McCouch SR (2000a) Mapping and genome organization of microsatellite sequences in rice (*Oryza sativa* L.). *Theor. Appl. Genet.,* **100** : 697-712

Theil EC (1990) Regulation of ferritin and transfer in receptor mRNAs. *J. Biol. Chem.,* **265** : 4771-4774.

Thomzik JE, Stenzel K, Stoecker R, Schreier PH, Hain R, Stahl DJ (1997) Synthesis of a grapewine phytoalexin in transgenic tomatoes (*Lycopersicon esculentum* Mill.) conditions resistance against *Phytophthora infestans. Physiol. Mol. Plant Pathol.,* **51** : 265-278.

Toenniessen GH (1991) Potentially useful genes for rice genetic engineering. In: *Rice Biotechnology* (Eds. Khush GS, Toenniessen GH),CAB-International, IRRI, pp 253-281.

Toenniessen GH (1999) A physiologist's perspective on plants and population. *Trends Plant Sci.,* **4** : 242.

Toki S (1997) Rapid and efficient *Agrobacterium*-mediated transformation in rice. *Plant Mol. Biol. Rep.,* **15** : 16-27.

Tu J, Datta K, Khush GS, Zhang Q, Datta SK (2000) Field performance of Xa21 transgenic indica rice (*Oryza sativa* L.), IR72. *Theor. Appl. Genet.,* **101** : 15-20.

Tu J, Ona I, Zhang Q, Mew TW, Khush GS, Datta SK (1998) Transgenic rice variety 'IR72' with *Xa21* is resistant to bacterial blight. *Theor. Appl. Genet.,* **97** : 31-36.

Tyagi AK, Mohanty A, Bajaj S, Chaudhury A, Maheshwari SC (1999) Transgenic rice: a valuable monocot system for crop improvement and gene research. *Critical Rev. Biotechnol.,* **19** : 41-79.

Vain P, Worland B, Clarke MC, Richard G, Beavis M, Liu H, Kohli A, Leech M, Snape J, Christou P, Atkinson H (1998) Expression of an engineered cysteine proteinase inhibitor (Oryzacystatin-1AD86) for nematode resistance in transgenic rice plants. *Theor. Appl. Genet.,* **96** : 266-271.

Vaucheret H, Béclin C, Elmayan T, Feuerbach F, Godon C, Morel J-B, Mourrain P, Palauqui J-C, Vernhettes S (1998) Transgene-induced gene silencing in plants. *Plant J.,* **16** : 651-659.

Vijaychandra K, Palanichelvam K, Veluthambi K (1995) Rice scutellum induces *Agrobacterium tumefaciens vir* genes and T- strand generation. *Plant. Mol. Biol.,* **29** : 125-133.

Wang HZ, Huang DN, Lu RF, Liu JJ, Qian Q, Peng XX (2000) Salt tolerance of transgenic rice (Oryza sativa L.) with *mtlD* gene and *gutD* gene. *Chinese Sci. Bull.,* **45** : 1685-1690.

Wang MB, Li ZY, Upadhyaya NM, Brettell RIS, Waterhouse PM (1997) Intron-mediated improvement of a selectable marker gene for plant transformation using *Agrobacterium tumefaciens. J. Genet. Plant Breed.,* **51** : 325-334.

Wang G-L, Song W-L, Ruan D-L, Sideris S, Ronald PC (1996) The cloned gene, *xa21*, confers resistance to multiple *Xanthomonas oryzae pv. Oryzae* isolates in transgenic plants. *Mol. Plant-Microbe Interact,* **9** : 850-855.

Wenck A, Czako M, Kanevski I, Marton L (1997) Frequent collinear long transfer of DNA inclusive of the whole binary vector during *Agrobacterium*-mediated transformation. *Plant Mol. Biol.,* **34** : 913-922.

WFC (World Food Council) (1992) The global state of hunger and malnutrition: 1992 Report. Agenda Item 2(A), Nairobi (Kenya).

Wu C, Fan Y, Zhang C, Oliva N, Datta SK (1997) Transgenic fertile japonica rice plants expressing a modified *cry1A(b)* gene resistant to yellow stem borer. *Plant Cell Rep.,* **17** : 129-132.

Wu C-Y, Suzuki A, Washida H, Takaiwa F (1998) The GCN4 motif in a rice glutelin gene is essential for endosperm-specific gene expression and is activated by Opaque-2 in transgenic rice plants. *Plant J.,* **14** : 673-683.

Wünn J, Klöti A, Burkhardt PK, Ghosh Biswas GC, Launis K, Iglesias VA, Potrykus I (1996) Transgenic Indica rice breeding line IR58 expressing a synthetic *cry1A(b)* gene from *Bacillus thuringiensis* provides effective insect pest control. *Bio/technology,* **14** : 171-176.

Xu D, Duan X, Wang B, Hong B, Ho T-H D, Wu R (1996a) Expression of a late embryogenesis abundant protein gene, *HVA1*, from barley confers tolerance to water deficit and salt stress in transgenic rice. *Plant Physiol.,* **110** : 249-257.

Xu D, McElroy D, Thornburg RW, Wu R (1993) Systemic induction of a potato *pin2* promter by wounding, methyl jasmonate, and abscisic acid in transgenic rice plants. *Plant Mol. Biol.,* **22** : 573-588.

Xu D, Xue Q, McElroy D, Mawal Y, Hilder VA, Wu R (1996b) Constitutive expression of a cowpea trypsin inhibitor gene, *CpTi*, in transgenic plants confers resistance to two major insect pests. *Mol. Breed.,* **2** : 167-173.

Yamaguchi-Shinozaki K, Shinozaki K (1999) Improving drought, salt, and freezing stress tolerance using a single gene for a stress-inducible transcription factor in transgenic plants. In: *Genetic Improvement of Rice for Water-Limited Environments* (Eds Ito O, O'Toole J, Hardy B), IRRI, Los Baòos, Laguna, Philippines, pp 173-179.

Yamashita S, Ichikawa H, Itoh Y, Ohashi Y (1995) Development of new binary vectors stably maintained in *Agrobacterium* and their utilization to plant transformation. *Breed. Sci.,* **45** : 1-56.

Ye X, Al-Babili S, Klöti A, Zhang J, Lucca P, Beyer P, Potrykus I (2000) Engineering the provitamin A (ß-carotene) biosynthetic pathway into (carotenoid-free) rice endosperm. *Science,* **287** : 303-305.

Yin Z, Wang G-L (2000) Evidence of multiple complex patterns of T-DNA integration into the rice genome. *Theor. Appl. Genet.,* **100** : 461-470.

Yoder JI, Goldsbrough AP (1994) Transformation systems for generating marker free transgenic plants. *Bio/Technology,* **12** : 263-267.

Yokoi S, Higashi S-I, Kishitani S, Murata N and Toriyama K (1998) Introduction of the cDNA for *Arabidopsis* glycerol-3-phosphate acyltransferase (GPAT) confers unsaturation of fatty acids and chilling tolerance of photosynthesis on rice. *Mol. Breed.,* **4** : 269-275.

Yokoi S, Tsuchiya T, Toriyama K, Hinata K (1997) Tapetum-specific expression of the *Osg6B* promoter-b-glucuronidase gene in transgenic rice. *Plant Cell Rep.,* **16** : 363-367.

Yoo J, Jung G (1995) DNA uptake by imbibition and expression of a foreighn gene in rice. *Physiol Plant,* **94** : 453-459.

Zhang Y, Chantler SE, Gupta S, Zhao Y, Hannah LC, Meyer C, Weston J, Wu M-X, Preiss J, Okita TW (1996) Molecular approaches to enhance rice productivity through manipulations of starch metabolism during seed development. *Rice Genetics III,* Proceedings of the third International Rice Genetics Symposium (Ed. G.S. Khush), IRRI, Manila, Philippines, pp 809-813.

Zhang S, Song W-Y, Chen L, Ruan D, Taylor N, Ronald P, Beachy R, Fauquet C (1998) Transgenic elite Indica rice varieties, resistant to *Xanthomonas oryzae* pv. *Oryzae. Mol. Breed.,* **4** : 551-558.

Zhang W, Wu R (1988) Efficient regeneration of transgenic rice plants from rice protoplasts and correctly regulated expression of the foreign gene in the plants. *Theor. Appl. Genet.,* **6** : 835-840.

Zhang J, Xu RJ, Eliott MC, Chen DF (1997) *Agrobacterium*-mediated transformation of elite indica and japonica rice cultivars. *Mol. Biotechnol.,* **8** : 223-231.

Zheng HH, Li Y, Yu ZH, Li W, Chen MY, Ming XT, Casper R, Chen ZL (1997) Recovery of transgenic rice plants expressing the rice dwarf virus outer coat protein gene (S8). *Theor. Appl. Genet.,* **94** : 522-527.

Zhu B, Su J, Chang MC, Verma DPS, Fan YL, Wu R (1998) Overexpression of a D^1-pyrroline-5-carboxylate synthetase gene and analysis of tolerance to water- and salt-stress in transgenic rice. *Plant Sci.,* **139** : 41-48.

Zhu J-K, Hasegawa PM, Bressan RA (1997) Molecular aspects of osmotic stress in plants. *Critical Rev. Plant Sci.,* **16** : 253-277.

Zubko E, Scutt C, Meyer P (2000) Intrachromosomal recombination between attP regions as a tool to remove selectable marker genes from tobacco transgenes. *Nat. Biotechnol.,* **18** : 442-445.

Chapter 3

BIOTECHNOLOGICAL APPROACHES FOR MODIFICATION OF NITROGEN ASSIMILATION IN RICE

Tomoyuki Yamaya★

Department of Applied Plant Science, Graduate School of Agricultural Science, Tohoku University, 1-1Tsutsumidori-Amamiyamachi, Aoba-ku, Sendai 981-8555, Japan and Plant Science Center, RIKEN, 2-1 Hirosawa, Wako 351-0198, Japan

Summary

We have proposed from our immunolocalization studies that cytosolic glutamine synthetase (GS1) is important in the export of nitrogen via phloem in senescing leaves, whereas in developing leaf blade and spikelets, NADH-glutamate synthase (GOGAT) is a key enzyme in the utilization of glutamine transported from senescing organs. To evaluate our hypothesis, transgenic rice plants expressing antisense RNA for either GS1 or NADH-GOGAT were generated after Agrobacterium-mediated transformation. Characteristics of those transformants are discussed in relation to the mechanism of nitrogen utilization in rice plants.

Keywords : Antisense inhibition, cytosolic glutamine synthetase, localization, NADH-glutamate synthase, nitrogen utilization, rice

Abbreviations : GS1, cytosolic glutamine synthetase: GOGAT, glutamate synthase, Fd, ferredoxin, GUS, β-glucuronidase

★Corresponding author : E-mail : tyamaya@biochem.tohoku.ac.jp

1. INTRODUCTION

In Sasanishiki, a leading cultivar of japonica rice (*Oryza sativa* L.) in northern Japan, approximately 80% of total nitrogen in the panicle is remobilized through the phloem from older, senescing organs (Mae *et al.*, 1981). Senescing leaf blades contribute about 50% of total nitrogen in the panicle. Thus, the process of nitrogen recycling is very important in determining both the productivity and quality of rice. However, little has been known on the mechanisms of nitrogen remobilization from the senescing organs and the re-utilization of the remobilized nitrogen for biosynthetic reactions in developing organs. Major forms of nitrogen in the phloem sap of rice plants are glutamine (Gln) and asparagine (Asn) (Hayashi and Chino, 1990). The Asn is probably synthesized from Gln (Ireland and Lea, 1999; Lea *et al.*, 1990; Sechley *et al.*, 1992). Thus, the synthesis of Gln in senescing organs, as well as the utilization of Gln in developing organs are the key steps for nitrogen recycling in rice plants. Gln synthetase (GS; EC 6.3.1.2) and glutamate (Glu) synthase (GOGAT) are the key enzymes in the biosynthesis of Gln and Glu (Ireland and Lea, 1999; Lea *et al.*, 1990; Sechley *et al.*, 1992).

Nitrogen is available in the soil as either NO_3^- or NH_4^+. If the source of nitrogen is NO_3^-, as in the case of most crop plants, some of the NO_3^- could be reduced in the roots, but the majority is transferred to root-xylem and then shoots where it can be processed. In case of rice plants grown in paddy field, on the other hand, major form of nitrogen is NH_4^+. The NH_4^+ taken up by roots is rapidly assimilated within the roots to yield the amide group of Gln and transported via xylem (Lewis *et al.*, 1983; Oaks, 1992). However, the compartment involved in the assimilation of NH_4^+ in rice roots, as well as the mechanisms for the assimilation, have not been understood.

GS catalyzes the ATP dependent conversion of Glu to Gln utilizing NH_4^+ as a substrate. There are two groups of GS isoenzymes, i.e. chloroplastic/ plastidic enzyme (GS2) and cytosolic enzyme (GS1 in leaves and GSr in roots). Corresponding cDNAs have been obtained from a number of plant species (Ireland and Lea, 1999). GS2 is predominant in green tissues and is encoded by a single nuclear gene (Ireland and Lea, 1999). GOGAT catalyzes the reductive transfer of the amide group of Gln to 2-oxoglutarate to form two Glu molecules (Ireland and Lea, 1999). This is the major route to synthesize Glu in plants and the GOGAT reaction is tightly coupled with the GS reaction. GOGAT exists as two molecular species that use either reduced ferredoxin (Fd) or NADH as a reductant (Ireland and Lea, 1999). Fd-GOGAT (EC 1.4.7.1) is the major GOGAT species in green leaves of plants, including rice (Yamaya *et al.*, 1992). Studies with mutants lacking either GS2 in barley (Blackwell *et al.*, 1987; Wallsgrove *et al.*, 1987) or Fd-GOGAT in *Arabidopsis* (Somerville and Ogren, 1980) and in barley (Kendall *et al.*, 1986) clearly show that GS2 and Fd-GOGAT are essential in the reassimilation of NH_4^+ derived from photorespiration. These mutants contain GS1 or NADH-GOGAT (EC 1.4.1.14) comparable to the wild type and grow normally under non-photorespiratory conditions. Therefore, GS1 and NADH-GOGAT are probably important in the synthesis of Gln and Glu for normal growth and development. GS1/GSr and NADH-GOGAT are predominant in non-chlorophyllous organs or tissues, such as roots and vascular bundles of leaves and spikelets (Ireland and Lea, 1999). One approach to consider a role for these enzymes in nitrogen metabolism is to localize corresponding enzyme protein in situ. To obtain a conclusive evidence, studies with mutant lacking either enzyme are required. However, because any mutant lacking either GS1 or NADH-GOGAT has not been isolated from plant kingdom including *Arabidopsis* or barley, it should be very difficult to obtain such mutant from rice. Alternatively, we decide to develop strategies for : 1) cellular localization to identify the compartment and 2) an expression of an antisense RNA to decrease the accumulation of GS1 or NADH-GOGAT protein in rice. The antisense strategy has been successfully employed to inhibit the expression of genes involved in a variety of metabolic pathways, such as nitrate reduction (Vaucheret *et al.*, 1992), ammonium assimilation (Temple *et al.*, 1998; Brugière *et al.*, 1999), and carbon assimilation (Quick *et al.*, 1991). In this chapter, results obtained from my laboratory are mainly described, because studies with rice plants on GS/GOGAT are not actively carried out by other groups.

2. CELLULAR AND SUBCELLULAR LOCALIZATION OF GS1 AND NADH-GOGAT IN RICE

Both GS2 and Fd-GOGAT are major molecular species in green leaves, and their localization and molecular properties are well characterized in many plants (Ireland and Lea, 1999). In rice leaves, GS2 is located mainly in mesophyll cells and parenchyma sheath cells of leaf blade (Sakurai *et al.*, 1996). Fd-GOGAT is also detected in these compartments in leaf blades and in chloroplast-containing cross-cells of pericarp of the grains (Hayakawa *et al.*, 1994). These results are consistent with results from mutants that show that both enzymes are active in photorespiratory nitrogen metabolism (Blackwell *et al.*, 1987; Kendall *et al.*, 1986; Somerville and Ogren, 1980; Wallsgrove *et al.*, 1987).

Sequences of amino acids for rice GS1 and GS2 are homologous (Sakamoto *et al.*, 1989) and polyclonal antibody raised against GS2 purified from rice leaf blades recognize both GS1 and GS2 proteins (Kamachi *et al.*, 1991). Because mono-specific and highly purified antibodies are required to localize small amounts of antigens on the surface of plant segments, rice GS1 immunoglobulin G (IgG) raised against the synthetic peptide, which was less homologous region to GS2 deduced from cDNAs, was affinity-purified with the antigen (Yamaya *et al.*, 1992). Polyclonal IgG raised against the rice cell NADH-GOGAT (Hayakawa *et al.*, 1992) was also affinity-purified. These IgGs mono-specifically recognize the corresponding antigen (Hayakawa *et al.*, 1992; Sakurai *et al.*, 1996). Immunoblotting showed that high content of GS1 protein was detected in senescing leaf blade (Kamachi *et al.*, 1991), whereas that of NADH-GOGAT protein was in developing young organs, such as unexpanded non-green leaves (Yamaya *et al.*, 1992), spikelets at the early stage of ripening (Hayakawa *et al.*, 1993), and root tips (Ishiyama *et al.*, 1998) of rice plants. Distribution of these enzymes was complementary to that of GS2 and Fd-GOGAT, suggesting that GS isoenzymes or GOGAT species have a distinct role in nitrogen metabolisms in rice plants.

In rice roots, the mRNA and protein for NADH-GOGAT accumulated markedly within a few hours of supplying low concentration of NH_4^+ (Yamaya *et al.*, 1995; Hirose *et al.*, 1997). This accumulation occurred in two cell layers of the root surface, the epidermis and exodermis (Ishiyama *et al.*, 1998). Gln is probably a signal substance for the accumulation of NADH-GOGAT mRNA (Hirose *et al.*, 1997) and okadaic acid-sensitive protein phosphatase could be involved in the regulation of NADH-GOGAT gene expression and the signal transduction (Hirose and Yamaya, 1999). Although the supply of NH_4^+ had less effect on the expression of GSr, this enzyme was also found to be located in the epidermis and exodermis, as well as other cell-types in rice roots (Ishiyama *et al.*, 1998). This is consistent with subcellular localization studies with barley roots (Peat and Tobin, 1996). This finding indicates that NADH-GOGAT, coupled with the GSr reaction, is important for the primary assimilation of NH_4^+ in two cell layers of the root surface.

Our immunocytological studies with rice leaves at the reproductive stage showed that GS1 protein was detected in companion cells and vascular parenchyma cells of senescing leaf blade (Sakurai *et al.*, 1996). Localization of GS1 protein in the phloem companion cells have also been shown with other plant species (Carvalho *et al.*, 1992; Pereira *et al.*, 1992; Dubois *et al.*, 1996). Localization of the enzyme in phloem companion cells suggests that GS1 is important for the export of leaf nitrogen from senescing leaves, because companion cells are important for phloem loading of solutes (Van Bel, 1993).

NADH-GOGAT is located in the plastids of non-photosynthetic tissues such as roots (Hayakawa *et al.*, 1999) and root-nodules (Trepp *et al.*, 1999). NADH-GOGAT is also active in vascular tissues of developing leaf blade and grains of rice plants (Hayakawa *et al.*, 1993) and 5'-upstream region of gene for this enzyme expresses a reporter gene specifically in vascular bundles of these organs (Kojima *et al.*, 2000). Our immunolocalization studies showed that this enzyme protein was detected in vascular-parenchyma cells and mestome-sheath cells of developing leaf blade and in vascular-parenchyma cells of both dorsal and lateral vascular bundles and in the nucellar projection, nucellar epidermis, and aleaurone cells of developing grains (Hayakawa *et al.*, 1994). All these cells containing NADH-GOGAT protein are important for solute transport from phloem and xylem, suggesting that NADH-GOGAT is responsible in young organs for the synthesis of Glu from Gln that is transported from senescing organs or roots via phloem and xylem vessel elements (Hayakawa *et al.*, 1994).

3. ANTISENSE STRATEGY

Because any mutant lacking GS1 or NADH-GOGAT has not been isolated from plant kingdom including *Arabidopsis* or barley, conclusive evidence to support our hypothesis is not presented. This is probably because GS1 is encoded by a family of genes and each member is differentially expressed in a particular organ or tissue (Ireland and Lea, 1999). Alternatively, lack of GS1 or NADH-GOGAT can result in a lethal mutation. To investigate the role of these enzymes, we have attempted to silence either GS1 or NADH-GOGAT using RNA antisense strategy, in rice plants.

Nucleotide sequences for GS1, GSr, and GS2 cDNAs are highly homologous in rice (Sakamoto *et al.*, 1989). To reduce expression specifically in companion cells and vascular parenchyma cells of rice leaves, a 131-bp fragment of 3'-untranslated region of rice GS1 cDNA, corresponding to a less homologous region of rice GS2 and GSr, was placed in the antisense orientation and this chimeric gene was fused with a rice thioredoxin h promoter. This promoter directs companion cell-specific accumulation of a mRNA in situ in rice (Ishiwatari *et al.*, 1998). Own promoter (Kojima *et al.*, 2000a,b) was also used for the antisense-transformation. The chimeric gene was then introduced into rice (*Oryza sativa* cv. Sasanishiki) via *Agrobacterium tumefaciens*-mediated transformation (Goto *et al.*, 1997). Analyses of the regenerated antisense lines are now in progress from many aspects. Brugière *et al.* (1999) successfully decreased the expression of GS1 in vascular tissues of *Nicotiana plummbaginifolia* using antisense RNA strategy. Their conclusion was that GS1 in phloem tissues of *N. plummbaginifolia* plays a major role in regulating proline production, consistent with the function of proline as a key metabolite synthesized in response to water stress.

To reduce the expression of *NADH-GOGAT* gene, a 856-bp fragment of cDNA for rice NADH-GOGAT (Goto *et al.*, 1998) was placed in the antisense orientation under the control of cauliflower mosaic virus 35S promoter. T_1 lines containing low copy number were analysed (K Ishiyama *et al.*, unpublished data). The transformants expressed the antisense RNA, decreased accumulation of mRNA for the enzyme, and contained less amounts of NADH-GOGAT protein in developing leaf blade and spikelets. The antisense plants showed comparable growth rate and plant height to the untransformed control. However, significant reduction in weight of 1,000 seeds was observed in the antisense plants, although total spikelet number per panicle were similar. Thus, the grain-filling of the antisense plants was severely inhibited. We conclude that NADH-GOGAT in vascular bundle of spikelets plays a major role in regulating reutilization of glutamine transported from other organs.

4. CONCLUSIONS AND FUTURE PROSPECTS

Unlike the works with mutants, the antisense effect causes an incomplete inhibition of a target gene expression. Therefore, sufficient background for the target molecule is required to evaluate its physiological role. Our results using antisense strategy as well as immunolocalization studies successfully show that GS1 and NADH-GOGAT are important in nitrogen utilization in rice plants. The current results were obtained from the transformants either at T_0 or T_1 generation which possesses heterozygous transgene. Analyses of nitrogen use efficiency in details are further required using homozygous progenies from those antisense plants.

An alternative approach for improving nitrogen utilization could be the use of genetic resources of rice plants. *Oryza sativa* is widespread and there are three types (or subspecies) identified for cultivated rice plants, i.e. japonica, indica, and javanica. Recent results from our laboratory show that a few cultivars of indica rice contained more GS1 protein in senescing leaf blade than Sasanishiki (japonica), whereas Sasanishiki contained more NADH-GOGAT in developing leaf blade than some of indica cultivars (Obara *et al.*, 2000). An overexpression of the target genes, i.e. *GS1* and *NADH-GOGAT*, in the corresponding compartment of japonica and indica cultivars, respectively, is expected to lead to higher efficiency of nitrogen use and higher yield. Recently, back-cross inbred lines between Nipponbare (japonica) and Kasalath (indica) have been generated and DNA makers were mapped on 12 chromosomes. Analyses of quantitative trait loci (QTL) using these inbred lines for determining contents of both enzymes and related parameters could be useful to understand whole process of metabolism and regulation on nitrogen utilization in rice plants. We have recently detected 7 QTLs for GS1 protein content and 6 for NADH-GOGAT (Obara *et al.*, 2001). Fine mapping is now in progress to identify individual genes.

ACKNOWLEDGEMENTS

This work was supported in part by a program of Research for the Future from Japan Society for the Promotion of Science (JSPS-RFTF96L00604).

REFERENCES

Blackwell RD, Murray AJS and Lea PJ (1987) Inhibition of photosynthesis in barley with decreased levels of chloroplastic glutamine synthetase activity. *J. Exp. Bot.*, **38** : 1799-1809.

Brugière N, Dubois F, Limami AM, Lelandais M, Roux Y, Sangwan RS and Hirel B (1999) Glutamine synthetase in the phloem plays a major role in controlling proline production. *Plant Cell*, **11**: 1995-2011.

Carvalho H, Pereira S, Sunkel C and Salema R (1992) Detection of a cytosolic glutamine synthetase in leaves of *Nicotiana tabacum* L. by immunocytochemical methods. *Plant Physiol.*, **100** : 1591-1594.

Dubois F, Brugiere N, Sangwan RS and Hirel B (1996) Localization of tobacco cytosolic glutamine synthetase enzymes and the corresponding transcripts shows organ- and cell-specific patterns of protein synthesis and gene expression. *Plant Mol. Biol.*, **31** : 803-817.

Goto S, Ishii Y, Hayakawa T and Yamaya T (1997). *Agrobacterium*-mediated transformation of Sasanishiki, a leading cultivar of rice (*Oryza sativa* L) in northern Japan. In : *Plant Nutrition - for Sustainable Food Production and Environment* (Ed. Ando T) Kluwer Academic Publisher, Dordrecht, The Netherlands, pp. 839-840.

Goto S, Akagawa T, Kojima S, Hayakawa T and Yamaya T (1998) Organization and structure of NADH-dependent glutamate synthase gene from rice plants. *Biochem. Biophys. Acta*, **1387** : 298-308.

Hayakawa T, Hopkins L, Peat LJ, Yamaya T and Tobin AK (1999) Quantitative intercellular localization of NADH-dependent glutamate synthase protein in different types of root cells in rice plants. *Plant Physiol.*, **119** : 409-416.

Hayakawa T, Yamaya T, Mae T and Ojima K (1993) Changes in the content of two glutamate synthase proteins in spikelets of rice (*Oryza sativa*) plants during ripening. *Plant Physiol.*, **101** : 1257-1262.

Hayakawa T, Yamaya T, Mae T and Ojima K (1992) Purification, characterization, and immunological properties of NADH-dependent glutamate synthase from rice cell cultures. *Plant Physiol.*, **98** : 1317-1322.

Hayakawa T, Nakamura T, Hattori F, Mae T, Ojima K and Yamaya T (1994) Cellular localization of NADH-dependent glutamate-synthase protein in vascular bundles of unexpanded leaf blades and young grains of rice plants. *Planta*, **193** : 455-460.

Hayashi H and Chino M (1990) Chemical composition of phloem sap from the upper most internode of the rice plant. *Plant Cell Physiol.*, **31** : 247-251.

Hirose N, Hayakawa T and Yamaya T (1997) Inducible accumulation of mRNA for NADH-dependent glutamate synthase in rice roots in response to ammonium ions. *Plant Cell Physiol.*, **38** : 1295-1297.

Hirose N and Yamaya T (1999) Okadaic acid mimics nitrogen-stimulated transcription of NADH-glutamate synthase gene in rice cell cultures. *Plant Physiol.*, **121** : 805-812.

Ireland RJ and Lea PJ (1999) The Enzymes of Glutamine, Glutamate, Asparagine, and Aspartate Metabolism. In: *Plant Amino Acids: Biochemistry and Biotechnology* (Ed. Singh BK), Marcel Dekker, Inc., New York, U.S.A. pp. 49-109.

Ishiwatari Y, Fujiwara T, McFarland KC, Nemoto K, Hayashi H, Chino M and Lucas WJ (1998) Rice phloem thioredoxin has the capacity to mediate its own cell-to-cell transport through plasmodesmata. *Planta*, **205** : 12-22.

Ishiyama K, Hayakawa T and Yamaya T (1998) Expression of NADH-dependent glutamate synthase protein in the epidermis and exodermis of rice roots in response to the supply of nitrogen. *Planta*, **204** : 288-294.

Kamachi K, Yamaya T, Mae T and Ojima K (1991) A role for glutamine synthetase in the remobilization of leaf nitrogen during natural senescence in rice leaves. *Plant Physiol.*, **96** : 411-417.

Kendall AC, Wallsgrove RM, Hall NP, Turner JC and Lea PJ (1986) Carbon and nitrogen metabolism in barley (*Hordeum vulgare* L.) mutants lacking ferredoxin-dependent glutamate synthase. *Planta*, **168** : 316-323.

Kojima S, Kimura M, Nozaki Y and Yamaya T (2000a) Analysis of a promoter for NADH-glutamate synthase gene in rice (*Oryza sativa*): Cell-type specific expression in developing organs of transgenic rice plants. *Aust. J. Plant Physiol.*, **27** : 787-793.

Kojima S, Hanzawa S, Hayakawa T, Hayashi M and Yamaya T (2000b) Nucleotide sequence of a genomic DNA (Accession No. AB037664) and a cDNA (Accession No. AB037595) encoding cytosolic glutamine synthetase in Sasanishiki, a leading cultivar of rice (*Oryza sativa* L) in northern Japan. (PGR 00-048). *Plant Physiol.,* **122** : 1459.

Lea PJ, Robinson SA and Stewart GR (1990) The enzymology and metabolism of glutamine, glutamate and asparagine. In: *The biochemistry of plants, Intermediary nitrogen metabolism* (Eds Miflin B J and Lea P J), Academic Press, San Diego, CA pp. 121-157.

Mae T, Makino A and Ohira K (1981) The remobilization of nitrogen related to leaf growth and senescence in rice plants (*Oryza sativa* L.). *Plant Cell Physiol.*, **22** : 1067-1074.

Obara M, Sato T and Yamaya T (2000) High content of cytosolic glutamine synthetase does not accompany with a high activity of the enzyme in rice (*Oryza sativa* L) leaves of indica cultivars during the life span. *Physiol. Plant,* **108** : 11-18.

Obara M, Kajiwara M, Fukuta Y, Yano M, Hayashi M, Yamaya T and Sato T (2001) Mapping of QTLs associated with contents of cytosolic glutamine synthetase and NADH-glutamate synthase in rice (*Oryza sativa* L.) *J. Exp. Bot.,* **52** : 1209-1217.

Peat LJ and Tobin AK (1996) The effect of nitrogen nutrition on the cellular localization of glutamine synthetase isoforms in barley roots. *Plant Physiol.*, **111** : 1109-1117.

Pereira S, Cavalho H, Sunkel C and Salema R (1992) Immunocytolocalization of glutamine synthetase in mesophyll and phloem of leaves of *Solanum tuberosum* L. *Protoplasma*, **167** : 66-73.

Quick WP, Schurr U, Fichtner K, Schulze ED, Rodermel SR, Bogorad L and Stitt M (1991) The impact of decreased Rubisco on photosynthesis, growth, allocation and storage in tobacco plants which have been transformed with antisense *rbc*S. *Plant J.*, **1** : 51-58.

Sakamoto A, Ogawa M, Masumura T, Shibata D, Takeba G, Tanaka K and Fujii S (1989) Three cDNA sequences coding for glutamine synthetase polypeptides in *Oryza sativa* L. *Plant Mol. Biol.*, **13** : 611-614.

Sakurai N, Hayakawa T, Nakamura T and Yamaya T (1996) Changes in the cellular localization of cytosolic glutamine synthetase protein in vascular bundles of rice leaves at various stages of development. *Planta*, **200** : 306-311.

Sechley KA, Yamaya T and Oaks A (1992) Compartmentation of nitrogen assimilation in higher plants. *Int. Rev. Cytol.*, **134** : 85-163.

Somerville CR and Ogren WL (1980) Inhibition of photosynthesis in *Arabidopsis* mutants lacking in leaf glutamate synthase activity. *Nature,* **286** : 257-259.

Trepp GB, Plank DW, Gantt JS and Vance CP (1999) NADH-glutamate synthase in alfalfa nodules: immunocytochemical localization. *Plant Physiol.,* **119** : 829-837

Van Bel AJE (1993) Strategies of phloem loading. *Annu. Rev. Plant Physiol. Mol. Biol.*, **44** : 253-281.

Vaucheret H, Kronenberger J, Lepingle A, Vilaine F, Boutin JP and Caboche M (1992) Inhibition of tobacco nitrite reductase activity by expression of antisense RNA. *Plant J.,* **2** : 559-569.

Temple SJ, Bagga S and Sengupta-Gopalan C (1998) Down-regulation of specific members of the glutamine synthetase gene family in alfalfa by antisense RNA technology. *Plant Mol. Biol.*, **37** : 535-547.

Wallsgrove RM, Turner JC, Hall NP, Kendall AC and Bright SWJ (1987) Barley mutants lacking chloroplast glutamate synthetase - biochemical and genetic analysis. *Plant Physiol.*, **83** : 155-158.

Yamaya T, Hayakawa T, Tanasawa K, Kamachi K, Mae T and Ojima K (1992) Tissue distribution of glutamate synthase and glutamine synthetase in rice leaves. Occurrence of NADH-dependent glutamate synthase protein and activity in the unexpanded non-green leaf blades. *Plant Physiol.*, **100** : 1427-1432.

Yamaya T, Tanno H, Hirose N, Watanabe S and Hayakawa T (1995) A supply of nitrogen causes increase in the level of NADH-dependent glutamate synthase protein and in the activity of the enzyme in roots of rice seedlings. *Plant Cell Physiol.*, **36** : 1197-1204.

Chapter 4

TISSUE CULTURE OF WHEAT

SL Kothari★ and Satish Kumar

Department of Botany, University of Rajasthan, Jaipur – 302 004, India

Summary

Wheat is one of most important cereal crops accounting for about one-third of the grain consumption in the world. In vitro regeneration of plants in wheat has been reported from different explants and callus cultures. The plant regeneration is strongly influenced by the genotype and plant growth regulators used in the medium. The in vitro plant regeneration in wheat occurs through somatic embryogenesis or shoot bud formation and very often both the processes also occur in the same culture. The present review will look into the various aspects of in vitro plant regeneration in wheat.

Keywords : *Triticum aestivum*, explant culture, somatic embryogenesis, organogenesis, plant regeneration.

1. INTRODUCTION

Wheat is one of the most important cereal crops in the world in terms of area under cultivation, yield and contribution to the human diet. It accounts for about one-third of the grain consumption in the world (Jha, 1998). The classical plant breeding has led to tremendous improvement in yield, quality and agronomic performance of crop plants. Biotechnology now offers a helping hand to classical plant breeding by providing additional and novel variations through foreign DNA insertions and DNA marker assisted selection of simple genes and the quantitative trait loci. *In vitro* regeneration of plants from single cells or tissues is a pre-requisite for

★Corresponding author : E-mail : SLKothari@lycos.com

obtaining genetically engineered (transgenic) plants. Until about late 1970s, tissue culture of cereal crops was considered highly difficult. It was in the 1970s that successful callus formation and plant regeneration was reported in cereals (Rangan, 1973, 1976; Dudits *et al.*, 1975; King *et al.*, 1978). Highly efficient and reproducible regeneration of plants from callus and suspension cultures was established in the 1980s largely through the efforts of Vasil and co-workers in Florida, Green and Phillips and Potrykus *et al.* (see Morrish *et al.*, 1987; Bhaskaran and Smith, 1990). The morphologic identification of embryogenic callus and its selective propagation and its use in raising totipotent suspension and protoplast cultures were the crucial steps in cereal biotechnology which paved the way for genetic transformation and generation of transgenic plants in cereal crops. The present review will highlight the important development in the area of tissue culture of wheat. Successful plant tissue culture depends upon selection of suitable explants, nutrient medium, plant growth regulators and other media supplements. Effect of genotype on regeneration, morphogenic pathways of regeneration and work on cell suspension and protoplast cultures will be discussed.

2. EXPLANTS

Selection of proper explant is crucial for successful plant regeneration via *in vitro* culture techniques. Not every cell or tissue is totipotent to give rise to regenerable callus and subsequent plant regeneration. Availability of explant for tissue culture is another problem because immature explants are not available throughout the year. A broad category of explant used for tissue culture of wheat has been discussed in the following section.

2.1. Caryopses and mature embryos

The benefit of using seeds and mature embryos as explant is that they are available throughout the year. In wheat seed and mature embryos have been used by a number of workers (Cure and Mott, 1978; MacKinnon *et al.*, 1987; Mohmand and Nabors, 1990, 1991; O'Hara and Street, 1978; Eapen and Rao, 1982b; Rao and Kothari, 1992). Scutellum and embryo axis are sources of callus when seeds were cultured on medium with different hormonal combinations (Cure and Mott, 1978; Eapen and Rao, 1985; Gosch-Wackerle *et al.*, 1979; Heyser *et al.*, 1985). However, the callus was non-regenerable but in 1985 Eapen and Rao reported shoot differentiation on IAA and Zeatin supplemented media. The embryogenic cultures from mature seeds exhibited three types of calli-smooth embryogenic, nodular embryogenic and non-embryogenic, on very high levels of 2,4-D (20 mg/1 (Heyser *et al.*, 1985).

In the earliest report on mature embryo culture in wheat, Chin and Scott (1977) reported 7% regeneration on a hormone free medium, from the callus obtained on 2,4-D supplemented medium. Varshney *et al.* (1999) recently reported plant regeneration from mature embryos of 17 cultivars of *Triticum aestivum* and 3 cultivars of *Triticum durum*. They also noted variation in regeneration response in different cultivars. Callus induction and maintenance required 2,4-D (2.5 mg/1). Plant regeneration was achieved on medium supplemented with BAP (1 mg/1) + NAA (0.2 mg/1) from the callus subcultured at least once on maintenance medium containing 2,4-D (2.5 mg/1). Plant regeneration frequency as high as 90% was reported by them. Ozgen *et al.* (1998) used mature embryos for induction of embryogenic callus in wheat on medium supplemented with higher concentration of 2,4-D (8 mg/1). They also noted a strong genotypic effect on the culture response.

2.2. Leaf, leaf bases, mesocotyl, coleoptile, nodes, internodes and apical meristem

Leaf explant which lacks meristem except at the base, provides an ideal material for obtaining totally undifferentiated callus. In fact, molecular basis of differentiation can best be investigated with leaf (Jamora and Scott, 1983). After 1980 a number of workers successfully attempted callus induction and plant regeneration from leaf and leaf base segments (Ahuja *et al.*, 1982; McHughen, 1983; Rajyalakshmi *et al.*, 1991; Tabaeizadeh *et al.*, 1990; Wernicke and Milkovits, 1984, 1986). It has been observed that in all cases, 2,4-D was the potent embryogenic callus induction hormone. However, Zamora and Scott (1983), added arginine and yeast extract to the medium.

The earlier work with leaf explant of wheat exhibited plant regeneration through organogenesis (Ahuja *et al.*, 1982; Zamora and Scott, 1983). Wernicke and Milkovits (1984, 1986) reported formation of callus with structures resembling to the somatic embryos. Plantlet regeneration via somatic embryogenesis has been reported on hormone free medium from the callus induced from basal segments of 2nd and 3rd leaves on a 2,4-D containing medium (Rajyalakshmi *et al.*, 1991).

From the young seedling, a number of different explants can be used for callus induction, *viz.*, mesocotyl segments, stem axes, nodal and internodal regions of stem, apical meristems etc. From mesocotyl, efficient plantlet regeneration has been reported to be about 30-40%. Bartok and Sagi (1990) developed a new method of culturing mesocotyl of mature wheat embryo from which plant regeneration was obtained via organogenesis. For this purpose, coleoptile and coleorrhiza were surgically removed and callus was induced from mesocotyl, on medium containing 2,4-D. Varshney *et al.* (1998) reported callus induction from coleoptiles on MS medium containing 2,4-D and subsequent plantlet formation on NAA, BAP containing medium, nodal and internodel regions or stems of seedlings have also been reported to be callus inductive, and the later could be maintained for about 3 years by periodical subculture (O'Hara and Street, 1978).

Induction of callus and plant regeneration can best be reported from apical meristems (Ishii, 1982). For the first time, Wernicke and Milkovits, 1986, reported regeneration via somatic embryogenesis from shoot meristem culture. They also exhibited an enhanced effect of 2,4-D on morphogenesis. Regeneration of apical explants upto the formation of fertile plants was shown by Simmonds *et al.* (1992).

2.3. Inflorescence

In wheat, immature inflorescence has also been as an important source of callus inducible tissue. The calli obtained from the immature inflorescence produce more morphogenetic cultures. It has been proved that immature inflorescences have a number of suppressed meristematic zones, which propagate on contact with the nutrient medium. The earlier work with immature inflorescence reviews that callus response was poor and there was no sign of differentiation (Chin and Scott, 1997). Ozia-Akins and Vasil, 1982, reported compact callus and complete plant regeneration. Maddock and Semple, 1986 and Rajyalakshmi *et al.*, 1988, reported high frequency plantlet regeneration from immature inflorescence induced callus on medium supplemented with coconut milk and casein hydrolysate, respectively. Regeneration was *via* somatic embryogenesis. Hunsinger and Schauz, 1987, reported callus induction and plant regeneration on LM medium supplemented with Dicamba (Li and Dong, 1994; Maddock *et al.*, 1983; Menki-Milczarek and Zimny, 1992; Redway *et al.*, 1990b).

Size of cultured inflorescence plays an important role in callusing response. Sharma *et al.* (1995); Izabela *et al.* (1995) and Karen *et al.* (2000) cultured different sizes of immature inflorescence and reported that 0.5 and 1.0 cm long inflorescences provided higher regeneration frequency and efficiency.

2.4. Immature embryo

Unlike mature embryos which remain available throughout the year and can be utilized when required, the immature embryos are the explants which are available only for a limited period of the season are the most responsive explants in tissue culture. Shimada 1978, first regenerated the complete plantlets from immature embryos on 'LS' medium supplemented with 2,4-D. Later, Shimada and Yamada (1979) studied various factors affecting tissue culture. Younger embryos about 14-d old found to be best for callus induction. Gosch-Wacerle *et al.* (1979) used 14-18 d old immature embryos and obtained shoot regeneration using IAA and Zeatin.

A number of studies have been done on immature embryo culture of wheat and describing a number of different protocols. Bohorova *et al.* (1994) reported induction of callus on 'N6' medium using Dicamba. On MMS, SD1 media supplemented with IAA, NAA, and 2,4-D or Dicamba. Carman *et al.* (1987a,b); Chawla (1988) reported induction of callus on NAA supplemented MS medium and subsequent plant regeneration on BAP. Ahmad and Sayed-Shahram (1999); Fennel *et al.* (1996) observed callus induction on MS medium supplemented with 2,4-D. In an extensive study, Comeau *et al.* (1992) used Norstog II, B5 and MS media for callus induction and plantlet differentiation with IAA and GA_3. Similarly, Eapen and Rao (1982a) reported induction of callus and plant regeneration on B_5, MS, Lin and Staba (1961) media using a broad range of auxins and Kn. Potato tuber extract supplemented with 2,4-D has also been proved to be favourable for callus induction (Lazer *et al.*, 1987). Redway *et al.* (1990b) used a variety of media and reported callus induction when supplemented with 2,4-D and Dicamba, and subsequent plant regeneration on hormone supplemented basal medium.

Although the nutrient medium provided to the explant is so formulated that it can favour the maximum proliferation, however, a few other growth adjuvants like CW, CH, tryptophan etc. significantly improve the *in vitro* response of the explants. Maddock *et al.* 1983 and Mathias and Simpson 1986 observed a promotory effect of coconut water on callus induction, embryoid formation and plant regeneration. Carman *et al.* (1987a) and Nabors *et al.* (1983) reported a significantly increased embryogenic callus formation.

In wheat, plant regeneration occurs mainly through somatic embryogenesis. Histological observations of the tissue cultured embryos exhibited that the scutellum is the main tissue from which the somatic embryos arise. The zygotic embryogeny and embryogenic development is similar in somatic embryogenesis (Magnusson and Bornmann, 1985). It has been well established that in wheat both embryogenic and organogenic mode of regeneration take place through scutellum and epiblast derived callus (He *et al*., 1990 and Kirsten *et al*., 1987).

The embryogenic potential of an explant is not only explored through the composition of nutrient medium, but can also be improved by selecting an immature explant at suitable age. Various studies confirm that younger embryos produce comparatively more somatic embryos (He *et al*., 1990 and Hunsinger and Schauz, 1987; Varshney *et al*., 1996). The position of immature embryo on medium is also critical. Eapen and Rao (1985), He *et al*. (1988) and Heyser *et al*. (1985) suggested that if scutellum is faced away from the medium, the frequency of plant regeneration could be increased.

2.5. Roots

Roots, root tips containing root apices have also been explored to be a favourable explant for prolific callus induction and plant regeneration via tissue culture. However, a number of culture media and hormones with growth additives were tried but there was no successful plant regeneration (Chin and Scott, 1977; Dudits *et al*., 1975; Ishii, 1982; Mascarenhas *et al*., 1975; O'Hara and Street, 1978; Trione *et al*., 1968). Adachi and Katayawa (1969) and Shimada *et al*. (1969) must be credited for reporting successful plantlet regeneration from root cultures. Bhojwani and Hayward 1977 added asparagine to the medium and enhanced regeneration of plantlets. Nabors *et al*. (1983) obtained rather improved regeneration from root derived cultures. They reported that embryogenic calli showed 33 times more regeneration than non-embryogenic calli.

3. PLANT GROWTH REGULATORS AND OTHER ADJUVANTS IN CALLUS INDUCTION AND REGENERATION

Formulation of a favourable nutrient medium is the first requirement of successful regeneration in tissue culture. The nutrient medium which is a composition of various organic and inorganic compounds is further supplemented with plant growth regulators and other adjuvants. The general requirement of PGR for most of the callus cultures of wheat are auxins and cytokinins. Auxins – a class of compounds that stimulates callus induction, maintenance and shoot cell elongation, resemble IAA in their roles. Cytokinins promote cell proliferation tissue culture. Besides a number of auxins, both natural and synthetic, 2,4-D is the principal auxin which is mainly responsible for callus induction, maintenance and proliferation of somatic embryos (Abe and Futsuhara, 1985, 1986b; Bregitzer *et al.*, 1989; Fatokun and Yamada, 1984; He *et al.*, 1986; Mikami and Kinoshita, 1988; Ozava and Komamine, 1989; Raghava Ram and Nabors, 1984; Sears and Deckard, 1982; Vasil, 1987; Wernicke and Brittell, 1982; Yao and Krikorian, 1981).

With almost all of the explants of wheat, 2,4-D exhibits remarkable performance in induction of callus. However, its omission from the medium results in the regeneration *via* both somatic embryogenesis and organogenesis.

A few other auxins like Dicamba, pCPA, 2,4,5-T, NAA, IAA, PAA have also been reported to be lesser, equal or greater efficient in callus induction and plant regeneration (Bohorova *et al.*, 1994; Carman *et al.*, 1987b; Dudits *et al.*, 1975. Hunsinger and Schauz, 1987; Redway *et al.*, 1990b; Chin and Scott, 1977; Eapen and Rao, 1982; Hu and Kasha, 1995 and Mascarenhas *et al.*, 1975). Hunsinger and Schauz (1987) and Ishii (1982) observed that dicamba resulted in initiation of larger calli and large number of shoots from cultures maintained in dark when supplemented as substitution of 2,4-D.

Precocious germination of embryos sometimes reduces frequency of somatic embryogenesis in wheat. To suppress precocious germination and increase the somatic embryogenesis frequency, various protocols have been developed. Bapat *et al.* (1988) suggested the use of high concentration of 2,4-D. Carman *et al.* (1987a,b) and Papenfuss and Carman (1987) reported positivie role of dicamba to overcome precocious germination. Sorbitol (0.3 M) is also able to suppress the precocious germination (Ryschka *et al.*, 1991).

The cytokinins alone have been supposed to discourage the induction of somatic embryogenesis. However, when supplemented with 2,4-D, dicamba, and 2,4,5-T, Kn exhibits a promotory effect (Mathias *et al.*, 1986; Lazer *et al.*, 1988; Carman *et al.*, 1987b, 1988b; Chu *et al.*, 1990; Huang, 1987; Liang *et al.*, 1982; Lu *et al.*, 1991; Nabors *et al.*, 1983). Lazer *et al.*, 1983 noted a depressive effect of Kn on callus proliferation. Low concentration of ABA enhanced embryogenesis in callus cultures (Brown *et al.*, 1989). A few reports on BAP are also available but these exhibit a negligible role in induction of callus (Dudits *et al.*, 1975; Ozias-

Akins and Vasil, 1982). They reported callus induction from immature embryo when 2,4-D was used with 2-iP.

Although, various protocols for efficient plant regeneration have been developed by a number of workers, but to maximize the efficiency of regeneration, investigators also manipulated the composition of basal medium as well as included complex substances to the medium. There is a variety of growth promoting adjuvants which have been utilized in tissue culture of wheat. These are – coconut water, casein hydrolysate, amino acids, vitamins, antibiotics etc. Ozias-Akins and Vasil (1983b) made an extensive study on the role of different growth adjuvants and reported that vitamins, casein hydrolysate supported 30% increase in induction of embryogenesis. Maddock *et al.* (1983) and Mathias and Simpson (1986) concluded that coconut water has a promotive effect on callus induction, embryoid formation and subsequent plant regeneration. Carman *et al.* (1987b) and Nabors *et al.* (1983) added tryptophan and increased embryogenic callus formation, significantly.

Mathias and Boyd (1986) advised addition of cefotaxime and carbenicillin for improved callus induction and plant regeneration. Carman *et al.* (1987a) used GA_3 and obtained plantlet formation while Ahloowalia (1982) and Redway *et al.* (1990b) practiced with zeatin and reported plant regeneration.

4. GENOTYPIC EFFECT

It has now generally been accepted that conditions optimal for plant regeneration in one cultivar fail to produce plants in another cultivar of the same species. It indicates that various *in vitro* responses are genotype dependent. Induction of callus, somatic embryogenesis, somaclonal variations, regeneration etc. are controlled by nuclear genes (Hodges *et al.*, 1986; Tomes and Smith, 1985). Willman *et al.* (1989) suggested that at least one gene or a block of genes controls the expression of somatic embryogenesis in tissue cultures. Close and Gallgher-Ludeman (1989) have observed the role of nuclear genes in the control of responsiveness to plant growth regulator type and concentration.

The differences in performance from cultivar to cultivar may be attributed to variations in the endogenous hormone levels. Moreover, explants from a single genotype do not respond identically under culture conditions, most likely due to varying gradiants of endogenous hormones (Wernicke and Brettell, 1982).

Wheat is also affected by genotypic variations in callus induction and regeneration from different explants (Ahloowalia, 1982; Carman *et al.*, 1987a,b; Mathias *et al.*, 1987; Sears and Deckard, 1982; Wang and Nguyen, 1990; Machii *et al.*, 1998 and Karen *et al.*, 2000). Maddock and Semple (1983) examined 25 cultivars and found a variation in culture responses from 12 to 96%. Production of plants from embryogenic and non-embryogenic calli and callus induction from epiblast have been found to be strictly genotype dependent (He *et al.*, 1986, 1988; Heyser *et al.*, 1985). A comprehensive study of the matter indicates that something within an explant is also as critical as its genotype. Many of the genotypic differences can be minimized by growing the plants under optimal conditions, and also by varying nutrients and hormonal complement in the culture medium (Duncan *et al.*, 1985).

5. MORPHOGENIC PATHWAYS

Plant regeneration in tissue follows the following pathways: (i) Shoot morphogenesis from the axillary buds or adventitious formation of shoot buds (ii) *de novo* formation of shoot buds in callus cultures and (iii) *de novo* or adventitious formation of somatic embryos. In wheat, formation of axillary meristems is not very common but addition of cytokinins and auxins in the nutrient media sometimes results in the formation of axillary buds from the germinating mature embryos or precociously germinating somatic embryos resulting in the formation of multiple shoots which has also been described as microtillering (Rao and Kothari, 1992). The increase in number of basal shoots from the culms regenerated in tissue cultures resembles basal tillering characteristic of grasses. Shoots regenerated in this way are well suited for clonal propagation as they are genetically uniform (Kothari and Chandara, 1995).

Somatic embryogenesis has been proposed as the most common pathway of plant regeneration in cereals (Vasil, 1987) including wheat (Maheshwari *et al.*, 1995). But plant regeneration in wheat has also been shown to occur through organogenesis or both organogenesis and somatic embryogenesis (Nabors *et al.*, 1985; He *et al.*, 1988, 1990; Kothari and Varshney, 1998). There are several reports also where both embryogenesis and shoot bud formation have been shown to occur in the same cultures in other cereals (Chen *et al.*, 1988; Lorz *et al.*, 1988; Bhaskaran and Smith, 1990; Kothari and Chandra, 1995). Evidence for single cell origin of embryoids has been shown in wheat (Magnusson and Bornmann, 1985).

Histological studies have been made on the development of callus in wheat from the scutellum and epiblast callus (Ozias-Akins and Vasil, 1982, 1983; He *et al.*, 1990), showed that both epiblast and scutellum forms callus in which both embryogenesis and organogenesis occurred.

They also observed shoot forming protuberances from the primary callus and subcultured callus which were broadly based and were of multiple cell origin. Such structures resembled the previously described atypical embryoids (Ozias-Akins and Vasil, 1983). Regenerating callus of wheat has also been shown to have structures that were considered shoot buds (Kothari and Varshney, 1998). They observed meristematic domes on the surface of nodular compact callus. The domes enclosed by two leaf primordia were common. Such shoot meristems were devoid of any embryonal structures such as coleoptile, coleorhiza scutellum or epiblast. Kothari and Varshney (1998) provided histological evidence for the occurrence of both somatic embryogenesis and shoot bud formation in aged, repeatedly subcultured callus. Presence of globular structures and structures with lateral notch and presence of completely developed embryoids with embryonal structures such as coleoptile, scutellum, epiblast and coleorhiza were taken as evidence for somatic embryogenesis. But at the same time in the same cultures the formation of *de novo* shoot buds were also common. Presence of shoot apex surrounded by leaf primordia with vascular connections to the basal callus and the absence of embryonal structures was taken as clear evidence for shoot organogenesis. In conclusion it can be said that somatic embryogenesis is a predominant node of plant regeneration in wheat but the formation of *de novo* shoot buds is also of common occurrence. Both the processes can occur in the same callus (Chen *et al.*, 1985; Lorz *et al.*, 1988; He *et al.*, 1990; Kothari and Chandra, 1995; Kothari and Varshney, 1980).

6. PROTOPLAST CULTURES

Diaz (1994) isolated protoplasts from endosperm, aleurone,, leaves and roots of wheat for use in the DNA uptake experiments and study of transient *gus* gene expression. It was observed that each tissue had a different requirement of specific combination of enzymes and their concentrations for maximum yield of protoplasts. The growth conditions of the donor tissue also influenced the yield and quality of protoplasts.

Regeneration of plants from cereal protoplasts has never been easy particularly with, mesophyll protoplasts. Totipotent protoplasts in cereals have always used the embryogenic suspension cultures (Vasil *et al.*,

1990). Redway *et al.* (1990) described the conditions for obtaining embryogenic suspension cultures in wheat from which plants were readily regenerated. Using the suspension cultures, Vasil *et al.* (1990) reported regeneration of plants from protoplasts of wheat cultured on Kao and Michyaluk or MS based nutrient media.

Earlier reports on cell suspension cultures of wheat (Ahuja *et al.*, 1982; Chen *et al.*, 1985; Maddock, 1987; Harris *et al.*, 1989) did not show regeneration of plants. The earliest reports on protoplast to plant regeneration where one regenerated plant was transferred to soil includes Vasil *et al.* (1990) and Chang *et al.* (1991). He *et al.* (1992) reported highly reproducible regeneration of normal green plants from protoplasts of Australian plant with normal chromosome complement.

7. CONCLUSIONS AND FUTURE PROSPECTS

Wheat is one of major staple food of the world which need urgent attention in wake of the population pressure. *In vitro* plant regeneration, a pre-requisite for genetic transformation has been achieved using different explants and callus cultures, however, the regeneration depends on genotype and hormonal composition in the media. Plant regeneration from protoplast and cell suspension cultures has been reported only in a few cases. I*n vitro* morphogenesis has been either through somatic embryogenesis or shoot bud formation and very often both the processes also occur in the same culture. Improvements in nutrient utilization efficiency and partitioning of photosynthates between vegetative and reproductive parts, resistance/tolerance to biotic and abiotic stresses and modification in nutritional and industrial qualities of wheat grains through genetic engineering require an efficient regeneration system compatible to *Agrobacterium* mediated transformation in the large number of genotype.

REFERENCES

Abe T and Futsuhara Y (1985) Efficient plant regeneration by somatic embryogenesis from root callus tissues of rice. *J. Plant Physiol.,* **121** : 111-118.

Abe T and Futsuhara Y (1986b) Genotypic variability for callus formation and plant regeneration in rice (*Oryza sativa* L.) *Thero. Appl. Genet.,* **72** : 3-10.

Adachi T and Katayama Y (1969) Callus formation and shoot differentiation in wheat tissue culture. *Bull. Fac. Agric. Univ. Miyajaki,* **16** : 77.

Ahloowalia BS (1982) Plant regeneration from callus culture in wheat. *Crop. Sci.,* **23** : 405-410.

Ahmad A and Sayed-Shahram M (1999) Response of durum wheat cultivars to immature embryo culture, callus induction and *in vitro* salt stress. *Plant Cell Tiss. Org. Cult.,* **58** : 67-72.

Ahuja PS and Pental D and Cocking EC (1982) Plant regeneration from leaf base callus and cell suspensions of *Triticum aestivum. Z. Pflanzenziichtg.*, **89** : 139-144.

Bapat SA, Rawal SK and Mascarenhas AF (1988) Occurrence and frequency of precocious germination of somatic embryos is a genotype dependent phenomenon in wheat. *Plant Cell Rep.*, **7** : 538-545.

Bastok T and Sagi F (1990) A new endosperm supported callus induction method for wheat (*Triticum aestivum* L.) *Plant Cell Tiss. Org. Cult.*, **22** : 37.

Bhaskaran S and Smith RH (1990) Regeneration in cereal tissue culture: A review. *Crop Sci.*, **30** : 1328-1337.

Bhojwani SS and Hayward C (1977) Some observations and comments on tissue culture of wheat. *Z. Pflanzenphysiol*, **85** : 341.

Bohorova ME, Wanginkel, Rajaram S and Hoisington DA (1994) Tissue culture response of CIMMYT elite bread wheat varieties and evaluation of regenerated plants. *In vitro Plant*, **30** : 229-235.

Bregitzer P, Somers DA and Rines HW (1989) Development and characterization of friable embryogenic oat callus. *Crop Sci.*, **29** : 798-803.

Brown C, Broks FJ, Pearson D and Mathias RJ (1989) Control of embryogenesis and organogenesis in immature wheat embryo callus using increased medium osmolarity and abscissic acid. *J. Plant Physiol.*, *133* : 727.

Carman JE. Jeffersson NE and Campbell WF (1987a) Induction of embryogenic *Triticum aestivum* calli. I. Quantification of genotype and culture medium effects. *Plant Cell Tiss. Org. Cult.*, **10** : 101-113.

Carman JE, Jeffersson NE and Campbell WF (1987b) Induction of embryogenic *Triticum aestivum* L., calli II. Quantification of organic addenda and other culture variable effects. *Plant Cell Tiss. Org. Cult.*, **12** : 97-110.

Chang YF, Wang WC, Warfield CY, Nguyen HT and Wong JR (1991) Plant regeneration from protoplasts isolated from long term cultures of wheat (*Triticum aestivum* L.) *Plant Cell Rep.*, **9** : 611.

Chawala HS (1988) Isozyme modifications during morphogenesis of callus from barley and wheat. *Plant Cell Tiss. Org. Cult.*, **12** : 299-304.

Chen Thh, Kartha KK and Gusta LV (1985) Cryopreservation of wheat suspension culture and regenerable callus. *Plant Cell Tiss. Org. Cult.*, **4** : 101.

Chin JC and Scott KJ (1977) Studies on the formation of roots and shoots in wheat callus cultures. *Ann. Bot.*, **41** : 473-481.

Chu CC (1978) The N6 medium and its applications to anther culture of cereal crops. In: *Proc. Symp. Plant Tissue Culture, Science* Press, Beizing. 43-50.

Chu CC, Hill RD and Brule-Babel AL (1990) High frequency of pollen embryoid formation and plant regeneration in *Triticum aestivum* L. on *monosachharide* containing media. *Plant Sci.*, **66** : 255-262.

Close KR and Gallagher – Ludman LA (1989) Structure-activity relationship of auxin like plant growth regulators and genetic influences on the culture induction responses in maize (*Zea mays* L.) *Plant Sci.*, **61** : 245-252.

Comeau A, Nadeau P, Plourde A, Simard R, Mais C, Colky S, Harper L, Lettre J, Landry B and Pearre St. PA (1992) Media for the ovule culture of proembryo of wheat derived interspecific hybrids on haploids. *Plant Sci.*, **81** : 117-125.

Cure WW and Mott RL (1978) A comparative anatomical study of organogenesis in cultured tissues of maize, wheat and oats. *Physiol. Plant*, **42** : 91-96.

Diaz I (1994) Optimization of conditions for DNA uptake and transient GUS expression in protoplasts for different tissues of wheat and barley. *Plant Sci.,* **96** : 179-187.

Dudits D, Nemet G and Hayks Z (1975) Study of callus growth and organ formation in wheat (*Triticum aestivum*) tissue cultures. *Can. J. Bot.,* **53** : 957-963.

Duncan DR, Williams BE, Zehr BE and Widholm JM (1985) The production of callus capable of plant regeneration from immature embryos of numerous *Zea mays* genotypes. *Planta,* **165** : 322-332.

Eapen S and Rao PS (1982a) Organogenesis and plantlet formation from callus cultures of different cultivars of bread wheat (*Triticum aestivum*) *Proc. Indian. Nat. Sci. Acad.,* **48** : 371-377.

Eapen S and Rao PS (1982b) Callus induction and plant regeneration from immature embryos of rye and triticale. *Plant Cell Tiss. Org. Cult.,* **1** : 221-227.

Eapen S and Rao PS (1983) Factors controlling pollen embryogenesis in triticale and wheat. *Proc. Indian Natl. Sci. Acad.,* **51** : 352-355.

Fatokun CA and Yamada Y (1984) Variation in callus formation and plant regeneration in African rice (*Oryza glaberriana* Steud.) *J. Plant Physiol.,* **117** : 179-183.

Fennel S, Bohorava N, Ginkel van M, Crossa J and Hoisingtan D (1996) Plant regeneration from immature embryos of 48 elite CIMMYT bread wheat.

Gamborg OL, Miller RA, Ozima K (1968) Nutrient requirements of suspension cultures of soybean root cells. *Exp Cell Res.,* **50** : 151-158.

Gosch-Wackerle G, Arivi L and Galun E (1979) Induction, culture and differentiation of callus from immature rachises, seeds and embryos of *Triticum. Z. Pflanzen Physiol.,* **91** : 267-278.

Harris R, Wright M, Byrne M, Varnum J, Bright-Well B, Schubert K (1988) Callus formation and plantlet regeneration from protoplasts derived from suspension cultures of wheat (*Triticum aestivum* L.) *Plant Cell Rep.,* **7** : 337-340.

He DG, Tanner G and Scott KJ (1986) Somatic embryogenesis and morphogenesis in callus derived from the epiblast of immature embryos of wheat (*Triticum aestivum*) *Plant Sci.,* **45** : 119-124.

He DG, Yang YM and Scott KJ (1988) A comparison of scutellum callus and the epiblast callus induction in wheat: The effect of genotype, embryo age and medium. *Plant Sci.,* **57** : 225.

He DG, Yang YM and Scott KJ (1992) Plant regeneration from protoplasts of wheat (*Triticum aestivum* CV. Hortog) *Plant Cell Rep.,* **11** : 16.

He DG, Yang YM, Bertram J and Scott KJ (1990) The histological development of the regenerative tissue derived from cultured immature embryos of wheat (*Triticum aestivum* L.) *Plant Sci.,* **68** : 103-111.

Heyser JW, Nabors MW, Mac-Kinnon C, Dykes TA, Dimott KJ, Kauztman DC and Muzeebkazi A (1985) Long term high frequency plant regeneration and the induction of somatic embryogenesis in callus cultures of wheat (*Triticum aestivum* L.) *Pflanzenziichtg,* **94** : 218-233.

Hodges TK, Kamo KK, Imbrie CW and Beewar MR (1986) Genotype specificity of somatic embryogenesis and regeneration in maize. *Biotechnology,* **4** : 219-223.

Hu T and Kasha KJ (1995) Improved embryogenesis and plant regeneration from isolated microsopre culture of wheat (*Triticum aestivum* L.) *In Vitro* 31:pt.2:74A.

Huang B (1987) Effects of incubation temperature on microspore callus production in plant regeneration in wheat anther cultures. *Plant Cell Tiss. Org. Cult.,* **9** : 45-48.

Hunsinger H and Schauz K (1987) The influence of dicamba on somatic embryogenesis and frequency of plant regeneration from cultured immature embryos of wheat (*Triticum aestivum* L.) *Plant Breed.,* **98** : 119-123.

Ishii C (1982) Callus induction and shoot differentiation of wheat, oat and barely. *Proc. 5th Int. Congr. Plant Tissue Cell Culture,* 185.

Izabela M, Franciszek D and Biesaga-Koscielnak J (1995) Transfer of the ability to flower in winter wheat via callus tissue regenerated from immature inflorescences. *Plant Cell Tiss. Org. Cult.,* **41** : 285-288.

Jamora AB and Scott KJ (1983) Callus formation and plant regeneration from wheat leaves. *Plant Sci. Lett.,* **29** : 183-189.

Jha D (1998) Future wheat needs in South Asia region. In: *Wheat research needs beyond 2000 AD. Proc. Internatl. Group Meet.* Narosa Publishing House, India. 29-33.

Kao KN and Mickayluk MR (1975) Nutritional requirements for growth of *Vicia hajastana* cells and protoplast at very low population density in liquid media. *Planta.,* **126** : 105-110.

Karen LC, Nick LL and Chibbar RN (2000) An efficient method for *in vitro* regeneration from immature inflorescences explants of Candian wheat cultivars. *Plant Cell Tiss. Org. Cult.,* **60** : 69-73.

King PG, Potrykus I and Thomas E (1978) *In vitro* genetics of cereals: Problems and perspectives. *Physiol. Veg.,* **16** : 381-399.

Kirsten V, Angela S, Markus I and Dieter H (1998) Regeneration of German spring wheat varieties from embryogenic scutellar callus. *J. Plant Physiol.,* **152** : 167-172.

Kothari SL and Chandra N (1995) Advances in tissue culture and genetic transformation of cereals. *J. Indian Bot. Soc.,* **74A** : 323-342.

Kothari SL and Varshney A (1998) Morphogenesis in long-term maintained immature embryo derived callus of wheat (*Triticum aestivum* L.) – Histological evidence for both somatic embryogenesis and organogenesis. *J. Plant Biochem. Biotech.,* **7** : 93-98.

Lazer MD, Chen THH, Gusta LV and Kartha KK (1988) Somaclonal variation for freezing tolerance in a population derived from norstar winter wheat. *Theor. Appl. Genet.,* **75** : 480-484.

Lazer MD, Chen THH, Scoles GJ and Kartha KK (1987) Immature embryo and anther culture of chromosome addition lines of rye in Chinese spring wheat. *Plant Sci.,* **51** : 77-81.

Lazer MD, Collins GV and Vian WE (1983) Genetic and environmental effects on the growth and differentiation of wheat somatic cell cutlures. *J. Heredity,* **74** : 353.

Li LH and Dong YS (1994). Somaclonal variation in tissue culture of *Triticum aestivum* × *Agropyron desertorum* F_1 hybrid. *Plant Breed.,* **112** : 160-166.

Liang GHL, Sangduan N, Heyne EG and Sears RE (1982) Polyploid production through anther culture in common wheat. *J. Hered.,* **73** : 360-364.

Lin ML and Staba EJ (1961) Peppermint and spearmint tissue culture. I: Callus formation and submerged culture. *Lloydia.,* **24** : 139-145.

Linsmair EM and Skoog F (1985) Organic growth factor requirements of tobacco tissue cultures. *Physiol. Plant.,* **18** : 100-127.

Lorz H, Gobel E and Brown P (1988) Advances in tissue culture and progress towards genetic transformation of cereals. *Plant. Breed.,* **100** : 1-25.

Lu GS, Sharma HC and Ohm HW (1991) Wheat anther culture: Effect of genotypes and environmental conditions. *Plant Cell Tiss. Org. Cult.,* **24** : 233-236.

Machii H, Mizuno H, Hirabayashi T, Li H and Hagio T (1998) Screening wheat genotypes for high callus induction and regeneration capability from anther and immature embryo cultures. *Plant Cell Tiss. Org. Cult.,* **53** : 67-74.

Mac-Kinnon C, Gunderson G and Nabors MW (1987) High efficiency plant regeneration by somatic embryogenesis from callus of mature embryo explant of bread wheat (*Triticum aestivum*) and grain sorghum (*Sorghum bicolor*). *In Vitro Cell Dev. Biol.,* **23** : 443-448.

Maddok SE and Semple JT (1986) Field assessment of somaclonal variation in wheat. *J. Exp. Bot.,* **37** : 1063-1073.

Maddok SE, Lancaster VA, Risiott R and Franklin J (1983) Plant regeneration from cultured immature embryos and inflorescence of 25 cultivars of wheat (*Triticum aestivum*) *J. Exp. Bot.,* **34** : 915-926.

Magnusson I and Bornman CH (1985) Anatomical observations on somatic embryogenesis from scutellar tissues of immature zygotic embryos of *Triticum aestivum. Physiol. Plant,* **63** : 137-142.

Maheshwari N, Rajyalakshmi K, Baweja K, Dhir SK, Chowdhry CM and Maheshwari SC (1995) *In vitro* culture of wheat and genetic transformation-retrospect and prospect. *Crit. Rev. Plant Sci.,* **14** : 149-178.

Mascarenhas AF, Pathak M, Henbi RR and Jagannathan V (1975) Tissue cultures of maize, wheat, rice and sorghum. Part I. Initiation of viable callus and root cultures. *Indian J. Exp. Biol.,* **13** : 103-107.

Mathias RJ and Boyd LA (1986) Cefotaxime stimulates callus growth, emrbyogenesis and regeneration in hexaploid bread wheat (*Triticum aestivum* L. em. Thell). *Plant Sci.,* **46** : 217.

Mthias RJ and Simpson ES (1986) The interaction of genotype and culture medium on the tissue culture responses (*Triticum aestivum* L. em. Tehll) callus. *Plant Cell Tiss. Org. Cult.,* **7** : 31-37.

McHughen A (1983) Rapid regeneration of wheat *in vitro* (short communication) *Ann. Bot.,* **51** : 851-853.

Menke-Milczarek I and Zimny J (1982) The efficiency of somatic embryogenesis of wheat (*Triticum aestivum* L.) Embryogenic reaction of inflorescence. *Bullet. Polish. Acad. Sci. Bio. Sci.,* **40** : 181-188.

Mikami T and Kinoshita (1988) Genotypic effect on the callus formation from different explants of rice (*Oryza sativa* L.) *Plant Cell Tiss. Org. Cult.,* **12** : 311-314.

Mohmand AS and Nabors MW (1990) Somaclonal variant plants of wheat derived from mature embryo explants of 3 genotypes. *Plant Cell Rep.,* **8** : 558-560.

Mohmand AS and Nabors MW (1991) Comparison of two methods for callus culture and plant regeneration in wheat (*Triticum aestivum*) *Plant Cell Tiss. Org. Cult.,* **26** : 185-187.

Morrish F, Vasil V and Vasil IK (1987) Development, morphogenesis and genetic manipulation tissue and cell cutlures of Gramineae. *Advances in Genetics,* **24** : 431-499.

Murashige T and Skoog F (1962) A revised medium for rapid growth and bioassays with tobacco tissue cultures. *Physiol. Plant.,* **15** : 431-497.

Nabors MW, Heyser JW, Dykes TA and Demott KJ (1983) Long duration high frequency plant regeneration from cereal tissue cultures. *Planta.,* **157** : 385-391.

Norstog K (1973) New synthetic medium for the culture of premature barley embryos. *In Vitro,* **8** : 307-308.

O'Hara JF and Street HF (1978) Wheat callus culture: the initiation, growth and organogenesis of callus derived from various explant sources. *Ann. Bot.,* **42** : 1029-1038.

Ozava K and Komamine A (1989) Establishment of a system of high frequency embryogenesis from long term cell suspension cultures of rice (*Oryza sativa* L.) *Theor. Appl. Genet.,* **77** : 205-211.

Ozgen M, Turet M, Altinok S and Savak C (1998) Efficient callus induction and plant regeneration from mature embryo culture of winter wheat (*Triticum aestivum* L.) genotypes. *Plant Cell Rep.,* **18** : 331-335.

Ozias-Akins A and Vasil IK (1983b) Improved efficiency and normalization of somatic embryogenesis in *Triticum aestivum* (wheat). *Protoplasma,* **117** : 40-44.

Ozias-Akins P and Vasil IK (1982) Plant regeneration from cultured immature embryos and inflorescences of *Triticum aestivum* L. (wheat): Evidence for somatic embryogenesis. *Protoplasma,* **110** : 95-105.

Papenfuss JM and Carman JG (1987) Enhanced regeneration from wheat callus cultures using Dicamba and Kinetin. *Crop. Sci.,* **27** : 588-593.

Raghava Ram NY and Nabors MW (1984) Cytokinin mediated long term high frequency plant regeneration in rice tissue cultures. *Z. Pflanzenphysiol,* **113** : 315-323.

Rajyalakshmi K, Dhir SK, Maheshwari N and Maheshwari SC (1988) Callusing and regeneration of plantlets via somatic embryogenesis from inflorescence cultures of *Triticum aestivum* L. Role of genotype and long term retention of morphogenic potential. *Plant Breed.,* **101** : 80-85.

Rajyalakshmi K, Grover A, Maheshwari N, Tyagi AK and Maheshwari SC (1991) High frequency regeneration of plantlets from the leaf bases via somatic embryogenesis and comparison of polypeptide profiles from morphogenic and non-morphogenic calli in wheat (*Triticum aestivum*) *Physiol. Plant.,* **82** : 617-623.

Rangan TS (1973) Morphogenetic investigations on the tissue cultures of *Panicum miliaceum. Z. Pflanzenphysiol,* **72** : 456-469.

Rangan TS (1976) Growth and plantlet regeneration in tissue cultures of some Indian millets: *Paspalum scrobiculatum* L., *Eleusine coracana* Gaertn. and *Pennisetum typhoideum* P. *Z. Pflanzenphysiol,* **78** : 208-216.

Rao A and Kothari SL (1992) Micropropagation on wheat (*Triticum aestivum* Desf.) through microtillering. *J. Indian Bot. Soc.,* **71** : 47-49.

Redway FA, Vasil V, Lu D and Vasil IK (1990b) Identification of callus types for long term maintenance and regeneration from commercial cultivars of wheat (*Triticum aestivum* L.) *Theor. Appl. Genet.,* **79** : 609-617.

Ryschka S, Ryschka U and Schulze J (1991) Anatomical studies on the development of somatic embryos in wheat and barely explants. *Biochem. Physiol. Pflanz.,* **187** : 31.

Sears RG and Deckard EL (1982) Tissue culture variability in wheat: Callus induction and plant regeneration. *Crop Sci.,* **22** : 546-550.

Sharma VK, Rao A, Varsheny A and Kothari SL (1995) Comparison of developmental stages of inflorescence for high frequency plant regeneration in *Triticum aestivum* L. and *Triticum durum*. Desf. *Plant Cell Rep.,* **15** : 227-231.

Shimada T (1978) Plant regeneration from the callus induced from wheat embryo. *Jpn. J. Genet.,* **53** : 371.

Shimada T and Yamada Y (1979) Wheat plants regenerated from embryo cell cultures. *Jpn. J. Genet.,* **54** : 379-385.

Shimada T, Sasakuma T and Tsunewaki K (1969) *In vitro* culture of wheat tissue I. Callus formation, organ redifferentiation and single cell culture. *Can. J. Genet. Cytol.,* **11** : 294.

Simmonds J, Stewart P and Simmonds D (1992) Regeneration of *Triticum aestivum* apical explants after microinjection of germline progeniter cells with DNA. *Physiol. Plant.,* **85** : 197.

Tabaeizadeh Z, Plourde A and Comeau A (1990) Somatic embryogenesis and plant regeneration in *Triticum aestivum* × *Leymus augustus* F_1 hybrids and the parental lines. *Plant Cell Rep.,* **9** : 204-206.

Thomas DT and Smith OS (1985) The effect of parental genotype on initiation of embryogenic callus from elite maize (*Zea mays* L.) germplasm. *Theor. Appl. Genet.,* **70** : 505-509.

Trione EJ. Jones LE and Metzgere RJ (1968). *In vitro* culture of somatic wheat callus tissue. *Am. J. Bot.,* **55** : 529.

Varsheny A, Swati J and Kothari SL (1999) Plant regeneration from mature embryos of 20 cultivars of wheat (*Triticum aestivum* L. and *Triticum durum* Desf.) *Cereal Res. Comm.,* **27** : 163-170.

Varshney A, Kant T and Kothari SL (1998) Plant regeneration from coleoptile tissue of wheat (*Triticum aestivum* L.) *Biol. Plant.,* **40** : 137-141.

Vasil IK (1987) Developing cell and tissue culture systems for the improvement of cereal and grass crops. *J. Plant Physiol.,* **128** : 193-218.

Wang WC and Nguyen HT (1990) A novel approach for efficient plant regeneration from long term suspension culture of wheat. *Plant Cell Rep.,* **8** : 539-642.

Wernicke W and Brittell R (1982) Morphogenesis from cultured leaf tissues of *Sorghum bicolor* (L.) culture initiation. *Protoplasma,* **3** : 19-27.

Wernicke W and Milkovits L (1984) Development gradients in wheat leaves response of leaf segments in different genotypes cultured *in vtiro. J. Plant Physiol.,* **115** : 49-58.

Wernicke W and Milkovits L (1986) The regeneration potential of wheat shoot meristem in the presence and absence of 2,4-dichlorophenoxyacetic acid. *Protoplasma,* **131** : 131-141.

Wilman MR, Schroll SM and Hodges TK (1989) Inheritance of somatic embryogenesis and plant regeneration from primary (Type I) callus in maize. *In Vitro Cell Dev. Biol.,* **25** : 95-100.

Yao DY and Krikorian AD (1981) Multiplication of rice (*Oryza sativa* L.) from aseptically cultured nodes. *Ann. Bot.,* **48** : 255-259.

Chapter 5

TRANSGENIC WHEAT

ME Cannell* **and HD Jones**

Cereal Transformation Group, Crop Performance and Improvement Division, IACR-Rothamsted, Harpenden, Hertfordshire, AL5 2JQ, United Kingdom

Summary

In this chapter we present a review of the current status of wheat improvement through biotechnology. The opening sections (2, 3 and 4) focus on aspects of wheat transformation technology, whilst section 5 covers trangene expression. The penultimate section 6 concentrates on current targets for wheat improvement, covering mainly end-use quality but including some detail on agronomic and other traits. Finally section 7 draws conclusions and discusses a few key areas for future research.

Keywords : Wheat transformation, wheat improvement, end-use quality, agronomic trait, review

1. INTRODUCTION

Food products derived from wheat are one of the most important sources of calorific intake worldwide. Around 215 million hectares of wheat were grown in 1997, producing over 580 million tonnes of grain (FAO, 1999). Despite this, wheat still lags behind the other cereals with respect to progress in transformation research, due largely to its recalcitrance to *in vitro* culture, and was the last of the major crops to be transformed (Vasil *et al.*, 1992). Since those early days, wheat transformation has become a routine undertaking in many laboratories, although frequencies remain relatively low and the process is very labour-

*Corresponding author : E-mail : martin.cannell@bbsrc.ac.uk

intensive. In this chapter we present a review of the principal wheat transformation methods, molecular and genetic analyses so far conducted and targets for crop improvement. The focus is mainly on bread wheat, and although pasta wheat (durum wheat) has been successfully transformed (Lamacchia *et al.*, 2001; He *et al.*, 1999; Bommineni *et al.*, 1997), it remains similarly recalcitrant. Targets for pasta wheat improvement are discussed in section 6. Regarding methods and analyses, we have chosen to concentrate exclusively on the regeneration of plants containing stably integrated transgenes, and on data that have been published. However, because of the commercial relevance of wheat, a great deal of research is being carried out in the private sector, the results of which are not generally available. Some of the limitations of current transformation methodologies are described and the potential of developing ones outlined and expounded. Because research in the other cereals (such as rice and maize) is often more advanced than in wheat, we have occasionally referred to these studies. However, due to differences in biology, genome size and ploidy, extrapolations to the wheat situation must be carefully considered.

For ease of description, we have divided target traits for wheat improvement into end-use quality, agronomic and 'other'. Clearly a major focus for wheat improvement is grain quality and we concentrate on this area in section 6. However, there is increasing interest in the manipulation of agronomic and other traits, such as pest/disease resistance, herbicide tolerance etc., which are summarised in Tables 4 A and B.

2. GENE DELIVERY

2.1. Direct gene transfer (DGT)

2.1.1. Electroporation

He *et al.* (1994), generated stably transformed wheat plants via protoplast electroporation but since then few reports documenting this approach have been published. The main limitation is that prolonged cell treatments are required and whole plant regeneration from wheat protoplasts is difficult. To avoid protoplast preparation and culture, systems for electroporating intact somatic tissues have been developed. Sorokin *et al.* (2000) reported stable transformation via tissue electroporation of immature embryos and He *et al.* (2001) transformed a hexaploid relative of bread wheat, namely tritordeum, via electroporation of immature inflorescences. However, the reported transformation efficiencies were generally low compared to particle bombardment. In addition, target explants required careful preparation and treatment prior to and after electrical discharge. The use and development of tissue electroporation for wheat transformation has therefore been limited but the method could probably be improved in terms of effectiveness and extent of application.

2.1.2. Particle bombardment

The most widely used method of direct gene delivery is microprojectile or particle bombardment. The BioRad PDS 1000/He is the most commonly used delivery device (gun). Variable parameters when using particle bombardment include microprojectile type, size and quantity, DNA quantity and method of precipitation and 'gun' parameters such as propellant force, Helium pressures and target distances. All parameters can influence the efficiency of DNA delivery and some reports have described their systematic investigation and optimisation (Perl *et al.*, 1992; Altpeter *et al.*, 1996; Harwood *et al.*, 2000; Ingram *et al.*, 1999; Rasco-Gaunt *et al.*, 1999). Recent descriptions of bombardment procedures leading to stable wheat transformation cover a range of values for some of these parameters. Some of the lower and higher values are shown in Table 1 in which the range of values shown indicate the upper and lower extremes that are commonly used. Values lying between those shown are used most frequently.

A few limitations can be levelled at both electroporation and at particle bombardment. Both methods require the purchase and assembly of

Table 1. Some higher and lower values reported for particle 'gun' delivery parameters.

Parameter	Low value (source)	High value (source)
Microprojectile size	0.6 µm - 1.2 µm (Jordan, 2000)	1-3 µm (Bliffeld *et al.*, 1999)
Microprojectile quantity per shot	30-70 µg (Stoger *et al.*, 1999b)	580 µg (Weeks *et al.*, 2000)
µg DNA per gold preparation	0.5 - 1 µg (Uze *et al.*, 1999)	25 µg (Weeks *et al.*, 2000)
Propellant force	650-1100 psi (Rasco-Gaunt *et al.*, 1999)	1550 psi (De Block *et al.*, 1997)
Target distances	3-9 cm (Jordan, 2000)	7-10 cm (WP Chen *et al.*, 1998)

expensive and specialist equipment. In addition, the operation of such equipment requires skills that are not often taught at undergraduate level. A further disadvantage of these methods is that whole plasmid DNA molecules are usually delivered and that during integration by illegitimate recombination, plasmid breakage occurs essentially at random. Thus, not only are vector backbone sequences always assumed to be present in transformants produced by DGT of whole plasmids, but transgene sequences are often also truncated or rearranged unpredictably, which can sometimes induce transgene silencing. For commercial applications the presence of vector sequences is considered undesirable so that in many laboratories DNA fragments are used, which have been digested and purified from the plasmid DNA and that contain only the expression cassette.

2.2. *Agrobacterium*-mediated transformation

Rice (Dai *et al.*, 2001; Azhakanandam *et al.*, 2000; Hiei *et al.*, 1994) and maize (Negrotto *et al.*, 2000; Ishida *et al.*, 1996) are now routinely transformable at high frequencies using *Agrobacterium*. Whilst transformation of wheat has been demonstrated (Cheng *et al.*, 1997), efficiencies were lower, averaging at 1.6%. Further success using *Agrobacterium* for stable wheat transformation has remained very limited since that first report, despite the efforts of a number of groups working in this area (Amoah *et al.*, 2001; Uze *et al.*, 2000; Singh and Chawla, 1999; Peters *et al.*, 1999; Guo *et al.*, 1998). In maize, barley and wheat, the factor with the greatest influence on transformation is the variation in response and regenerability between individual genotypes in tissue culture. It is no coincidence that successful *Agrobacterium*-mediated transformations of maize (Negrotto *et al.*, 2000), barley (Tingay *et al.*, 1997) and wheat (Cheng *et al.*, 1997) were all carried out using genotypes generally considered to the best of the tissue culture model varieties for those species (A188, Golden Promise and Bobwhite respectively). Other factors influencing *Agrobacterium*-mediated transformation include the type and developmental stage of the targeted tissue, the bacterial strain and Ti plasmid combination, bacterial growth and co-cultivation conditions. Continued investigation of these parameters may lead to improved transformation protocols and efficiencies. However, since the major limitation appears to be tissue culture response, the most progress may ultimately be made through the optimisation and identification of culture systems with increased regeneration capacities and breadth of genotype applicability. *Agrobacterium* transformation offers a number of advantages over direct gene transfer, such as lower cost and increased simplicity of the methodology, more predictable transgene integration/organization, higher frequencies of single transgene copies inserted and the potential for integrating large, intact fragments of DNA. However, a potential disadvantage stems from the discovery that sequences outside the T-DNA borders can sometimes become integrated (De Buck *et al.*, 2000; Wenck *et al.*, 1997).

3. TISSUE CULTURE FOR TRANSGENIC WHEAT PLANTS

3.1. Explant source

A number of tissue types have been investigated as target explants for cereal transformation, including microspores, shoot meristems, mature seed embryos, immature embryos and immature inflorescences. Of these, only the latter two have been broadly successful for wheat. These tissues contain cells with the potential to form somatic embryos that may be further induced to regenerate into a whole plant.

Immature embryos (more specifically the scutellar tissue) are the most commonly used source of explant. The optimal stage of caryopsis develoment for embryo isolation can differ between genotypes but is around 11-16 days post anthesis. An improvement in transformation was observed when scutella were isolated from donor plants that had themselves been regenerated via embryogenesis and tissue culture (Harvey *et al.*, 1999) A major limitation of immature embryos is that their isolation is highly labor-intensive. Thus mature seed embryos have received interest as an alternative source of explant as they are easier to isolate and preclude the need for expensive donor growth rooms. Although success with mature seed embryos of rice has been reported (Jiange *et al.*, 2000; Sudhakar *et al.*, 1998), the methodology requires improvement in wheat.

Immature inflorescences provide a useful alternative source of explant, as they are easier to isolate and are harvested from much younger plants. This allows a more efficient use of growth room space for donor material. Stable transformation of durum wheat (He *et al.*, 1999), tritordeum (Barro *et al.*, 1998) and bread wheat (cv Baldus) (Sparks *et al.*, 2001) has been reported using immature inflorescence explants. However, their tissue culture response is highly genotype specific, with some varieties being particularly unresponsive (Rasco-Gaunt and Barcelo, 1998).

3.2. Culture conditions

Most schemes for wheat transformation involve induction of embryogenesis from somatic tissues, followed by the regeneration of whole plants from individual somatic embryos. Somatic cells with embryogenic potential are good targets for chromosomal integration of transgene DNA, since each has the potential to become an independent transgenic plant.

The basic salt formulation used in most wheat tissue culture protocols is solidified MS (Murashige and Skoog, 1962) supplemented with vitamins, hormones and sugars. The auxin 2,4-D (2,4-dichlorophenoxyacetic acid) is frequently used to induce embryogenesis, although the use of picloram has also been described, (see Barro *et al.*, 1998; Campbell *et al.*, 2000), either with or without low levels of cytokinins such as zeatin and benzyladenine purine (BAP). In shoot regeneration, higher levels of cytokinins are employed either with or without lower auxin levels.

Osmotic treatment is thought to offer protection to bombarded material by minimising cytoplasm leakage from target cells (Vain *et al.*, 1993). Various osmotic treatments induced by different concentrations and combinations of sugar-alcohols and sugars have been employed. The use of sugar-alcohols usually involves a temporary incubation of several hours on media supplemented with them, prior to bombardment. Mannitol and sorbitol are frequently used (Altpeter *et al.*, 1996; Ortiz *et al.*, 1996; Blechl and Anderson, 1996; Stoger *et al.*, 1999a; Jordan, 2000; Weeks *et al.*, 2000; Brinch-Pedersen *et al.*, 2000), and a raffinose/mannitol combination has also been employed (Campbell *et al.*, 2000). One of the sugars most commonly included is sucrose, often at 2-3% concentration (Nehra *et al.*, 1994; De Block *et al.*, 1997; W.P. Chen *et al.*, 1998; Wirtzens *et al.*, 1998) and the effect of a range of sucrose concentrations has recently been investigated (Rasco-Gaunt *et al.*, 2001). An additional approach for osmotic shock, involving a temporary transfer of material from 2% maltose to 20% maltose, has also been used (Uze *et al.*, 1999). The length of pre-culture of explant material is usually balanced against medium composition to optimise the stage of cellular differentiation, cell cycle and osmotic conditioning. Examples of different pre-culture periods that have been used are shown in Table 2.

3.3. Selection systems

A variety of selection systems are employed in wheat transformation procedures. The most common uses phosphinothricin (PPT)-based

Table 2. Reported explant pre-culture periods

Explant preculture period	Source
1-2 days	Barro *et al.*, 1997; WP Chen *et al.*, 1998
4 days	Campbell *et al.*, 2000
5-7 days	Altpeter *et al.*, 1996; Brinch-Pedersen, 2000; Jordan, 2000; Weeks *et al.*, 2000
7-10 days	Harvey *et al.*, 1999; Ortiz *et al.*, 1996; Takumi *et al.*, 1999; Uze *et al.*, 1999
3-5 weeks	Nehra *et al.*, 1994; Karunaratne *et al.*, 1996; De Block *et al.*, 1997

herbicides (such as Bialaphos, BASTA and Challenge) as the selective agent, and the *bar* or *pat* gene (encoding phosphinothricin acetyl transferase) as the resistance-conferring gene. Glyphosate has also been used as a selective agent, with the *epsps* (enolpyruvylshikimate phosphate synthase) and *gox* (glyphosate oxidoreductase) genes conferring resistance (Zhou *et al.*, 1995). However, its use is much less widespread than the PPT-based herbicides, perhaps because observed transformation efficiencies were lower compared to those obtained with other genes and agents. Also used is the *neo* or *nptII* gene (encoding neomycin phosphotransferase), that confers resistance to the antibiotic kanamycin and its analogues such as paromomycin and geneticin sulphate (G418). In addition, the antibiotic hygromycin has been employed, with resistance conferred by the gene *hph,* encoding hygromycin phosphotransferase (Ortiz *et al.*, 1996). Whilst PPT-based systems are generally successful in terms of transformation frequencies, in our hands the frequency of 'escapes' (selected regenerants that are non-transformed) generated is sometimes higher than observed with the *neo* system. The visualisation of green fluorescent protein (GFP) expression in regenerating calli and plantlets has also demonstrated utility as a selection system (Jordan, 2000) although efficiencies may be improved if used in combination with antibiotic resistance markers.

The selection systems described above have different qualities and uses and choice may be dependent on the scientific context of particular projects. However, the limitations of them all are manifold. As well as being largely inefficient at selecting transformed cells their use is often subject to proprietary status. Some environmentalists are concerned over the possibility of herbicide resistance genes 'flowing' to weed populations and public reaction to the presence of antibiotic resistance genes has prohibited their use in GM wheat improvement programs.

Thus alternative selectable markers are continually sought which are either more efficient and/or address one or more of the other concerns. For example, Reed *et al.* (1999) described a system that relies on the efficient utilisation of mannose as a carbon source by cells transformed with the mannose-6-phosphate isomerase gene (*manA*). This scheme differs from conventional ones in that mannose is not toxic to plant cells and that expression of the selectable marker gene provides a metabolic advantage over non-transformed cells. Referred to as a 'positive' selection system, *manA* addresses environmental concerns over gene flow and does not require the inclusion of toxic or expensive chemicals in the culture medium. It has been successfully applied in rice and maize, and in wheat it appears to facilitate some of the highest transformation frequencies observed. Marketed by Syngenta as Positech™, the constructs are now freely available to academic researchers worldwide. The maize glutathion S-transferase subunit 27 (GST-27) was recently shown by Milligan *et al.* (1999) to confer tolerance in transgenic wheat seedlings to the chloroacetanilide herbicide alachlor, thus showing potential utility as a selectable marker and Weeks *et al.* (2000) demonstrated the use of the fungal gene cyanamide hydratase (*Cah*) which confers resistance to the compound cyanamide. This approach addresses concerns over the use of selectable antibiotic resistance genes, and whilst it may confer an 'unwelcome' environmental advantage (due to herbicide tolerance), it may also confer secondary whole plant benefits to transformants (such as the possibility of allowing more efficient fertiliser and fungicide usage).

4. TRANSGENE INTEGRATION

Transgene integration patterns show ranges of plasmid insertion numbers and complexities, which may be determined by multifarious factors. From current literature, insertion numbers in wheat generally range from 1 to about 15, with the estimated average being around 5. Although there is evidence that the relationship between transgene copy number and expression level may have a linear quality in wheat (Stoger *et al.*, 1998), it is not generally considered to be the case in plants, due to position effects (Mlynarova *et al.*, 1994, 1995, 1996) and transgene silencing (Iyer *et al.*, 2000). For most applications, fewer copy insertions are more desirable as these are easier to characterise at the molecular level and are less likely to be targeted by silencing mechanisms.

Unfortunately, the frequency of single copy insertions in particle bombardment-generated plants is rare, and there are very little data on the factors influencing integration patterns and copy number. [Notable exceptions in wheat include De Block *et al.* (1997) who demonstrated that the inclusion of niacinamide in the culture medium was effective in reducing copy number, and Uze *et al.* (1999), who showed that linear DNA (both single and double stranded) gave better transformation efficiencies than circular DNA under the conditions described]. Extrapolations must therefore be made from studies in the other cereals, but it should be remembered that wheat, having the largest and most complex genome of all crop species, may not always be comparable.

LL Chen and co-workers (1998) demonstrated increased co-transformation frequencies in rice following an increase in the relative molar quantity of plasmids encoding the gene of interest, and also provided evidence that the nature of the DNA sequence did not influence integration efficiencies. Interestingly, Fu *et al.* (2000) showed, also in rice, that delivery of a transgene cassette on whole plasmid DNA (supercoiled *or* linear) resulted in more complex integration patterns compared to when the same cassettes were delivered as DNA fragments that had been purified from the plasmid backbone. These data suggest that the presence of bacterial vector sequences may influence the integration process. Moreover, studies in tobacco (Muller *et al.*, 1999) and rice (Kohli *et al.*, 1999) have shown that the conformation of the DNA sequence may also affect integration, and that secondary DNA structures such as stem-loops and cruciforms are often highly recombinogenic. Insertion numbers produced by *Agrobacterium*-mediated transformation of barley (Tingay *et al.*, 1997), maize (Ishida *et al.*, 1996) and rice (Hiei *et al.*, 1994; Dai *et al.*, 2001) demonstrate the potential of this system for generating higher frequencies of low and single copy insertions in the cereals, although as mentioned in Chapter 1, routine *Agrobacterium*-mediated transformation of wheat remains elusive (with the notable exception of Cheng *et al.* (1997).

Regarding the distribution of transgenes across the wheat genome, particle bombardment-mediated integration occurs most commonly at a single genetic locus (even though multiple transgene copies become integrated). Accurate frequencies are difficult to calculate, although from the limited published data and our own observations we estimate that 10-15% of transformants generated contain more than one transgene locus. The occurrence of multiple transgene loci is desirable as this may allow segregation of the ‘target transgene’ away from the selectable marker (although the frequency of 'clean' target transgene loci will be even lower than 10-15%, due to high co-transformation frequencies). From the limited number of genetic studies undertaken on cereals transformed by *Agrobacterium,* the frequency of insertion at more than one locus does not appear to be significantly different. However, the use of super-binary vectors has the potential to increase the frequency as shown in rice and tobacco (Komari *et al.*, 1996).

Additional studies on the spatial organisation of transgenes delivered by particle bombardment in oat (Pawlowski and Somers *et al.*, 1998), rice (Kohli *et al.*, 1998) and wheat (Abranches *et al.*, 2000), have revealed that tracts of genomic DNA often lie between multiple plasmid insertions within a single locus. There is also evidence to suggest that gene-rich areas of the rice genome are preferentially targeted during *Agrobacterium*-mediated transformation (Barakat *et al.*, 2000).

5. TRANSGENE EXPRESSION

5.1. Promoters

A list of promoters reported to be active in stably transformed wheat is shown in Table 3. Sample references of recent publications where these promoters were used are given. The scarcity of wheat derived promoters, means that those from other species are frequently employed. However, such heterologous promoters do not always maintain the activity originally observed in the species of origin. For example, an endosperm-specific wheat promoter (high molecular weight subunit 1Dx5) has been observed to drive GUS expression in roots, leaves and pollen of barley (Y Zhang *et al.* personal communication). The CaMV 35S promoter is one of the most commonly used in dicots but without enhancer elements in wheat has shown poor activity and a tendency to be silenced (WP Chen *et al.*, 1998, 1999). The promoter that appears to give the highest and most stable constitutive expression in wheat is the maize ubiquitin promoter and intron (Ubi1). However, although considered a constitutive promoter, recent studies to characterise its specificity in detail have shown that position effect, developmental stage (Rooke *et al.*, 2000) and stress (Stroger *et al.*, 1999b) may all affect its activity.

The best-studied and most widely available promoters in the public sector, are the constitutive ones but there is a paucity of those with proven specificity that may be used in strategic research. Although expression was detectable from them, the rice sucrose synthase (*Rss1*) and the grapevine stilbene synthase (*Vst1*) promoters used by Stoger *et*

Table 3. Promoters active in stably transformed wheat

Source	Promoter	Enhancer element	Intended activity	Sample reference
Maize	ubiquitin	*Ubi* intron	constitutive	Rooke *et al.*, 2000
Rice	actin	Actin intron	constitutive	Jordan, 2000
CaMV	35S		constitutive	Chen *et al.*, 1999
CaMV	35S-Sh ex/int	maize *shrunken* exon and intron	constitutive	Karunaratne *et al.*, 1996
CaMV	35S-35S		constitutive	Zhou *et al.*, 1995
CaMV	35S-*adh1*	maize *adh1* intron	constitutive	Barro *et al.*, 1998
Maize	35S-RTBV	rice tungro bacilliform virus intron		Bieri *et al.*, 2000
Maize	*adh1*/pEmu	Maize AREs and *Adh1* intron. Octopine synthase	constitutive	Karunaratne *et al.*, 1996
Maize	alcohol dehydrogenase		constitutive	Vasil *et al.*, 1992
Wheat	high molecular weight glutenin (HMWG) subunit 1Dx5		endosperm	Lamacchia *et al.*, 2001
Wheat	HMWG subunit 1Dy10		endosperm	Blechl and Anderson, 1996
Wheat	low molecular weight glutenin (LMWG1D1)		endosperm	Stoger *et al.*, 1999a

Table 3. Continued

source	Promoter	Enhancer element	Intended activity	Sample reference
Maize	T-72 tapetum specific		tapetum	De Block *et al.*, 1997
Rice	E1 tapetum specific		tapetum	De Block *et al.*, 1997
Rice	T72 tapetum specific		tapetum	De Block *et al.*, 1997
Rice	sucrose synthase		phloem	Stoger *et al.*, 1999b
Maize	rubisco small subunit		green tissue	Sparks *et al.*, 2001
Grapevine	native stilbene synthase promoter – along and with 4 × 35S enhancer	4 × 35S enhancer	pathogen induced	Leckband and Lörz, 1998

al. (1999a) and Leckband and Lörz (1998) respectively, showed limited success in generating the desired phenotypes in wheat (even though it was observed in barley in the latter study). Clearly the target for crop improvement determines the choice of promoter. A principal target for wheat modification is grain quality and although effective endosperm promoters are available (Lamacchia *et al.*, 2001; Stoger *et al.*, 1999b), other grain specific promoters with different developmental stage specificities are continuously sought (see section 6 also). Further targets for promoter specificities include green tissue (for engineering of photosynthetic components, disease/pest resistance), root tissue (nutrient uptake and disease/pest resistance), flowers (for male sterility), phloem (aphid resistance) and mesophyll (*eg.* for engineering C_3/C_4 metabolism). Promoters that are active under a range of biotic and abiotic stresses are also sought. So far, except for the limited success of the wound-inducible *Vst1* promoter (Lekband and Lörz, 1998), no inducible promoters have been demonstrated in wheat. However, the glutathion-S-transferase 27 (GST 27) promoter from maize was able to drive marker gene expression upon application of a herbicide safener to transgenic maize plants (Greenland *et al.* 1996). The rice basic chitinase (RC24) and rice glycine-rich cell-wall protein (Osgrp1) promoters have shown wound-inducibility in rice (Xu *et al.*, 1996), and Caddick *et al.* (1998), demonstrated the successful ethanol induction of the *alc* regulon (from *Aspergillus nidulans*) in transgenic tobacco.

5.2. Stability of transgene expression

In recent years there have been a growing number of studies giving significant details of the inheritance of transgene expression and phenotypes over several generations in wheat (Stoger *et al.*, 1998, 1999b; Cannell *et al.*, 1999; Takumi *et al.*, 1999; Bliffeld *et al.*, 1999; Sivamani *et al.*, 2000a; Bieri *et al.*, 2000; Brinch-Pederson *et al.*, 2000; Clausen *et al.*, 2000). In the majority of cases phenotypes were stably inherited although analyses in most studies was limited to the T_3 generation.

Some incidences of transgene instability have been well documented. Demeke *et al.* (1999) detailed the expression of *uidA* and *nptII* (regulated by the rice actin promoter) in the T_4 and T_5 generations of two independent wheat transformants (containing multiple transgene copies). In many progeny plants, decreased transgene activity was observed which was demonstrated to be associated with transgene methylation. In addition, Cannell *et al.* (1999), in a study of six lines over three generations, identified one transformant (copy number ~5) that gave non-Mendelian inheritance ratios for *uidA* and *bar* gene expression from plasmid pAHC25 (Ubi1-*uidA*::Ubi1-*bar*). Silencing was not observed until the T_2 generation and was subsequently found to result from promoter methylation (J. Howarth, personal communication).

Two incidences of silencing following transformation with high molecular weight glutenin subunits (HMW-GS) have been reported in wheat. Six genes encode this family of storage proteins, only 3-5 of which are expressed in any given genotype. Alvarez *et al.* (2000) transformed the wheat variety 'Federal', which expresses subunits 1Ax2, 1Dx5, 1Bx7, 1By9 and 1Dy10. In two lines containing the 1Dx5 transgene (approximately 3 copies), increased HMW glutenin content was observed. Two further lines transformed with the 1Ax1 gene (approximately 3 copies) were shown to express the transgene but caused loss of expression of the endogenous 1Ax2 gene. Finally, two lines containing multiple (10-20) copies of 1Ax1 and 1Dx5 showed either transgene expression or partial transgene silencing whereas all 5 of the endogenous genes were silenced. In one of the latter lines this silencing became reversed in T_3 generation. A potentially related phenomenon was observed by Blechl and Anderson (1996) who described a population of transformants containing a chimeric high molecular weight glutenin construct consisting of the 1Dy10 promoter and partial coding sequence fused to a larger portion of the 1Dx5 gene (plus 1Dx5 terminator). In one of the transformants, containing 5-6 copies, weak expression of the transgene was accompanied by a 70% reduction in expression of at least three related endogenous genes (co-suppression). A detailed review of transgene silencing in monocots is presented by Iyer *et al.* (2000). A frequent cause is the insertion of multiple, rearranged transgenes – both common features of direct gene delivery. Interestingly however, Stoger *et al.* (1998) analysed a sub-population of transformants containing the Ubi::*uidA* transgene and found those with more copies expressed at higher levels.

6. TARGETS FOR WHEAT IMPROVEMENT

6.1. End-use quality

6.1.1. Bread-making quality

Wheat grain is composed mainly of protein and starch, which account for approximately 10-15% and 70-80% dry weight of the grain, respectively, and provides significant calorific and nutritional input in many human diets. The unique ability of wheat flour to make leavened bread is determined by the visco-elasticity of dough conferred by gluten proteins. Variation in the amount and composition of the HMW subunits of gluten are associated with differences in bread-making qualities of different wheat varieties (Payne, 1987). The role of the various gluten protein groups in the biotechnology of bread-making was reviewed by Shewry *et al.* (1995), who also commented on the potential for improvement of end-use quality of wheat by two main strategies: i) the addition of extra HMW subunit genes copies in an attempt to increase the glutenin proteins known to be associated with good bread-making, and ii) the modification of structural features likely to contribute to the functionality of gluten, such as the number and distribution of cysteine residues or the length of the repetitive domain in the HMW glutenin subunits. These approaches have been successfully used in several laboratories and the highlights of this work are reviewed below.

Genes encoding a number of proteins known to be associated with good bread-making quality have been introduced into hexaploid wheat (*Triticum aestivum*), durum wheat (*Triticum durum*) and tritordeum, and some studies have shown altered functionality. For instance, a hybrid Dy10/Dx5 protein was over-expressed in the model wheat cultivar 'Bobwhite' using the native Dy10 promoter (Blechl and Anderson, 1996). A genomic clone of the 1Ax1 gene, including its native promoter (Halford *et al.*, 1992) was introduced into Bobwhite, resulting in a marked increase in total HMW subunit protein (Altpeter *et al.*, 1996). Mixograph studies on dough from transgenic lines expressing the 1Ax1 and 1Ax1 + 1Dx5 proteins under the control of their native promoters, showed a step-wise increase in both the maximum resistance to dough mixing (peak dough resistance), and the time to maximum resistance (mixing time) (Barro *et al.*, 1997). Transgenic lines expressing the 1Ax1 and 1Dx5 transgene, resulted in significant increases in the proportion of these proteins in the grain, the combined proportion of HMW subunits was increased from 12% to over 20% of the total gluten protein fraction (Rooke *et al.*, 1999b). One particular 1Dx5 over-expressing line gave an over-strong dough with abnormal mixing behaviour. Dough from this line was too strong for conventional bread-making, but could prove valuable for blending with poor quality wheats and for novel end-uses for gluten where a high degree of elasticity is required (Rooke *et al.*, 1999b). elevated proportions of HMW subunits were also achieved in a commercial cultivar of bread wheat (Federal) already expressing five native HMW subunit genes (Alvarez *et al.*, 2000).

In a different approach, He *et al.* (2000) created a series of lines transformed with modified 1Dx5 subunit genes encoding HMW glutenin subunits with repetitive domains about 34% and 17% shorter, and 22% larger than the native 1Dx5 protein (D'Ovidio *et al.*, 1997). The expression of the mutant forms of subunit 1Dx5 was confirmed by SDS-PAGE of seed proteins over at least two generations. The same group also demonstrated expression, in commercial lines of bread wheat, of two HMW glutenin subunits encoded by gene fragments from which vector backbone and ampicillin resistance gene had been removed (Pastori *et al.*, 2000). There has been some opposition to the inclusion of the ampicillin gene in field trials and commercial releases of GM plants, and this work demonstrated that the accumulation of 1Ax1 and 1Dx5 subunit proteins can be achieved by utilising minimal DNA fragments comprising an expression cassette only.

The majority of reports to date, are based on wheat transgenics grown under glasshouse or constant environment conditions. However, an initial analysis of field-grown plants over-expressing either subunit 1Ax1 or 1Dx5, from duplicate plots at IACR-Rothamsted and IACR-Long Ashton supported data from glasshouse-grown material showing modified glutenin aggregation, visco-elasticity and mixing properties of doughs especially in the 1Dx5 transgenics (Popineau *et al.*, 2001). In addition, there was no effect of the transgene on agronomic performance (Fido *et al.*, 2000).

6.1.2. Improvements in durum wheat and tritordeum

Durum wheat (*Triticum durum*) is a tetraploid species used to make pasta, flat breads, bulgar and couscous in areas of southern Europe and North Africa. It is not suitable for making leavened bread, because it lacks the D genome associated with high gluten visco-elasticity, present in hexaploid bread wheat. The over-expression of 1Ax1 and 1Dx5 genes in three varieties of durum wheat resulted in elevated levels of the corresponding proteins. Mixograph studies also revealed an increase in dough strength in the transgenic lines, demonstrating the potential for using transformation to improve the bread-making properties of this species (He *et al.*, 1999). The quality of durum wheat for pasta making is also associated with the composition of low molecular weight (LMW) subunits (Pogna *et al.*, 1990). Three LMW subunit genes, incorporating a c-myc, epitope-tag, have been over-expressed in transgenic durum wheat. The presence of the epitope-tag will facilitate the use of a specific antibody against the novel transgenic protein and shed light on the still obscure mechanisms of trafficking and deposition of wheat prolamins (Tosi *et al.*, 2000).

Tritordeum is an amphiploid cereal derived from crossing the South American wild barley (*Hordeum chilense*) with either durum or bread wheat, resulting in hexaploid and octaploid forms, respectively (Martin *et al.*, 1995). It has good agronomic performance, with similar yields and protein contents to the parental wheat varieties, and is a crop of potential importance to the food industry (Alvarez *et al.*, 1992). Tritordeum varieties have proved to be particularly amenable to genetic transformation (Barcelo *et al.* 1994), and modification with 1Ax1 and 1Dx5 genes resulted in increased protein levels and dough strength (Rooke *et al.*, 1999a).

6.1.3. Starch quality

The success of transgenic approaches to modification of wheat HMW seed proteins, suggests that it may be possible to produce starches with a wide range of structures and properties. In potato and rice, transformation has been used to modify starch functionality (*e.g.* Visser *et al.*, 1991; Flipse *et al.*, 1996a/b; Itoh *et al.*, 1997), and a search of the patent databases reveals significant industrial interest in genetic modification of starch quality in wheat. Potential applications include wheat starches that mimic starch mutants in maize, phosphorylated starch (currently obtained from potato) and thermoplastic and biodegradable starches for packaging. Obvious target genes are those that encode the various isoforms and classes of soluble and granule-bound starch synthases, which could fine-tune the proportion of amylose and amylopectin, and produce 'designer' starches with novel functional properties (Miflin *et al.*, 1999).

One of the strategies to modify wheat starch structure involves identification of germplasm with null alleles for starch biosynthetic genes, followed by exchange of functional alleles with the identified null alleles through classical plant breeding. This technique has successfully been used to combine the three null alleles for granule-bound starch synthase I (GBSSI) to develop a wheat line that produces amylopectin-rich (>95%) starch (waxy starch) (Nakamura *et al.*, 1995; Yasui *et al.*, 1996; Baga *et al.*, 1999). Another strategy to alter expression levels of starch biosynthetic genes employs recent advances in molecular biology and genetic engineering of wheat. For this approach, various monocot vectors have been developed that drive expression of wheat starch branching enzyme I (SBEI) cDNA sequences in the anti-sense orientation. Baga *et al.* (1999) report that several of the wheat lines transformed with the antisense vectors express branching enzyme (BE) activity at a significantly lower level than non-transformed cells. One transgenic wheat plant expressing the anti-sense SBEI RNA produces a ten-fold lower level of BE activity in kernels than wild-type wheat. A similar approach was taken by Woplin *et al.* (1998) who attempted to down-regulate starch synthase activity by generating sense and antisense wheat lines with maize GBSSI and SSI cDNAs. The expression of transgenes was screened at the RNA, protein and enzyme activity level, however, no significant change was found in any of the plants analysed.

6.1.4. Grain hardness

Endosperm texture, *i.e.* the hardness or softness of the grain, is one of the most important quality criteria in cereals. It determines the milling performance, as well as affecting properties for food production. Grain hardness is genetically determined and controlled by the Ha locus on the short arm of chromosome 5D in hexaploid bread wheat (Law *et al.*, 1978). Genes for the proteins puroindoline a (Pin a), puroindoline b (Pin b) and grain softness proteins (GSP) are lined to Ha locus, and area associated with grain hardness. Friabilin, a marker for grain hardness, is composed of Pin a, Pin b and other proteins (Morris *et al.*, 1994; Oda and Scholfield, 1997). Hard wheats may lack Pin a protein and/or contain mutant forms of pin b, notably with a glycine-to-serine mutation (Giroux and Morris, 1997, 1998). However, such mutations are not universally present in hard wheats, and although there is strong correlative data associating puroindolines with hardness, genuine 'cause and effect' is yet to be established. A transgenic approach, over-expressing these genes in null (*e.g.* durum) wheat backgrounds is currently underway at IACR UK, and may help to resolve some of the uncertainties. There are currently no reports in the literature of genetic modification in wheat with genes encoding grain hardness proteins. However, over-expression of Pin a and Pin b in rice, a cereal in which Pin a and Pin b are normally absent, has been achieved, and demonstrated reductions in hardness and in starch damage (Krishnamurthy and Giroux, 2001).

6.1.5. Alpha amylase content

Pre-harvest sprouting (PHS) of wheat is the germination of immature grains whilst on the crop ear, and is one cause of high alpha amylase in the mature grain. This is a major factor in determining end-use, particularly bread-making quality in most wheat-growing countries worldwide. Alpha amylase degrades the starch present in flour producing poor quality, sticky bread that clogs slicing machines. PHS has been linked to the lack of dormancy induction during embryo development (Gale and Lenton, 1987) and transgenic approaches to enhance dormancy in developing wheat grain has potential to reduce the significance of this problem. One strategy has been to use the wild oat homologue (*AfVP1*) of the maize Viviparous 1 (*VP1*) gene. Expression of this gene is strongly correlated with the level of embryo dormancy in wild oat (Jones *et al.*, 1997), and over-expression of the *AfVP1* gene in wheat might be expected to increase resistance to PHS. This approach has been taken by Holdsworth *et al.* (2001) who are currently analysing wheat lines genetically modified with this gene.

6.2. Agronomic

There is much evidence of transgenic approaches being used to investigate and improve a wide range of agronomic wheat traits including pest/disease resistance, tolerance to abiotic stresses, herbicide tolerance, alteration of plant architecture, etc. These are summarised in Tables 4 a and b.

7. CONCLUSIONS AND FUTURE PROSPECTS

Wheat transformation is a now a routine undertaking in many laboratories, but transformation frequencies remain relatively low and the process is still labour-intensive. Advances in the basic transformation procedure are still being made, for example, culture conditions (Rasco-Gaunt *et al.*, 2001) and the influence of donor material age (Pastori *et al.*, 2001), have been investigated. The results of such studies constitute an improvement towards the broad application of transformation in elite wheat genotypes. However, there remains considerable scope to further improve efficiencies and remove sources of variation in parts of the transformation process.

The focus of much transformation research is now moving towards the development of strategies to improve transgene integration, expression and stability. Work in these areas of basic research is often limited to model species, and although advancements are sometimes transferable between models and crops, new strategies must ultimately be proven in the wheat background (especially given its complex genetic composition). The numbers of such studies conducted in wheat are therefore increasing. For example, Srivastava *et al.* (1999) demonstrated use of the

Tables 4. Transgenic approaches for the modification of agronomic and other traits in wheat.

Table 4 (a) : Pest/disease resistance

Target species	Promoter/Gene	Wheat variety (s)	Observed phenotype	Reference
		Fungal disease resistance		
Erysiphe graminis	Barley seed class II chitinase. Maize ubiquitin promoter and intron (*Ubi1*)	Bobwhite	Chitinase localised in apoplast. Increased resistance to powdery-mildew	Bliffeld *et al.* (1999)
Erysiphe graminis	Barley seed ribosome-inactivating protein. CaMV35S promoter plus RTBV intron	Frisal	Moderate or no protection shown to *Erysiphe graminis*	Bieri *et al.* (2000)
General	Grape (*Vitis vinifera*) stilbene synthase cDNA. Promoter *Ubi1*.	Hanno, Combi	Expression of sts shown by RT-PCR and accumulation of reservatol by HPLC & MS.	Fettig and Hess, 1999
Fusarium	Rice thaumatin-like protein (*tlp*). Promoter *Ubi1*.	Bobwhite	Enhanced resistance to wheat scab (*Fusarium graminaerum*)	Chen *et al.* (1999)
General	Rice chitinase. CaMV35S promoter.	Bobwhite	Accumulation of rice chitinase protein demonstrated by Western blot analysis	Chen *et al.* (1998)
Ustilago/Tilletia	Antifungal protein KP4 from *Ustilago maydis*-infecting virus. Promoter *Ubi1*.	Golin, Greina, Frisal	Some lines showed antifungal activity against *U. maydis* and increased resistance to stinking smut (*T. tritici*).	Clausen *et al.* (2000)

Table 4 (a). Continued

Target species	Promoter/Gene	Wheat variety (s)	Observed phenotype	Reference
		Viral disease resistance		
Wheat streak mosaic virus	Wheat streak somaic virus replicase gene (*NIb*). Promoter *Ubi1*	Hi-Line	Milder symptoms or a delay in onset of symptoms after mechanical inoculation of WSMV.	Sivamani *et al.* (2000b)
Barley stripe mosaic virus	Mutant bacterial ribonuclease III (*rnc70*). Promoter *Ubi1*	Bobwhite	Reduction in virus symptoms and reduced accumulation of virons in plants challenged with barley stripe mosaic virus	Zhang *et al.* (2001)
Barley yellow mosaic virus	Barley yellow mosaic virus coat protein. 35S promoter plus maize *shrunken* exon and intron (*sh* ex/int)	Hartog	Barley yellow mosaic virus coat protein was detected	Karunaratne *et al.* (1996)
		Insect resistance		
Grain moth *Sitotroga cerealella*	Barley trypsin inhibitor (*BTI-CMe*). Promoter *Ubi1*	Bobwihte	Integrity of BTI-Cme was confirmed by trypsin inhibitor assay. A reduction in survival rate of the grain moth (*Sitotroga cerealella*) was observed	Altpeter *et al.* (1999)
Grain aphid *Sitobion avenae*	Snowdrop lectin (*Galanthus nivalis agglutinin* GNA). Rice sucrose synthase promoter	Bobwhite	Decreased fecundity of grain aphid (*Sitobion avenae*) on feeding on some lines.	Stoger *et al.* (1999b)

Table 4 (b). Other triats

Target species	Promoter/Gene	Wheat variety (s)	Observed phenotype	Reference
		Other traits		
Abiotic stress/ Polyamine metabolism	Sense and antisense tritordeum and tomato S-adenosylmethionine decarboxylase (*SAMDC*). Maize ubiquitin promoter and intron (*Ubi1*)	Avans, Cadenza, Rialto, Riband	Antisense lines had a reduced level of *samdc* mRNA, a decreased polyamine titre and a dwarf stature	Rasco-Gaunt *et al.* (1999b)
Drought tolerance	Barley *HVA1* (a group 3 Late Embryogenesis Abundant) gene. Promoter *Ubi1*	Hi-Line	Improved water-use efficiency and growth characteristics under moderate deficit conditions	Sivamani *et al.* (2000a)
Yield/Carbon metabolism	Tritordeum glycine decarboxylase (*GDC*). Rice tungrovirus promoter	Baldus	A decrease in GDC activity correlated with reduced growth characteristics, in some plants.	Sparks *et al.* (2001)
Yield/Plant architecture	Oat phytochrome A. *A. thaliana* PHY A, B and C. Promoter *Ubi1*	Florida	Far-red light grown seedlings showed inhibition of coleoptile extension compared to wild-type seedlings	Shlumukov *et al.* (2001) Reid *et al.* (2001)
Pre-harvest sprouting/dwarfing	Bean (*Phaseolus coccineus*) GA2 oxidase cDNA. Promoter *Ubi1*.	Canon, Cadenza	Altered growth characteristics, severe dwarf morphology.	Stone *et al.* (2001); Hedden and Phillips (2000)

Table 4 (b). Continued

Target species	Promoter/Gene	Wheat variety (s)	Observed phenotype	Reference
Lipid profile	*Pisum sativum* glycerol-3-phosphate acyltransferase gene and *A. thaliana* acyl-ACP thioesterase gene. Promtoer *Ubi1*.	Avans, Canon, Riband Cadenza	Morphological (growth, organelle development) and metabolic changes (fatty acid labelling of chloroplast and non-chloroplast lipids)	Edin *et al.* (2000)
Herbicide tolerance	Maize glutathione S-transferase (*GST-27*). Promoter *Ubi1*.	Florida	Tolerance to the chloroacetanilide herbicide 'alachlor' was correlated with GST-27 expression levels	Milligan *et al.* (1999)
Herbicide tolerance	*Agrobacterium* strain CP4 enolpyruvylshikimate phosphate synthase gene (*EPSPS*) and a bacterial glyphosate oxidoreductase (*GOX*). Promoter *Ubi1* and duplicated 35S promoters	Bobwhite	Transgenic plants tolerant to glyphosate were recovered	Zhou *et al.* (1995)
Cytoplasmic male sterility	*Barnase* driven by maize and rice tapetum promoter	Pavon	Premature degeneration of the tapetum	De Block *et al.* (1997)

bacteriophage *Cre-lox* recombinase system for the resolution of multiple copy to single copy integrations. In addition, Bliffeld *et al.* (1999) and Bieri *et al.* (2000) used transformation vectors for wheat studies which contained matrix attachment regions (MARs). Although these have been shown to stabilise transgene expression in plants (Holmes-Davis and Comai, 1998), previous studies have been limited to a narrow range of species (mostly models) and heterologous MARs and transgenes. New approaches for the evaluation of cereal MARs in wheat which will allow the comparison of unselected events with selected events of equal copy number are currently being developed in our laboratory.

The development of reproducible methodologies for stable *Agrobacterium*-mediated wheat transformation is an important goal, which may address some of the problems associated with transgene instability, unwanted DNA integration and low frequencies of multi-locus integration. However, the associated problems of non-T-DNA integration must also be addressed. Another key area is the identification and characterisation of tissue-specific and inducible promoters and those with activities of well-defined developmental regulation. An additional component of this subject includes gaining a deeper understanding of the activities of commonly used promoters in wheat, as these can vary between species and individual transformants. Regarding the post-genomics era, transposon-based methods have shown potential for the tagging and trapping of genes and promoters in rice (Izawa *et al.*, 1997) and maize (Walbot, 2000; Tacke *et al.*, 1995). Functionality of the Ac/Ds system has also been demonstrated in transgenic barley (Koprek *et al.*, 2000) and wheat (Takumi *et al.*, 1999) plants, and is being evaluated for similar applications in these species by a number of different groups. However, their large genome sizes may make complete genome saturation a difficult undertaking, given that Ds transpositions occur preferentially within linked genomic regions. In addition it is believed that 'knockout' tagging events in hexaploid wheat may be masked by complementation between homeologous genes. For this reason there is increased interest in transformation of diploid wheat and the first successes in this area are likely to be reported soon.

Chloroplast transformation of wheat, although a very ambitious goal, is receiving some interest due to the many potential advantages for crop improvement offered by the engineering of these organelles – such as increased expression levels, integration of transgenes through homologous recombination and sequestration of foreign proteins (Heifetz, 2000). Another advantage of chloroplast transformation is that transgene

Table 5. Applications for field-trials of GM wheat in USA and Canada

Year	USA	Canada
2000	67	72
1999	36	123
1998	19	13
1997	18	19
1996	8	2
<1996	6	5

sources:
http://www.isb.vt.edu/cfdocs/fieldtests1.cfm
http://wwwxcfia-acia.agr.ca/english/plaveg/pbo/pbobbve.shtml

inheritance would be uniparental, thus helping to prevent pollen transmission of foreign DNA to wild relatives. The relevance of this is put into perspective given that the number of field trials of wheat with genetically modified trials in USA and Canada is increasing markedly (see Table 5). In addition to the traits already discussed, 'horizon-scanning' reveals commercial interest in i) increasing the 'health-food value' of wheat grain for example, through the expression of antioxidants, the alteration of polyamine content and the reduction in allergenicity of the gluten proteins, ii) the expression of antibodies for pharmaceutical, diagnostic and therapeutic applications and iii) improving nutritional composition, for example by increasing iron content in the grain (Drakaki *et al.*, 2000). However, the commercial future of these, or other transgenic approaches to crop improvement depends on a high level of consumer acceptance which in some countries, particularly Europe and Japan, is not evident at present.

ACKNOWLEDGEMENTS

IACR receives grant-aided support from the Biotechnology and Biological Sciences Research Council (BBSRC) of the UK. The authors thank Professor Peter R Shewry for his input to the end-use quality section, and Caroline Sparks for proof-reading the manuscript.

REFERENCES

Abranches R, Santos AP, Wegel E, Williams S, Castilho A, Christou P, Shaw P and Stoger E (2000) Widely separated multiple transgene integration sites in wheat chromosomes are brought together at interphase. *Plant J.*, **24** : 713-723.

Altpeter F, Vasil V, Srivastava V, Stoger E, Vasil IK (1996) Accelerated production of transgenic wheat (*Triticum aestivum* L) plants. *Plant Cell Rep.,* **16** : 12-17.

Altpeter F, Diaz I, McAuslane H, Gaddour K, Carbonero P and Vasil IK (1999) Increased insect resistance in transgenic wheat stably expressing trypsin inhibitor *CMe. Mol. Breed.,* **5** : 53-63.

Alvarez JB, Ballesteros J, Sillero JA and Martin LM (1992) Tritordeum: a new crop of potential importance in the food industry. *Hereditas,* **116** : 193-197.

Alvarez ML, Guelman S, Halford NG, Lusting S, Reggiardo MI, Ryabushkina N, Shewry P, Stein J and Vallejos RH (2000) Silencing of HMW glutenins in transgenic wheat expressing extra HMW subunits. *Theor. Appl. Genet.,* **100** : 319-327.

Amoah BK, Wu H, Sparks C and Jones HD (2001) Factors influencing *Agrobacterium*-mediated transient expression of *uidA* in wheat inflorescence tissue. *J. Exp. Bot.,* **52** : 1-8.

Azhakanandam K, McCabe MS, Power JB, Lowe KC, Clocking EC and Davey MR (2000) T-DNA transfer, integration, expression and inheritance in rice: effects of plant genotype and *Agrobacterium* super-virulence. *J. Plant Physiol.,* **157** : 429-439.

Baga M, Repellin A, Demeke T, Caswell K, Leung N, Abdel-Aal ES, Hucl P and Chibbar RN (1999) Wheat starch modification through biotechnology. *Starch-Starke,* **51** : 111-116.

Barakat A, Gallois P, Raynal M, Mestre-Ortega D, Sallaud C, Guiderdoni E, Delseny M and Bernardi G (2000) The distribution of T-DNA in the genome of transgenic *Arabidopsis* and rice. *FEBS Lett.,* **471** : 161-164.

Barcelo P, Hagel C, Becker D, Martin A and Lörz H (1994) Transgenic cereal (tritordeum) plants obtained at high efficiency by microprojectile bombardment of inflorescence tissue. *Plant J.,* **5** : 583-592.

Barro F, Rooke L, Békés F, Gras P, Tatham AS, Fido R, Lazzeri PA, Shewry PR and Barcelo P (1997) Transformation of wheat with high molecular weight subunit genes results in improved functional properties. *Nat. Biotechnol.,* **15** : 1295-1299.

Barro F, Cannell ME, Lazzeri PA and Barcelo P (1998) The influence of auxins on transformation in wheat and tritordeum and analysis of transgene integration patterns in transformants. *Theor. Appl. Genet.,* **97** : 684-695.

Bieri S, Potrykus I and Futterer J (2000) Expression of active barley seed ribosome-inactivating protein in transgenic wheat. *Theor. Appl. Genet.,* **100** : 755-763.

Blechl AE and Anderson OD (1996) Expression of a novel high-molecular-weight glutenin subunit gene in transgenic wheat. *Nat. Biotechnol.,* **14** : 875-879.

Bliffeld M, Mundy J, Potrykus I and Futterer J (1999) Genetic engineering of wheat for increased resistance to powdery mildew disease. *Theor. Appl. Genet.,* **98** : 1079-1086.

Bommineni VR, Jauhar PP and Peterson TS (1997) Transgenic durum wheat by microprojectile bombardment of isolated scutella. *J. Hered.,* **88** : 475-481.

Brinch-Pedersen H, Olesen A, Rasmussen SK and Holm PB (2000) Generation of transgenic wheat (*Triticum aestivum* L.) for constitutive accumulation of an *Aspergillus phytase. Mol. Breed.,* **6** : 195-206.

Caddick MX, Greenland AJ, Jepson I, Krause KP, Qu N, Riddell KV, Salter MG, Schuch W, Sonnewald U and Tomsett AB (1998) An ethanol inducible gene switch for plants used to manipulate carbon metabolism. *Nat. Biotechnol.,* **16** : 177-180.

Campbell BT, Baenziger PS, Sato AMS and Clemente T (2000) Inheritance of multiple transgene in wheat. *Crop Sci.,* **40** : 1133-1141.

Cannell ME, Doherty A, Lazzeri PA and Barcelo P (1999) A population of wheat and tritordeum transformants showing a high degree of marker gene stability and heritability. *Theor. Appl. Genet.,* **99** : 772-784.

Chen WP, Gu X, Liang GH, Muthukrishnan S, Chen PD, Liu DJ and Gill BS (1998) Introduction and constitutive expression of a rice chitinase gene in bread wheat using biolistic bombardment and the *bar* gene as a selectable marker. *Theor. Appl. Genet.,* **97** : 1296-1306.

Chen LL, Marmey P, Taylor NJ, Brizard JP, Espinoza C, D'Cruz P, Huet H, Zhang SP, de Kochko A, Beachy RN and Fauquet CM (1998) Expression and inheritance of multiple trasngenes in rice plants. *Nat. Biotechnol.,* **16** : 1060-1064.

Chen WP, Chen PD, Liu DJ, Kynast R, Friebe B, Velazhahan R, Muthukrishnan S and Gill BS (1999) Development of wheat scab symptoms is delayed in trasngenic wheat plants that constitutively express a rice thaumatin-like protein gene. *Theor. Appl. Genet.,* **99** : 755-560.

Cheng M, Fry JE, Pang SZ, Zhou HP, Hironaka CM, Duncan DR, Conner TW and Wan YC (1997) Genetic transformation of wheat mediated by *Agrobacterium tumefaciens. Plant Physiol.,* **115** : 971-980.

Clausen M, Krauter R, Schachermayr G, Potrykus I and Sautter C (2000) Antifungal activity of a virally encoded gene in transgenic wheat. *Nat. Biotechnol.,* **18** : 446-449.

Dai SH, Zheng P, Marmey P, Zhang SP, Tian WZ, Chen SY, Beachy RN and Fauquet C (2001) Comparative analysis of transgenic rice plants obtained by *Agrobacterium*-mediated transformation and particle bombardment. *Mol. Breed.,* **7** : 35-33.

De Buck S, De Wilde C, Van Montagu M and Depicker A (2000) T-DNA vector backbone sequences are frequently integrated into the genome of transgenic plants obtained by *Agrobacterium*-mediated transformation. *Mol. Breed.,* **6** : 459-468.

De Block M, Debrouwer D and Moens T (1997) The development of a nuclear male sterility system in wheat. Expression of the barnase gene under the control of tapetum specific promoters. *Theor. Appl. Genet.,* **95** : 125-131.

Demeke T, Hucl P, Baga M, Caswell K, Leung N and Chibbar RN (1999) Transgene inheritance and silencing in hexaploid spring wheat. *Theor. Appl. Genet.,* **99** : 947-953.

D'Ovidio R, Anderson OD, Masci S, Skerritt J and Porceddu E (1997) Consturction of novel wheat high-M.W. glutenin subunit gene variability: Modification of the repetitive domain and expression in *E. coli. J. Cereal Sci.,* **25** : 1-8.

Drakaki G, Christou P and Stoger E (2000) Constitutive expression of soybean ferritin cDNA in transgenic wheat and rice results in increased iron levels in vegetative tissues but not in seeds. *Transgenic Res.,* **9** : 445-452.

Edlin DAN, Kille P, Wilkinson MD, Jones HD and Harwood JL (2000) Morphological and metabolic changes in transgenic wheat with altered glycerol-3-phosphate acyltransferase or acyl-acyl carrier protein (ACP) thioesterase activities. *Biochem. Soc. Trans. Trans.,* **28** : 682-683.

Fetting S and Hess D (1999) Expression of a chimeric stilbene synthase gene in transgenic wheat lines. *Transgenic Res.,* **8** : 179-189.

Food and Agriculture Organisation of the United Nations (1999) *FAO Production Yearbook 1998* (FAO Statistics Series No. 148). Rome. pp. 272.

Fido RJ, Darlington HF, Cannell ME, Jones HD, Tatham AS, Bekes F and Shewry PR (2000) Expression of HMW glutenin subunits in field grown transgenic wheat. In: *Wheat Gluten* (Eds. Shewry PR and Tatham AS) ISBN 0854048650. Royal Society of Chemistry, pp 77-79.

Flipse E, Keetals CJAM, Jacobsen E and Visser RGF (1996a) The dosage effect of the wild-type GBSS allele is linear for GBSS activity but not for amylose content: absence of amylose has distinct influence on the physio-chemical properties of starch. *Theor. Appl. Genet.,* **92** : 121-127.

Flipse E, Suurs L, Keetals CJAM, Kossermann J, Jacobsen E and Visser RGF (1996b) Introduction of sense and anitisense cDNA for branching enzyme in the amylose-free potato mutant leads to physico-chemical changes in the starch. *Planta.,* **198** : 340-347.

Fu XD, Duc LT, Gontana S, Bong BB, Tinjuangjun P, Sudhakar D, Twyman RM, Christou R and Kohli A (2000) Linear transgene constructs lacking vector backbone sequences generate low-copy-number transgenic plants with simple integration patterns. *Transgenic Res.,* **9** : 11-19.

Gale MD and Lenton JR (1987) Preharvest sprouting in wheat – a complex genetic and physiological problem affecting breadmaking quality of UK wheats. *Aspects of Applied Biology,* **15** : 115-124.

Giroux MJ and Morris CF (1997) A glycine to serine change in puroindoline b is associated with wheat grain hardness and low levels of starch-surface friabilin. *Theor. Appl. Genet.,* **95** : 857-864.

Giroux MJ and Morris CF (1998) What grain hardness results from highly conserved mutations in the friabilin components puroindoline a and b. *Proc. Natl. Acad. Sci. USA,* **95** : 6262-6266.

Greenland AJ, Bell PJ, Jepson I, Wright SY, Nevshemal T and Register III JC (1996) Chemically switched male sterility in maize. *J. Exp. Bot.,* **47** : 12.

Guo GQ, Maiwald F, Lorenzen P and Steinbiss HH (1998) Factors influencing T-DNA transfer into wheat and barley cells by *Agrobacterium tumefaciens. Cereal Res. Commun.,* **26** : 15-22.

Halford NG, Field JM, Blair H, Urwin P, Moore K, Robert L, Thompson R, Flavell RB, Tatham AS and Shewry PR (1992) Analysis of HMW glutenin subunits encoded by chromosome 1A bread wheat (*Triticum aestivum* L.) indicates quantitative effects on grain quality. *Theor. Appl. Genet.,* **83** : 373-378.

Harvey A, Moisan L, Lindup S and Lonsdale D (1999) Wheat regenerated from scutellum callus as a source of material for transformation. *Plant Cell Tiss. Org. Cult.,* **57** : 153-156.

Harwood WA, Ross SM, Cilento P and Snape JW (2000) The effect of DNA/gold particle preparation technique, and particle bombardment device, on the transformation of barley (*Hordeum vulgare*). *Euphytica,* **111** : 67-76.

He DG, Mouradov A, Yang YM, Mouradova E and Scott KJ (1994) Transformation of wheat (*Triticum aestivum* L). through electroporation of protoplasts. *Plant Cell Rep.,* **14** : 192-196.

He GY, Rooke L, Steele S, Bekes F, Gras P, Tatham AS, Fido R, Barcelo P, Shewry PR and Lazzeri PA (1999) Transformation of pasta wheat (*Triticum turgidum* L var. *durum*) with high-molecular-weight glutenin subunit genes and modification of dough functionality. *Mol. Breed.,* **5** : 377-386.

He GY, Ovidio R, Anderson OD, Fido RJ, Tatham AS, Jones HD, Lazzeri PA and Shewry PR (2000) Modification of storage protein composition in transgenic bread wheat. In: *Wheat Gluten* (Eds. Shewry P.R. and Tatham A.S.) ISBN 0854048650. Royal Society of Chemistry.

He GY, Lazzeri PA and Cannell ME (2001) Fertile transgenic plans obtained from tritordeum infloresences by tissue electroporation. *Plant Cell Rep.,* **20** : 67-72.

Hedden P and Phillips AL (2000) Gibberellin metabolism: new insights reveals by the genes. *Trends in Plant Sci.,* **5** : 523-530.

Heifetz PB (2000) Genetic engineering of the chloroplast. *Biochimie,* **82** : 655-666.

Hiei Y, Ohta S, Komari T and Kumashiro T (1994) Efficient transformation of rice (*Oryza sativa* L) mediated by *Agrobacterium* and sequence-analysis of the boundaries of the T-DNA. *Plant Physiol.,* **158** : 439-445.

Holdsworth MJ, Lenton J, Flintham J, Gale M, Kurup S, McKibbin R, Larner V and Russell L (2001) Genetic control mechanisms regulating the initiation of germination. *J. Plant Physiol.,* **158** : 439-445.

Holmes-Davis R and Comai L (1998) Nuclear matrix attachment regions and plant gene expression. *Trends in Plant Sci.,* **3** : 91-97.

Ingram HM, Power JB, Lowe KC and Davey MR (1999) Optimisation of procedures for microprojectile bombardment of microspore-derived embryos in wheat. *Plant Cell Tiss. Org. Cult.,* **57** : 207-210.

Ishida Y, Saito H, Ohta S, Hieie Y, Komari T and Kumashiro T (1996) High efficiency transfromation of maize (*Zee mays* L) mediated by *Agrobacterium tumefaciens. Nat. Biotechnol.,* **14** : 745-750.

Itoh K, Nakajima M and Schimamoto K (1997) Silencing of waxy genes in rice containing *wx* transgenes. *Mol. Gen. Genet.,* **225** : 351-358.

Iyer LM, Kumpatla SP, Chandrasekharan MB and Hall TC (2000) Transgene silencing in monocots. *Plant Mol. Biol.,* **43** : 323-346.

Izawa T, Ohnishi T, Nakano T, Ishida N, Enoki H, Hashimoto H, Itoh K, Terada R, Wu CY, Miyazaki C, Endo T, Iida S and Shimamoto K (1997) Transposon tagging in rice. *Plant Mol. Biol.,* **35** : 219-229.

Jiang JD, Linscombe SD, Wang JJ and Oard JH (2000) High efficiency transformation of US rice lines from mature seed-derived calli and segregation of glufosinate resistance under field conditions. *Crop Sci.,* **40** : 1729-1741.

Jones HD, Peters NCB and Holdsworth M (1997) Genotype and environment interact to control dormancy and differential expression of the *Viviparous 1* homologue in embryos of *Avena fatua. Plant J.,* **129** : 911-921.

Jordan MC (2000) Green fluorescent protein as a visual marker for wheat transformation. *Plant Cell Rep.,* **19** : 1069-1075.

Karunaratne S, Sohn A, Mouradov A, Scott J, Steinbiss HH and Scott KJ (1996) Transformation of wheat with the gene encoding the coat protein of barley yellow mosaic virus. *Aust. J. Plant Physiol.,* **23** : 429-435.

Ke XY, Chen DF, Huang Y, Shi HP, Elliott MC and Li BJ (1997) Commercial wheat transformed by electroporation of immature embryos. *Biotechnol. Biotechnol. Equip.,* **11** : 28-31.

Kohli A, Leech M, Vain P, Laurie DA and Christou P (1998) Transgene organization in rice engineered through direct DNA transfer supports a two-phase integration mechanism mediated by the establishment of integration hot spots. *Proc. Natl. Acad. Sci. USA,* **95** : 7203-7208.

Kohli A, Griffiths S, palacios N, Twyman RM, Vain P, Laurie DA and Christou P (1999) Molecular characterization of transforming plasmid rearrangements in transgenic rice reveals a recombination hotspot in the CaMV 35S promoter and confirms the predominance of microhomology mediated recombination. *Plant J.,* **17** : 591-601.

Komari T, Hiei Y, Saito Y, Murai N and Kumashiro T (1996) Vectors carrying two separate T-DNAs for co-transformation of higher plants mediated by *Agrobacterium tumefaciens* and segregation of transformants free from selection markers. *Plant J.,* **19** : 165-174.

Koprek T, McElroy D, Louwerse J, Williams-Carrier R and Lemaux PG (2000) An efficient method for dispersing Ds elements in the barley genome as a tool for determining gene function. *Plant J.,* **24** : 253-263.

Krishnamurthy K and Giroux MJ (2001) Expression of wheat puroindoline genes in transgenic rice enhances grain softness. *Nat. Biotechnol.,* **19 :** 162-166.

Lamacchia C, Shewry PR, Di Fonzo N, Forsyth JL, Harris N, Lazzeri PA, Napier JA, Halford NG and Barcelo P (2001) Endosperm-specific activity of a storage protein gene promoter in transgenic wheat seed. *J. Exp. Bot.,* **52** : 243-250.

Law CN, Young CF, Brown JWS, Snape JW and Worland AJ (1978) The study of grain protein control in wheat using whole chromosome substitution lines. In: *Seed Protein Improvement by Nuclear Techniques. Intl. Atomic Energy Agency, Vienna.* Pp. 483-502.

Leckband G and Lorz H (1998) Transformation and expression of a stilbene synthase gene of *Vitis vinifera* L. in barley and wheat for increased fungal resistance. *Theor. Appl. Genet.,* **96** : 1004-1012.

Martin A, Rubiales D, Rubio JM and Cabrera A (1995) Hybrids between *Hordeum vulgare* and tetra-, hexa-, and octoploid tritordeums (amphiploid *H. chilense* × *Triticum* spp). *Hereditas,* **123** : 175-182.

Miflin B, Napier J and Shewry P (1999) Improving plant product quality. *Nat. Biotechnol.,* **17 :** 13-14.

Milligan AS, Daly A, Jepson I and Lazzeri PA (1999) The expression of a maize glutathion *S*-transferase gene in transgenic wheat. *In Vitro Cell. Dev. Biol.,* 35: 39A. Abstract P1006.

Mlynarova L, Loonen A, Heldens J, Jansen RC, Keizer P, Stiekema WJ and Nap JP (1994) Reduced Position Effect in Mature Transgenic Plants Conferred by the Chicken Lysozyme matrix-Associated Region. *Plant Cell,* **6** : 417-426.

Mlynarova L, Jansen RC, Conner AJ, Stiekema WJ and Nap JP (1995) The Mar-Mediated Reduction in Position Effect Can Be Uncoupled from Copy Number-Dependent Expression in Transgenic Plants. *Plant Cell,* **7** : 599-609.

Mlynarova L, Keizer LCP, Stiekema WJ and Nap JP (1996) Approaching the lower limits of transgene variability. *Plant Cell,* **8** : 1589-1599.

Morris CF, Greenblatt GA, Bettge AD and Malkawi HI (1994) Isolation and characterization of multiple forms of Friabilin. *J. Cereal Sci.,* **21** : 167-174.

Muller AE, Kamisugi Y, Gruneberg R, Niedenhof I, Horold RJ and Meyer P (1999) Palindromic sequences and A. plus T-rich DNA elements promote illegitimate recombination in *Niocotiana tabacum. J. Mol. Biol.,* **291** : 29-46.

Murashige T and Skoog F (1962) A revised medium for rapid growth and bioassays with tobacco tissue cultures. *Physiol. Plant.,* **15** : 473-497.

Nakamura T, Yamamori M, Hirano H, Hidaka S and Nagamine T (1995) Production of waxy (amylose-free) wheats. *Mol. Gen. Genet.,* **248** : 253-259.

Negrotto D, Jolley M, Beer S, Wenck AR and Hansen G (2000) The use of phosphomannose-isomerase as a selectable marker to recover transgenic maize plants (*Zea mays* L.) via *Agrobacterium* transformation. *Plant Cell Rep.,* **19** : 798-803.

Nehra NS, Chibbar RN, Leung N, Caswell K, Mallard C, Steinhauer L, Baga M and Kartha KK (1994) Self-Fertile Transgenic Wheat Plant Regenerated from Isolated Scutellar Tissues Following Microprojectile Bombardment with Two Distinct Constructs. *Plant J.,* **5** : 285-297.

Oda S and Schofield JD (1997) Characterization of friabilin polypeptides. *J. Cereal Sci.,* **26** : 29-36.

Ortiz JPA, Reggiardo MI, Ravizzini RA, Altabe SG, Cervigni GDL, Spitteler MA, Morata MM, Elias FE and Vallejos RH (1996) Hygromycin resistance as an efficient selectable marker for wheat stable transformation. *Plant Cell Rep.,* **15** : 877-881.

Pastori GM, Steele SH, Jones HD and Shewry PR (2000) Transformation of commercial wheat varieties with high molecular weight glutenin subunit genes. In: *Wheat Gluten* (Eds. Shewry PR and Tatham AS) ISBN 0854048650. Royal Society of Chemistry, pp 88-92.

Pastori GM, Wilkinson M, Steele S, Sparks C, Jones HD and Parry M (2001) Transformation of elite wheat varieties at high frequencies. *J. Exp. Bot.,* **52** : 210-215.

Pawlowski WP and Somers DA (1998) Transgenic DNA integrated into the oat genome is frequently interspersed by host DNA. *Proc. Natl. Acad. Sci. USA,* **95** : 12106-12110.

Payne PI, Nightingale MA, Krattiger AF and Holt LM (1987) The relationship between HMW glutenin composition and the breadmaking quality of British grown wheat varieties. *J. Sci. Food Agr.,* **40** : 51-65.

Perl A, Kless H, Blumenthal A, Galili G and Galun E (1992) Improvement of Plant-Regeneration and Gus Expression in Scutellar Wheat Calli by Optimization of Culture Conditions and DNA-Microprojectile Delivery Procedures. *Mol. Gen. Genet.,* **235** : 279-284.

Peters NR, Ackerman S and Davis EA (1999) A modular vector for *Agrobacterium* mediated transformation of wheat. *Plant Mol. Biol. Rep.,* **17** : 323-331.

Pogna NE, Autran JC, Mellini F, Lafiandra D and Feillet P (1990) Chromosome 1B-encoded gliadins and gultenin subunits in durum wheat: genetics and relationship to gluten strength. *J. Cereal Sci.,* **11** : 15-34.

Popineau Y, Deshayes G, Lefebvre J, Fido R, Tatham AS and Shewry PR (2001) Prolamin aggregation, gluten viscoelasticity, and mixing properties of trasngenic wheat lines expressing 1Ax and 1Dx and high molecular weight glutenin subunit transgenes. *J. Agric. Food Chem.,* **49** : 395-40.

Rasco-Gaunt S and Barcelo P (1998) Immature inflorescence culture of cereals. A highly responsive system for regeneration and transformation. (Ed. Hall RD) In: *Method in Molecular Biology* vol 111: *Plant Cell Culture Protocols.*

Rasco-Gaunt S, Grierson D, Lazzeri PA and Barcelo P (1998) Genetic manipulation of S-adenosyl methionine decarboxylase (SAMDC) in wheat. (Ed. AE Slinkark). In *Proceedings of 9th International Wheat Genetics Symposium,* Saskatoon, Canada.

Rasco-Gaunt S, Riley A, Barcelo P and Lazzeri PA (1999) Analysis of particle bombardment parameters to optimise DNA delivery into wheat tissues. *Plant Cell Rep.,* **19** : 118-127.

Rasco-Gaunt S, Riley A, Cannell MC, Barcelo P and Lazzeri PA (2001) Development of procedures allowing the efficient transformation of a range of current wheat varieties *via* particle bombarment. *J. Exp. Bot.,* **52** : 220-227.

Reed JN, Chang YF, McNamara DD, Beer S and Miles PJ (1999) High frequency transformation of wheat with the selectable marker mannose-6-phosphate isomerase (PMI). *In Vitro Cell. Dev. Biol.,* **35** : 57A. Abstract P1079.

Reid R, Shlumukov L, Jones H and Smith H (2001) Genetic Engineering of Plant Architecture In *Wheat 19th Annual Missouri Symposium,* Plant Photobiology. Abstract.

Rooke L, Barro F, Tatham AS, Fido R, Steele S, Békés F, Gras P, Martin A, Lazzeri PA, Shewry PR and Barcelo P (1999a) Altered functional properties of tritordeum with HMW glutenin subunit genes. *Theor. Appl. Genet.,* **99** : 851-858.

Rooke L, Bekes F, Fido R, Barro F, Gras P, Tatham AS, Barcelo P, Lazzeri P and Shewry PR (1999b) Overexpression of a gluten protein in transgenic wheat results in greatly increased dough strength. *J. Cereal Sci.,* **30** : 115-120.

Rooke L, Byrne D and Salgueiro S (2000) Marker gene expression driven by the maize ubiquitin promoter in transgenic wheat. *Ann. Appl. Biol.,* **136** : 167-172.

Shewry PR, Tatham AS, Barro F, Barcelo P and Lazzeri PA (1995) Biotechnology of Breadmaking: Unraveling and manipulating the Multi-Protein Gluten Complex. *Biotechnology,* **13** : 1185-1190.

Shlumukov LR, Barro F, Barcelo P, Lazzeri P and Smith H (2001) Establishment of far-red high irradiance responses in wheat through transgenic expression of an oat phyochrome A. gene. *Plant Cell Eviron.* (in press)

Singh N and Chawla HS (1999) Use of silicon carbide fibers for *Agrobacteriam*-mediated transformation in wheat. *Curr. Sci.,* **76** : 1483-1485.

Sivamani E, Bahieldin A, Wraith JM, Al-Niemi T, Dyer WE, Ho THD and Qu RD (2000a) Improved biomass productivity and water use efficiency under water deficit conditions in transgenic wheat constitutively expressing the barely *HVA1* gene. *Plant Sci.,* **155** : 1-9.

Sivamani E, Brey CW, Dyer WE, Talbert LE and Qu RD (2000b) Resistance to wheat streak mosaic virus in transgenic wheat expressing the viral replicase (NIb) gene. *Mol. Breed.,* **6** : 469-477.

Sorokin AP, Ke XY, Chen DF and Elliott MC (2000) Production of fertile transgenic wheat plants via tissue electroporation. *Plant Sci.,* **156** : 227-233.

Sparks CA, Castleden CK, West J, Habash DZ, Madgwick PJ, Paul MJ, Noctor G, Garrison J, Wu R, Wildinson J, Quick WP, Parry MAJ, Foyer CJ and Miflin BJ (2001) Potential for manipulating carbon metabolism in wheat. *Ann. Appl. Biol.,* **138** : 33-45.

Srivastava V, Anderson OD and Ow DW (1999) Single-copy transgenic wheat generated through the resolution of complex integration patterns. *Proc. Natl. Acad. Sci. USA,* **96** : 11117-11121.

Stoger E, Williams S, Keen D and Christou P (1998) Molecular characteristics of transgenic wheat and the effect on transgene expression. *Transgenic Res.,* **7** : 463-471.

Stoger E, Williams S, Christou P, Down RE and Gatehouse JA (1999a) Expression of the insecticidal lectin from snowdrop (*Galanthus nivalis agglutinin*: GNA) in transgenic wheat plants: effects on predation by the grain aphid *Sitobion avenae. Mol. Breed.,* **5** : 65-73.

Sotger E, Williams S, Keen D and Christou P (1999b) Constitutive versus seed specific expression in transgenic wheat: tomporal and spatial control. *Transgenic Res.,* **8** : 73-82.

Stone MC, Wilkinson MD, Huttly AK, Lenton JR and Jones HD (2001) Development and hormonal regulation of alpha-amylase in transgenic wheat. *J. Exp. Bot.,* **52** (Suppl.) : 7.78.

Sudhakar D, Duc LT, Bong BB, Tinjuangjun P, Maqbool SB, Valdez M, Jefferson R and Christou P (1998) An efficient rice transformation system utilizing mature seed-derived explants and a portable, inexpensive particle bombardment device. *Transgenic Res.,* **7** : 289-294.

Tacke E, Korfhage C, Michel D, Maddaloni M, Motto M, Lanzini S, Salamini F and Doring HP (1995) Transposon tagging of the maize Glossy2 locus with the transposable element En/Spm. *Plant J.,* **8** : 907-917.

Takumi S, Murai K, Mori N and Nakamura C (1999) Trans-activation of a maize Ds transposable element in transgenic wheat plant expressing the Ac transposase gene. *Theor. Appl. Genet.,* **98** : 947-953.

Tingay S, McElroy D, Kalla R, Fieg S, Wang MB, Thornton S and Brettell R (1997) *Agrobacterium tumefaciens*-mediated barely transformation. *Plant J.,* **11** : 1369-1376.

Tosi P, Napier JA, D, Ovidio R, Jones HD and Shewry PR (2000) Modification of LMW Glutenin subunit composition of Durum Wheat by microprojectile-mediated transformation. In: *Wheat Gluten* (Eds. Shewry PR and Tatham AS) ISBN: 0854048650. Royal Society of Chemistry, pp 93-96.

Uze M, Potrykus I and Sautter C (1999) Single-stranded DNA in the genetic transformation of wheat (*Triticum aestivum* L.): transformation frequency and integration pattern. *Theor. Appl. Genet.,* **99** : 487-495.

Uze M, Potrykus I and Sautter C (2000) Factors influencing T-DNA transfer from *Agrobacterium* to pre-cultured immature wheat embryos (*Triticum aestivum* L.). *Cereal Res. Commun.,* **28** : 17-23.

Vain P, McMullen MD and Finer JJ (1993) Osmotic treatment enhances Particle Bombardment-Mediated Transient and Stable Transformation of Maize. *Plant Cell Rep.,* **12** : 84-88.

Vasil V, Castillo AM, Fromm ME and Vasil IK (1992) Herbicide Resistance Fertile Transgenic Wheat Plants Obtained by Microprojectile Bombardment of Regenerable Embryogenic Callus. *Biotechnology,* **10** : 667-674.

Visser RGF, Somhorst I, Kipers AGJ, Feenstra WJ and Jacobsen E (1991) Inhibition of the expression of the gene for granule-bound starch sysnthase in potato by antisense constructs. *Mol. Gen. Genet.,* **225** : 289-296.

Walbot V (2000) Saturation mutagenesis using maize transposons. *Curr. Opin. Plant Biol.,* **3** : 103-107.

Weeks JT, Koshiyama KY, Maier-Greiner U, Schaeffner T and Anderson OD (2000) Wheat transformation using cyanamide as a new selective agent. *Crop Sci.,* **40** : 1749-1754.

Wenck A, Czako M, Kanevski I and Marton L (1997) Frequent collinear long transfer of DNA inclusive of the whole binary vector during *Agrobacterium*-mediated transformation. *Plant Mol. Biol.,* **34** : 913-922.

Witrzens B, Brettell RIS, Murray FR, McElroy D, Li ZY and Dennis ES (1998) Comparison of three selectable marker genes for transformation of wheat by microprojectile bombardment. *Aust. J. Plant Physiol.,* **25** : 39-44.

Woplin R, Lazzeri PA, Tyson H and Cannell ME (1998) Genetic manipulation of wheat to produce novel starches. (J Moorby, Association of Applied Biologists) in abstracts: *Production and Uses of Starch,* AAB Edinurgh.

Xu Y, Zhu Q, Panbangred W, Shirasu K and Lamb C (1996) Amylose and lipid contents, amylopectin structure, and gelatinization properties of waxy wheat (*T. aestivum*) starch. *J. Cereal Sci.,* **24** : 131-137.

Zhang LY, French R, Langenber WG and Mitra A (2001) Accumulation of barley stripe mosaic virus is significantly reduced in transgenic wheat plants expressing a bacterial ribonuclease. *Transgenic Res.,* **10** : 13-19.

Zhou H, Arrowsmith JW, Fromm ME, Hironaka CM, Taylor ML, Rodriguez D, Pajeau ME, Brown SM, Santino CG and Fry JE (1995) Glyphosate-tolerant CP4 and GOX genes as a selectable marker in wheat transformation. *Plant Cell Rep.,* **15** : 159-163.

Chapter 6

GENETIC TRANSFORMATION OF OAT (*Avena sativa* L.)

David A Somers*, Kimberly A Torbert and Sergei K Svitashev

Department of Agronomy and Plant Genetics, University of Minnesota, 1991 Upper Buford Circle, St. Paul, MN - 55108, USA

Summary

Hexaploid oat (Avena sativa L.) was first genetically engineered in 1992 using microprojectile bombardment of embryogenic tissue cultures. Since that time substantial improvements in oat genetic engineering have been reported. These include efficient production of transgenic plants from a range of genotypes, reduced costs of genetically engineered plants and a range of selection systems for recovery of transgenic plants. Much also has been learned about the function and structure of transgene loci in oat. A high proportion of transgene loci in oat consist of multiple copies of the delivered DNA that are interspersed with genomic DNA. These complex transgene loci undoubtedly cause problems with transgene expression. This review presents and discusses the improvements in oat genetic engineering systems and investigations of transgene loci. Although the achievements in these areas are substantial, it is evident that continued improvement of several aspects of the extant oat genetic engineering systems is desirable. Possible strategies for these improvements are discussed.

Keywords : transformation, transgene, tissue culture, fluorescence *in situ* hybridization

---

*Corresponding author : E-mail : somers@biosci.cbs.umn.edu

1. INTRODUCTION

Oat (*Avena sativa* L.) was first genetically engineered in 1992 (Somers *et al.*, 1992). This accomplishment was built on substantial progress in development and improvement of regenerable oat tissue cultures (Bregitzer *et al.*, 1989; for reviews see Rines *et al.*, 1992; Bregitzer *et al.*, 1995). As in most cereals, progress in genetically engineering oat during that time period was hampered by the inability to routinely regenerate plants from protoplasts and to develop *Agrobacterium*-mediated DNA delivery methods. Microprojectile bombardment represented a significant breakthrough by providing a means of DNA delivery into cells eliminating the need for protoplast or *Agrobacterium* systems (Klein *et al.*, 1987; Sanford, 1988, 1990; Sanford *et al.*, 1987, 1991, 1993). Substantial recent improvements in oat genetic engineering systems have increased the efficiency of transgenic plant production and expanded the range of oat genotypes that can be genetically engineered (for previous reviews see Somers, 1999; Somers *et al.*, 1994; 1995). For discussion of the progress and prospects of applications of genetic engineering to oat improvement see Burrows and Altosaar (1995). Significant progress also has been achieved in characterization of transgene locus structure and function in transgenic oat plants. The aim of this review is to update progress in oat genetic engineering reported since 1992. Specific topics include 1) the advantages and limitations of current genetic engineering systems, 2) transgene locus function and structure, and 3) the requirements for creating an ideal oat genetic engineering system.

2. GENETIC ENGINEERING SYSTEMS

Oat genetic engineering systems as in all plants are based on three requisite components. These include 1) a source of totipotent cells as targets for DNA delivery, 2) a means of delivering DNA into the nuclear genome of the target cells, and 3) a system for selecting or identifying transgenic cells and regenerated plants. The flexibility and robustness of each component have significant impacts on several aspects of the transformation system. These include the labor and therefore the cost of producing transgenic plants, the range of genotypes that may be genetically engineered, the level of tissue culture-induced somaclonal variation in regenerated plants, the ecological or food safety issues posed by the selectable marker gene, and the capacity to produce transgenic plants with simple transgene inserts that exhibit reproducible transgene expression. Each component of oat genetic engineering systems will be reviewed in this section in relation to these considerations.

2.1. Target cells for DNA delivery

All oat genetic engineering systems reported use tissue culture cells as a source of totipotent target cells. The recent advances in oat genetic engineering systems achieved by a number of researchers largely stem from improvements in the tissue cultures used as target cells (Rines *et al.*, 1992; Bregitzer *et al.*, 1995; Somers, 1999). At least four different explants are used to initiate totipotent oat tissue cultures for genetic engineering. These are immature embryos (Cummings *et al.*, 1976; Lorz *et al.*, 1976; Bregitzer *et al.*, 1989; Somers *et al.*, 1992), mature embryos (Hassan *et al.*, 1999; Torbert *et al.*, 1998ab; Cho *et al.*, 1999), leaf base segments (Chen *et al.*, 1995ab; Gless *et al.*, 1998ab) and shoot meristems (Cho *et al.*, 1998; Zhang *et al.*, 1996, 1999). Culture media formulations have been reported that improve specific steps in tissue culture initiation and plant regeneration (Bregitzer *et al.*, 1995). For example, Cho *et al.* (1999) reported that callus induction from mature embryos performed under dim light conditions on medium containing 2,4-D, BAP and high cupric sulfate increased the regeneration capacity of transgenic tissue cultures.

For oat genetic engineering, it is important to minimize the duration in tissue culture because deleterious mutations and cytogenetic changes (somaclonal variation) accumulate as cultures age (McCoy *et al.*, 1982; Olhoft and Phillips, 1998). Moreover, the transformation process *per se* appears to increase the frequency of oat plants exhibiting aneuploidy and chromosome structural changes compared to the frequency in nontransgenic plants regenerated from tissue culture (Choi *et al.*, 2000). Accordingly, some culture systems offer the advantage of markedly shortening the period in tissue culture and thus potentially reducing the accumulation of somaclonal variation. For example, mature embryo-derived tissue cultures are maintained in culture for upto only 10 weeks before microprojectile bombardment (Torbert *et al.*, 1998c), a substantially shortened period compared to our original work with immature embryo-derived cultures (Somers *et al.*, 1992). Other systems such as the leaf base segments offer the additional advantage of being directly bombarded before tissue cultures are initiated and selection of transgenic cells is imposed (Gless *et al.*, 1998b). Progress toward development of non-tissue culture genetic engineering systems for oat has not been reported.

A major advantage common to the more recent oat tissue culture systems employed in oat genetic engineering is that they are useful for genetically engineering a number of different genotypes (Gless *et al.*, 1998ab; Torbert *et al.*, 1998bc; Zhang *et al.*, 1996, 1999). In our initial studies (Somers *et al.*, 1992; Torbert *et al.*, 1995), the genotype used (GAF/Park-1) was specifically bred and selected for high frequency establishment of regenerable tissue cultures from immature embryos. Investigations of media composition suggest that some elite genotypes may be induced to initiate regenerable tissue cultures from immature embryos (Bregitzer *et al.*, 1995; Gana *et al.*, 1995). However, the use of mature embryos for tissue culture initiation dramatically increased the number of elite genotypes from which regenerable and transformable tissue cultures can be initiated. We tested the tissue culture response of 16 North American spring oat genotypes and found that, although the frequency of response varied, all except one line initiated tissue cultures that appeared similar to our most transformable lines (Torbert *et al.*, 1998b). Transgenic plants have now been produced in four of these genotypes (H. Kaeppler, personal communication, K.A. Torbert and D.A. Somers, unpublished results). Similar progress has been reported for the leaf base system (Gless *et al.*, 1998ab) and meristem cultures (Cho *et al.*, 1998) indicating that genotype limitations to oat genetic engineering have been significantly reduced. The ability to genetically engineer leading cultivars dramatically enhances the application of genetic engineering to oat improvement.

2.2. DNA delivery

Extant oat genetic engineering systems are based on DNA delivery via microprojectile bombardment (Somers, 1999). While these systems efficiently produce fertile transgenic plants, the complex, multiple copy transgene loci resulting from microprojectile bombardment cause problems with transgene expression and will be further discussed in a later section. It seems likely that recent successes in regenerating plants from protoplasts of cereals such as maize (Wang *et al.*, 2000) may lead to similar protoplast culture systems in oat and eventually to transformation systems based on direct DNA delivery into protoplasts using either chemical methods or electroporation. However, it seems more likely that greater emphasis will be focused on the development of *Agrobacterium tumefaciens*-mediated transformation systems for oat based on the recent successes in *Agrobacterium*-mediated genetic engineering of rice (Hiei *et al.*, 1994), maize (Ishida *et al.*, 1996), wheat (Cheng *et al.*, 1997) and barley (Tingay *et al.*, 1997). While oat has not yet been genetically engineered using *Agrobacterium*, progress in *Agrobacterium*-mediated T-DNA delivery into various oat tissues and tissue cultures has been reported (Perret and Morris, 2000). These results indicate that further research on this system will lead to development of an *Agrobacterium*-mediated genetic engineering system for oat.

2.3. Selection of transgenic cells and plants

In the initial oat genetic engineering system, herbicide resistance was used for selection of transgenic tissue cultures (Somers *et al.,* 1992). Herbicide resistance as a selectable marker for selection of transgenic oat has distinct disadvantages related to the likelihood that the transgene could be transferred through out-crossing to wild *Avena* species. Thus, alternative selectable markers have been developed. The *E. coli* gene coding for neomycin phosphotransferase, *nptII*, in combination with paromomycin, a neomycin, was shown to be useful for selection of transgenic oat tissue cultures that regenerate fertile transgenic plants (Torbert *et al.,* 1995). More recently, Cho *et al.* (1999) reported development of an efficient transformation system that uses hygromycin phosphotransferase as a selectable marker gene for plant cell resistance to hygromycin. While antibiotic resistance genes are far less likely to pose ecological risks such as might occur when transgenic herbicide resistance genes are transferred to wild oats, there remains the possibility of a human or animal health risk associated with antibiotic resistance genes. Thus, there is continued need for new selection systems and to reduce the types and amount of unnecessary DNA transferred into plant genomes by microprojectile bombardment. Recently, positive selectable markers based on carbohydrate metabolism have been developed (Haldrup *et al*., 1998). Phosphomannose isomerase was shown to be effective for selection of transgenic maize plants (Negrotto *et al*., 2000; Wang *et al*., 2000). This type of selectable marker will likely be very useful for selection of transgenic oat plants.

A visual selectable marker has been developed for oat offering a new strategy for identifying transgenic lines. The green fluorescent protein gene (*gfp*) under the control of the maize ubiquitin promoter allows rapid and efficient visualization of transgenic oat callus sectors (Kaeppler *et al.,* 2000). Selective subcultures of these fluorescing sectors have produced the regeneration of transgenic, fertile, fluorescent plants. An emerging advantage of the *gfp* selection system is that transgenic callus sectors can be identified and isolated in approximately half the time that it takes for paromomycin selection (Torbert *et al.,* 1995).

Presumably, this is because there is less stress imposed on the cells in cultures using the visible marker compared to growth inhibition of both transgenic and nontransgenic tissue culture cells imparted by antibiotic or herbicide tissue culture selection systems. These and further improvements indicate that substantial progress has been realized in improving the efficiency of oat transformation and reducing ecological risk of transgenic plants (Somers *et al.,* 1994, 1995; Somers, 1999).

3. TRANSGENE LOCUS FUNCTION AND STRUCTURE

This section is focused on reviewing results conducted in oat to characterize the function and structure of transgene loci. Oat is an allohexaploid comprised of homoeologous A, C and D genomes. Polyploids with two or more sets of homoeologous chromosomes can presumably tolerate more severe aberrations than diploids (Sears, 1966). Considering that integration of large, complex transgene loci would likely create a significant mutational load for transgenic cells, the genetic buffering of the allohexaploid genome of oat makes it more likely that affected cells in oat would persist and participate in plant regeneration than in a diploid species. Thus, oat may be considered somewhat of a model system to characterize complex transgene loci. Although it seems likely that transgene loci in oat will be larger and more complex compared to the smaller size of transgene loci reported in diploid rice (Kohli *et al.*, 1998), many crop plants are polyploid or have duplicated regions in their genomes. Thus, the transgene loci in oat may be representative of those resulting from microprojectile bombardment in many crops.

The proportion of transgenic oat plants produced using microprojectile bombardment that exhibit normal Mendelian segregation of the transgene and its phenotype within an experiment is highly variable (Pawlowski *et al.*, 1998). Deviations from Mendelian segregation ratios in transgenic plants are due to transgene silencing (Pawlowski *et al.*, 1998) and aberrant transgene transmission (Svitashev *et al.*, 2000). Both phenomena are likely caused by the extensive rearrangements of transgene sequences and greater transgene copy numbers within transgene loci produced using microprojectile bombardment and are likely compounded by the high frequency of cytogenetic aberrations caused by the tissue culture process in oat (Choi *et al.*, 2000; McCoy *et al.*, 1982; Olhoft and Phillips, 1998).

Several co-transformation experiments in which different transgenes carried on two separate plasmids are co-introduced into oat cells have been reported using microprojectile bombardment. One of the plasmids usually carries a plant selectable marker gene while the other plasmid is not selectable and carries the transgene of interest. In oats as in other co-transformed plants, co-expression of both transgenes co-introduced on the separate plasmid DNAs is usually observed in more than 50 % of the transgenic plants (McGrath *et al.*, 1998; Torbert *et al.*, 1998a; Koev *et al.*, 1998; Tzafrir *et al.*, 1998). Typically in such experiments, genes from both plasmids also are found co-integrated into the same transgene locus (Chen *et al.*, 1998; Pawlowski and Somers, 1998; Leggett *et al.*, 2000).

The number of transgene copies integrated into a transgene locus produced by microprojectile bombardment varies from 1 to more than 20 copies (Klein and Jones, 1999; Pawlowski and Somers, 1996). Transgene loci are composed of multiple copies of transgenes that often co-segregate as a single transgene locus (Pawlowski and Somers, 1998; Pawlowski *et al.*, 1998). Molecular analyses of transgenes delivered by microprojectile bombardment frequently reveal extensive rearrangements of delivered DNA sequences (Somers *et al.*, 1992; Pawlowski and Somers, 1998; Svitashev *et al.*, 2000). The rearrangements are detected by the appearance of additional transgene-hybridizing restriction fragments of altered molecular weights in Southern blot analyses, when compared to the introduced plasmid DNA. Transgene concatemers, which are complexes of linearized plasmid copies linked in either head-to-head or head-to-tail arrays, are sometimes observed (Pawlowski and Somers, 1998; Svitashev *et al.*, 2000). Moreover, the presence of genomic interspersions among tightly linked, clustered transgenes integrated into a single locus is quite common in oat (Pawlowski and Somers, 1998; Svitashev *et al.*, 2000; Svitashev and Somers, 2002). These observations represent a greater magnitude of structural complexity than previously recognized in transgene loci produced using microprojectile bombardment.

Fluorescence *in situ hybridization* (FISH) has been extremely useful for investigating transgene loci localization and structure in transgenic plants and especially in oat (Svitashev and Somers, 2002). FISH was used to visualize transgene loci and to distinguish the oat C genome chromosomes in a population of 16 transgenic oat plants that exhibited a total of 26 transgene loci (Svitashev *et al.*, 2000). Multiple, telomeric translocations between the C and A/D genome chromosomes were detected using FISH in untransformed oat in agreement with previous reports (Jellen *et al.*, 1994; Leggett and Markhand, 1995). Eight transgene integration sites were observed on C genome chromosomes and 18 sites were on A/D genome chromosomes indicating random integration. Integration sites were observed in different locations along individual chromosomes; however, in 18 out of 26 integration events transgene integration occurred in telomeric and subtelomeric regions. These observations are probably related to the fact that in most plants distal regions of chromosomes are gene-rich whereas heterochromatic regions are gene-poor and are mainly located around the centromeres (Gill *et al.*, 1993, 1996). Transgene integration into heterochromatic regions may result in poor or unstable transgene expression (Iglesias *et al.*, 1997; Jakowitsch *et al.*, 1999) and, as a result, the transgenic cells would not be selected or regenerated into plants. Both Leggett *et al.* (2000) and Svitashev *et al.* (2000) observed that transgene loci are often located on the borders of C and A/D chromosome translocations. Since random genomic integration of delivered DNA would not be expected to repeatedly occur at apparently preexisting intergenomic translocation borders, these regions appear to be hotspots for transgene integration in oat. To date there is no published literature reporting this observation in other plants.

FISH on prophase and prometaphase chromosomes showed that in at least 9 out of the 26 transgene loci, clusters of two or three integration sites were separated with fragments of oat DNA (Svitashev *et al.*, 2000). These results provided visual evidence that some transgene loci were organized as clusters of transgenes interspersed with oat DNA. Considering that the size of an average oat chromosome is about 600 megabases and that the distance between transgenes within a cluster appeared to be greater than 1% of the total prometaphase chromosome length, it was estimated that the intervening oat DNA was at least a few megabases long indicating that the size of some transgene clusters may exceed several megabases. This size estimate was much greater than could be determined from Southern analyses of transgene loci (Pawlowski and Somers, 1998; Svitashev *et al.*, 2000).

Various chromosomal aberrations that result from chromosomal breakage are observed in plants regenerated from tissue culture (McCoy *et al.*, 1982; Olhoft and Phillips, 1998). Svitashev *et al.* (2000) observed that six transgene integration sites were associated with four rearranged chromosomes. This observation is much higher than expected for random transgene integration and random tissue culture-induced chromosome alterations. In three lines, transgene integration sites were detected on different arms of the same chromosomes. This frequency of double integration into a single chromosome also seems far higher than would be expected based on random integration. Moreover, there was no detectable recombination between the transgene integration sites suggesting chromosomal rearrangements in these chromosomes. We suggest that particle bombardment and/or the process of transgene integration into the plant genome may lead to host genomic DNA breakage and result in chromosomal aberrations.

The association of chromosome breakage with transgene loci suggests that damage to the host genome may be caused by delivery of DNA-coated microprojectiles to the nucleus (Hunold *et al.*, 1994). An extra-chromosomal pool of whole and sheared delivered transgene DNA mixed with host DNA fragments of variable lengths including subchromosome fragments would be created. How this DNA is integrated into the genome to form the transgene locus is not completely clear. Several possible mechanisms for transgene integration delivered by direct methods including microprojectile bombardment have been proposed (Gorbunova and Levy, 1999; Kohli *et al.*, 1998, 1999; Pawlowski and Somers, 1998; Svitashev *et al.*, 2000; Svitashev and Somers, 2002). This research is extremely important because understanding the mechanism(s) underlying transgene integration and locus formation will likely elucidate avenues of simplifying transgene locus structures and reducing genomic damage, which in turn should increase the proportion of transgenic events that exhibit robust transgene expression. Transgene loci produced by *Agrobacterium tumefaciens*-mediated transformation are less complex resulting in a greater proportion of transgene events that exhibit Mendelian segregation (Pawlowski and Somers, 1996). Consequently, since *Agrobacterium*-mediated transformation systems have been developed for most other cereal crops (Hiei *et al.*, 1994; Ishida *et al.*, 1996; Tingay *et al.*, 1997; Cheng *et al.*, 1997), there has been a shift away from using microprojectile bombardment to genetically engineer some crops. This trend will continue until the fidelity of integration of DNA delivered via microprojectile bombardment is improved.

4. CONCLUSION AND FUTURE PROSPECTS

The substantial progress realized in improving oat genetic engineering systems must be tempered with the realization that further improvements are required for the efficient application of genetic engineering to oat improvement. The required areas of improvement impact the three components of genetic engineering systems: totipotent target cells, DNA delivery and selection systems. In regard to totipotent target cells, it is apparent that the trend in improving oat tissue cultures is on the right track in that there is an attempt to shorten the duration cells are in tissue culture (Torbert *et al.*, 1998a; Gless *et al.*, 1998b). The next breakthrough will be realized when a whole plant genetic engineering system is developed for oat. Whole plant genetic engineering systems are well established in *Arabidopsis thaliana* (Bechtold *et al.*, 1993; Chang *et al.*, 1994; Clough and Bent, 1998; Feldmann and Marks, 1987) and more recently *Medicago truncatula* (Trieu *et al.*, 2000). Extension of these genetic engineering systems to a monocot has not been reported but once accomplished will likely lead to genotype-independent transformation systems that produce plants free of tissue culture-induced genetic and cytogenetic variation.

A higher proportion of transgene loci produced using *Agrobacterium tumefaciens* exhibit simple insertions of the T-DNA and lower transgene copy numbers (Komari and Kubo, 1999) compared to those produced via microprojectile bombardment. Because of this and the recent achievement of *Agrobacterium*-mediated genetic engineering of a number of cereals, more investigators are using *Agrobacterium* for cereal transformation. Progress in achieving *Agrobacterium*-mediated DNA delivery into oat cells has been reported (Perret and Morris, 2000), and consequently development of an *Agrobacterium*-mediated genetic engineering system for oat is imminent. The simple transgene loci presumably produced by an *Agrobacterium*-mediated oat genetic engineering system would be expected to exhibit reduced transgene silencing compared to the complex loci produced using microprojectile bombardment and thereby facilitate the use of genetic engineering in oat improvement.

An *Agrobacterium*-mediated genetic engineering system for oat would offer additional advantages. Recently, dual T-DNA strategies have been developed for *Agrobacterium*-mediated T-DNA delivery, in which two different T-DNAs, either on the same binary vector or separate vectors, are transferred into the plant cell (Daley *et al.*, 1998; Komari *et al.*, 1996). About 50 % of the transgenic plants produced from this DNA delivery strategy exhibit unlinked transgene loci. A common strategy is to place the selectable marker in one T-DNA while the transgene of interest is carried in the other T-DNA which allows elimination of the selectable marker gene by segregation in subsequent generations of the transgenic plant. Thus, selectable marker-free transgenic plants can be produced, which reduces potential ecological and food safety concerns associated with selectable marker genes in transgenic plants. Although the oat cells used initially for *Agrobacterium*-mediated DNA delivery will likely either be from a regenerable tissue culture or may be passaged through a tissue culture step, an *Agrobacterium* system in itself will be an important breakthrough for oat genetic engineering because of the aforementioned issues. The *Arabidopsis* and *Medicago truncatula* whole plant genetic engineering systems are based on *Agrobacterium*-mediated DNA delivery, thus development of an *Agrobacterium*-mediated genetic engineering system for oat tissue cultures will bring us one important step closer to the development of a whole plant transformation system for this crop.

ACKNOWLEDGEMENTS

The authors wish to express their gratitude to H.W. Rines, P.D. Matthews and N. Al-Saady for critical review of this manuscript. Publication of the Minnesota Agricultural Experiment Station.

REFERENCES

Bechtold N, Ellis J and Pelletier G (1993) In planta *Agrobacterium* mediated gene transfer by infiltration of adult *Arabidopsis thaliana* plants. *C.R. Acad. Sci. Paris, Life Sci.*, **316** : 1194-1199.

Bregitzer PP, Milach SKC, Rines HW and Somers DA (1995) Somatic embryogenesis in oat (*Avena sativa* L.). In *Biotechnology in Agriculture and Forestry*, Somatic Embryogenesis and Synthetic Seed Vol. 31 (Ed. Bajaj YPS), Springer-Verlag, Berlin, pp. 53-62.

Bregitzer PP, Somers DA and Rines HW (1989) Development and characterization of friable, embryogenic oat callus. *Crop Sci.*, **29** : 798-803.

Burrows VD and Altosaar I (1995) Biotechnology and oat improvement — progress and prospects. In : *The Oat Crop: Production and Utilization* (Ed Welch R W), Chapman & Hall, New York, pp. 533-560.

Chang SS, Park SK, Kim BC, Kang BJ, Kim DU and Nam HG (1994) Stable genetic transformation of *Arabidopsis thaliana* by *Agrobacterium* inoculation *in planta*. *Plant J.*, **5** : 551-558.

Chen L, Marmey P, Taylor NJ, Brizard J, Espinoza C, D'Cruz P, Huet H, Zhang S, de Kochko A, Beachy RN and Fauquet CM (1998) Expression and inheritance of multiple transgenes in rice plants. *Nat. Biotechnol.*, **16** : 1060-1064.

Chen H, Xu G, Loschke DC, Tomaska L and Rolfe BG (1995a) Efficient callus formation and plant regeneration from oat leaves (*Avena sativa* L.). *Plant Cell Rep.*, **14** : 393-397.

Chen Z, Zhuge Q and Sundqvist C (1995b) Oat leaf base: tissue with an efficient regeneration capacity. *Plant Cell Rep.*, **14** : 354-358.

Cheng M, Fry JE, Pang S, Zhou H, Hironaka CM, Duncan DR, Conner TW and Wan Y (1997) Genetic transformation of wheat mediated by *Agrobacterium tumefaciens*. *Plant Physiol.*, **115** : 971-980.

Cho MJ, Jiang W and Lemaux PG (1999) High frequency transformation of oat via microprojectile bombardment of seed derived highly regenerative cultures. *Plant Sci.*, **148** : 9-17.

Cho MJ, Zhang S and Lemaux PG (1998) Transformation of shoot meristem tissues of oat using three different selectable markers. *In Vitro Cell. Dev. Biol.*, **34** : 1012.

Choi H, Lemaux PG and Cho MJ (2000) High frequency of cytogenetic aberration in transgenic oat (*Avena sativa* L.) plants. *Plant Sci.*, **156** : 85-94.

Clough SJ and Bent AF (1998) Floral dip: a simplified method for *Agrobacterium*-mediated transformation of *Arabidopsis thaliana*. *Plant J.*, **16** : 735-743.

Cummings DP, Green CE and Stuthman DD (1976) Callus induction and plant regeneration in oats. *Crop Sci.*, **16** : 465-470.

Daley M, Knauf VC, Summerfelt KR and Turner JC (1998) Co-transformation with one *Agrobacterium tumefaciens* strain containing two binary plasmids as a method for producing marker-free transgenic plants. *Plant Cell Rep.*, **17** : 489-496.

Feldmann KA and Marks MD (1987) *Agrobacterium* mediated transformation of germinating seeds of *Arabidopsis thaliana*: a non-tissue culture approach. *Mol. Gen. Genet.*, **208** : 1-9.

Gana JA, Sharma GC, Zipf A, Saha S, Roberts J and Wesenberg DM (1995) Genotype effects on plant regeneration in callus and suspension cultures of *Avena*. *Plant Cell Tiss. Org. Cult.*, **40** : 217-224.

Gill KS, Gill BS and Endo TR (1993) A chromosome region-specific mapping strategy reveals gene-rich telomeric ends in wheat. *Chromosoma*, **102** : 374-381.

Gill KS, Gill BS, Endo TR and Boyko EV (1996) Identification and high-density mapping of gene-rich regions in chromosome group 5 of wheat. *Genetics*, **143** : 1001-1012.

Gless C, Lörz H and Jahne-Gartner A (1998a) Establishment of a highly efficient regeneration system from leaf base segments of oat (*Avena sativa* L.). *Plant Cell Rep.*, **17** : 441-445.

Gless C, Lörz H and Jahne-Gartner A (1998b) Transgenic oat plants obtained at high efficiency by microprojectile bombardment of leaf base segments. *J. Plant Physiol.*, **152** : 151-157.

Gorbunova V and Levy A (1999) How plants make ends meet: DNA double-strand break repair. *Trends Plant Sci.*, **4** : 263-269.

Haldrup A, Petersen SG and Okkels FT (1998) Positive selection: a plant selection principle based on xylose isomerase, an enzyme used in the food industry. *Plant Cell Rep.*, **18** : 76-81.

Hassan G, Zipf A, Sharma GC and Wesenberg D (1999) Plant regeneration from mature *Avena* tissue explants. *Cereal Res. Com.*, **27** : 25-32.

Hiei Y, Ohta A, Komari T and Kumashiro T (1994) Efficient transformation of rice (*Oryza sativa* L.) mediated by *Agrobacterium* and sequence analysis of the boundaries of the T-DNA. *Plant J.*, **6** : 271-282.

Hunold R, Bronner R and Hahne G (1994) Early events in microprojectile bombardment: cell viability and particle location. *Plant J.*, **5** : 593-604.

Iglesias VA, Moscone EA, Papp I, Neuhuber F, Michalowski S, Phelan T, Spiker S, Matzke M and Matzke AJM (1997) Molecular and cytogenetic analyses of stably and unstably expressed transgene loci in tobacco. *Plant Cell*, **9** : 1251-1264.

Ishida Y, Saito H, Ohta S, Hiei Y, Komari T and Kumashiro T (1996) High efficiency transformation of maize (*Zea mays* L.) mediated by *Agrobacterium tumefaciens*. *Nat. Biotechnol.*, **14** : 745-750.

Jakowitsch J, Papp I, Moscone EA, van der Winden J, Matzke M and Matzke AJM (1999) Molecular and cytogenetic characterization of a transgene locus that induces silencing and methylation of homologous promoters in *trans*. *Plant J.*, **17** : 131-140.

Jellen EN, Gill BS and Cox TS (1994) Genomic *in situ* hybridization differentiates between A/D- and C-genome chromatin and detects intergenomic translocations in polyploid oat species (genus *Avena*). *Genome*, **37** : 613-618.

Kaeppler HF, Menon GK, Skadsen RW, Nuutila AM and Carlson AR (2000) Transgenic oat plants via visual selection of cells expressing green fluorescent protein. *Plant Cell Rep.*, **19** : 661-666.

Klein TM and Jones TJ (1999) Methods of genetic transformation. In : *Molecular Improvement of Cereal Crops, Advances in Cellular and Molecular Biology of Plants*, Vol. 5, (Ed. Vasil IK), Kluwer Academic Publishers, Dordrecht, Netherlands, pp. 20-42.

Klein TM, Wolf ED, Wu R and Sanford JC (1987) High-velocity microprojectiles for delivering nucleic acids into living cells. *Nature*, **327** : 70-73.

Koev G, Mohan BR, Dinesh-Kumar SP, Torbert KA, Somers DA and Miller WA (1998) Extreme reduction of disease in oats transformed with the 5' half of the barley yellow dwarf virus-PAV genome. *Phytopath.*, **88** : 1013-1019.

Kohli A, Griffiths S, Palacios N, Twyman RM, Vain P, Laurie DA and Christou P (1999) Molecular characterization of transforming plasmid rearrangements in transgenic rice reveals a recombination hotspot in the CaMV 35S promoter and confirms the predominance of microhomology mediated recombination. *Plant J.*, **17** : 591-601.

Kohli A, Leech M, Vain F, Laurie D and Christou P (1998) Transgene organization in rice engineered through direct DNA transfer supports a two-phase integration mechanism mediated by the establishment of integration hot spots. *Proc. Natl. Acad. Sci. USA*, **95** : 7203-7208.

Komari T and Kubo T (1999) Methods of genetic transformation: *Agrobacterium tumefaciens*, In : *Molecular Improvement of Cereal Crops* (Ed. Vasil IK), Kluwer Academic Publishers, Dordrecht, Netherlands, pp. 43-82.

Komari T, Hiei Y, Saito Y, Murai N and Kumashiro T (1996) Vectors carrying two separate T-DNAs for co-transformation of higher plants mediated by *Agrobacterium tumefaciens* and segregation of transformants free from selection markers. *Plant J.*, **10** : 165-174.

Leggett JM, Perret SJ, Harper J and Morris P (2000) Chromosomal localization of cotransformed transgenes in the hexaploid cultivated oat *Avena sativa* L. using fluorescence *in situ* hybridization. *Heredity*, **84** : 46-53.

Leggett JM and Markhand GS (1995) The genomic identification of some monosomics of *Avena sativa* L. cv. Sun II using genomic *in situ* hybridization. *Genome*, **38** : 747-751.

Lörz H, Harms CT and Potrykus I (1976) Regeneration of plants from callus in *Avena sativa*, L. *Z. Pflanzenzuchtg.*, **77** : 257-259.

McCoy TJ, Phillips RL and Rines HW (1982) Cytogenetic analysis of plants regenerated from oat (*Avena sativa*) tissue culture; high frequency of partial chromosome loss. *Can. J. Genet. Cytol.*, **24** : 37-50.

McGrath PF, Vincent JR, Lei CH, Pawlowski WP, Torbert KA, Gu W, Kaeppler HF, Wan Y, Lemaux PG, Rines HW, Somers DA, Larkins BA and Lister RM (1998) Coat protein-mediated resistance to barley yellow dwarf in oats and barley. *European J. Plant Path.*, **103** : 695-710.

Negrotto D, Jolley M, Beer S, Wenck AR and Hansen G (2000) The use of phosphomannose-isomerase as a selectable marker to recover transgenic maize plants (*Zea mays* L.) via *Agrobacterium* transformation. *Plant Cell Rep.*, **19** : 798-803.

Olhoft PM and Phillips RL (1998) Genetic and epigenetic instability in tissue culture and regenerated progenies. In : *Plant response to environmental stresses: from phytohormones to genomic reorganization* (Ed. Lerner HR), M. Dekker Inc., New York, pp. 111-148.

Pawlowski WP and Somers DA (1996) Transgenic inheritance in plants genetically engineered using microprojectile bombardment. *Mol. Biotech.*, **6** : 17-30.

Pawlowski WP and Somers DA (1998) Transgenic DNA integrated into the oat genome is frequently interspersed by host DNA. *Proc. Natl. Acad. Sci. USA*, **95** : 12106-12110.

Pawlowski WP, Torbert KA, Rines HW and Somers DA (1998) Irregular patterns of transgene silencing in allohexaploid oat. *Plant Mol. Biol.*, **38** : 597-607.

Perret SJ and Morris P (2000) Factors influencing T-DNA transfer in oats. *Proceedings of the Sixth International Oat Conference,* 13-16 Nov 2000, Lincoln Univ., Canterbury, New Zealand (Ed. Cross RJ), New Zealand Institute for Crop and Food Research Limited, Christchurch, New Zealand, pp. 148-153.

Rines HW, Phillips RL and Somers DA (1992) Application of tissue cultures to oat improvement. In: *Oat Science and Technology,* American Society of Agronomy, Madison, WI. Agronomy Monograph No. 33 (Eds Marshall HG and Sorrells ME), pp. 777-792.

Sanford JC (1988) The biolistic process. *Trends Biotechnol.*, **6** : 299-302.

Sanford JC (1990) Biolistic plant transformation. *Physiol. Plant*, **79** : 206-209.

Sanford JC, DeVit MJ, Russell JA, Smith FD, Harpending PR, Roy MK and Johnston SA (1991) An improved, helium-driven biolistic device. *Techniques*, **3** : 3-16.

Sanford JC, Klein TM, Wolf ED and Allen N (1987) Delivery of substances into cells and tissues using a particle bombardment process. *Particle Sci. Technol.*, **5** : 27-37.

Sanford JC, Smith FD and Russell JA (1993) Optimizing the biolistic process for different biological applications. *Methods Enzymol.*, **217** : 483-509.

Sears ER (1966) Nullisomic-tetrasomic combinations in hexaploid wheat. In: *Chromosome manipulation and plant genetics* (Ed. Lewis DR). Oliver and Boyd, London, pp. 29-47.

Somers DA (1999) Transgenic Cereals: *Avena sativa* (oat). In: *Advances in Cellular and Molecular Biology of Plants*, Vol. 5 Molecular Improvement of Cereal Crops (Ed. Vasil IK), Kluwer Academic Press, Dordrecht, Netherlands, pp. 317-339.

Somers DA, Rines HW, Torbert KA, Pawlowski WP and Milach SRC (1995) Genetic engineering of *Avena sativa* L. (oat). In: *Biotechnology in Agriculture and Forestry, Plant Protoplasts and Genetic Engineering VII*, (Ed. Bajaj YPS), Springer, Berlin, Vol. 38, pp. 78-190.

Somers DA, Torbert KA, Pawlowski WP and Rines HW (1994) Genetic engineering of oat, In: *Improvement of Cereal Quality by Genetic Engineering*, (Ed. Henry RJ), Plenum Press, New York, pp. 37-46.

Somers DA, Rines HW, Gu W, Kaeppler HF and Bushnell WR (1992) Fertile, transgenic oat plants. *Bio/Technology*, **10** : 1589-1594.

Svitashev SK and Somers DA (2001) Genomic interspersions determine the size and complexity of transgene loci in transgenic plants produced by microprojectile bombardment. *Genome,* **44** : 691-697.

Svitashev SK and Somers DA (2002) Characterization of transgene loci in plants using FISH: Pictures worth a thousand words. *Plant Cell Tiss. Org. Cult.* (in press).

Svitashev S, Ananiev E, Pawlowski WP and Somers DA (2000) Association of transgene integration sites with chromosome rearrangements in hexaploid oat. *Theor. Appl. Genet.*, **100** : 872-880.

Tingay S, McElroy D, Kalla R, Fieg S, Wang M, Thornton S and Brettell R (1997) *Agrobacterium tumefaciens*-mediated barley transformation. *Plant J.*, **11** : 1369-1376.

Torbert KA, Rines HW and Somers DA (1995) Use of paromomycin as a selective agent for oat transformation. *Plant Cell Rep.*, **14** : 635-640.

Torbert KA, Gopalraj M, Medberry SL, Olszewski NE and Somers DA (1998a) Expression of the *Commelina* yellow mottle virus promoter in transgenic oat. *Plant Cell Rep.*, **17** : 284-287.

Torbert KA, Rines HW, Kaeppler HF, Menon GR and Somers DA (1998b) Genetically engineering elite oat cultivars. *Crop Sci.*, **38** : 1685-1687.

Torbert KA, Rines HW and Somers DA (1998c) Transformation of oat using mature embryo-derived tissue cultures. *Crop Sci.*, **38** : 226-231.

Trieu AT, Burleigh SH, Kardailsky IV, Maldonado-Mendoza IE, Versaw WK, Blaylock LA, Shin H, Chiou TJ, Katagi H, Dewbre GR, Weigel D and Harrison MJ (2000) Transformation of *Medicago truncatula* via infiltration of seedlings of flowering plants with *Agrobacterium*. *Plant J.*, **22** : 531-541.

Tzafrir I, Torbert KA, Lockhart BEL, Somers DA and Olszewski NE (1998) The sugarcane bacilliform badnavirus promoter is active in both monocots and dicots. *Plant Mol. Biol.*, **38** : 347-356.

Wang AS, Evans RA, Altendorf PR, Hanten JA, Doyle MC and Rosichan JL (2000) A mannose selection system for production of fertile transgenic maize plants from protoplasts. *Plant Cell Rep.*, **19** : 654-660.

Zhang S, Cho MJ, Koprek T, Yun R, Bregitzer P and Lemaux PG (1999) Genetic transformation of commercial cultivars of oat (*Avena sativa* L.) and barley (*Hordeum vulgare* L.) using *in vitro* shoot meristematic cultures derived from germinated seedlings. *Plant Cell Rep.*, **18** : 959-966.

Zhang S, Zhong H and Sticklen MB (1996) Production of multiple shoots from shoot apical meristems of oat (*Avena sativa* L.). *J. Plant Physiol.*, **148** : 667-671.

Chapter 7

GENETIC TRANSFORMATION OF BARLEY (*HORDEUM VULGARE* L.)

Makoto Kihara★ and Kazutoshi Ito

Plant Bioengineering Research Laboratories, Sapporo Breweries Ltd., 37-1, Kizaki, Nitta, Gunma 370-0393, Japan

Summary

Recently, methods for the production of transgenic cereal plants have been developed, and many successful productions have been reported in barley. Selectable marker genes, reporter genes or useful trait genes to increase disease resistant or to improve seed quality are introduced into barley using direct gene transfer to protoplasts, microprojectile bombardment and Agrobacterium-infection. These experiments indicate the transgenes are stably expressed and are stably transmitted to the progeny. This fact suggests that genetic engineering is a useful technique for the improvement of barley.

Keywords : Barley, transformation, protoplasts, bombardment, *Agrobacterium*

1. INTRODUCTION

In the past ten years, plant transformation has been successfully developed in cereals which include a number of agronomically important crops, and some methods have been widely used for gene introduction (Vasil, 1994). Barley is an important crop both for brewing and for animal feed. Therefore, great efforts have been made to produce

★Corresponding author : E-mail : makoto.kihara@sapporobeer.co.jp transgenic barley. While the gene transformation of barley is still difficult, efficient regeneration from cultured cells, combined with several methods for DNA delivery and the selection of transformed cells, has led to the production of fertile transgenic barley plants. This review chronicles these successful transformations of barley and evaluates the various approaches for genetic transformation.

2. SUCCESSFUL METHODS OF DNA DELIVERY

2.1. Direct DNA transfer to protoplasts

The use of cereal protoplasts was particularly attractive because of two important characteristics. First, protoplasts have no cell walls and are only surrounded by a plasma membrane, thus permitting the uptake of foreign DNA into cells, and second, they have totipotency. Most of the early attempts to produce transgenic cereals were restricted to the use of protoplasts, because none of the then existing methods permitted DNA delivery into intact cells (Vasil, 1994). Two methods, electric (electroporation) shock or chemical treatment with polyethylene glycol (PEG), have been used for the efficient delivery of DNA into cereal protoplasts, and for demonstrating the stable integration of the introduced genes (Krawtwig and Lörz, 1995). In barley, the transient expression of introduced genes was reported by both protoplast transformation methods (Junker *et al.*, 1987; Teeri *et al.*, 1989; Lee *et al.*, 1989), and Lazzeri *et al.* (1991) first reported the stable transformation in protoplast-derived calluses via PEG-induced direct gene transfer.

In contrast to most dicotyledonous species where even protoplasts isolated from mature tissues have the plant regeneration ability, only the embryogenic cell suspension is a major source of totipotent protoplasts in cereals. In 1991, the first successful plant regeneration from barley protoplasts was reported (Jähne *et al.*, 1991a). In this work, embryogenic suspension cells as a source of regenerable protoplasts were initiated from anther-derived calluses of cv. Igri, but these cells quickly lost their regeneration capacity and the efficiency of plant regeneration was not high enough for genetic manipulation (Jähne *et al.*, 1991b). Funatsuki *et al.* (1992) established suspension cells from immature embryo-derived calluses of cv. Dissa, and obtained fertile plants from protoplasts isolated from these suspension cells. Furhermore, Kihara and Funatsuki (1994) and Funatsuki and Kihara (1994) established embryogenic suspension cells from immature embryo-derived callses of cv. Igri, and showed high frequency plant regeneration from protoplasts. One year later, they succeeded in the recovery of fertile transgenic barley containing the *nptII* gene by PEG-induced direct gene transfer to embryogenic protoplasts (Funatsuki *et al.*, 1995) (Table 1). BaYMV and BaMMV coat protein genes (Kuroda *et al.*, 1996), the thermostable-amylase gene (Kihara *et al.*, 1997, 2000) or *uidA* gene (Okada *et al.*, 2000) with the *nptII* gene were also introduced into the transgenic barley plants using the protoplast transformation system. Compared with the PEG method, electroporation has not been widely used for barley transformation. Only one paper has reported that transgenic barley plants were obtained from microspore culture-derived protoplast electroporation (Salmenkallio-Marttila *et al.*, 1995).

2.2. Microprojectile bombardment

Microprojectile bombardment or biolistics has been widely used for the genetic transformation of cereals, although it is a relatively new method of transformation. The foreign DNA, coated on microscopic particles of tungsten or gold, is fired into the target tissues using special devices with pneumatic, helium, nitrogen, or an electronically triggered discharge. The principal advantage of the biolistics is its ability to deliver DNA into intact regenerable plant cells, thus eliminating the need for protoplasts. In addition, somaclonal variation, which causes abnormalities, is avoidable because DNA can be delivered into explants from which plants can be regenerated without a long-term culture period.

The first successful example of fertile transgenic plants induced by bombardment in cereals was obtained from maize in 1990 (Fromm *et al.*, 1990). Subsequently, fertile transgenic rice plants and transgenic wheat plants were reported (Christou *et al.*, 1991; Vasil *et al.*, 1992). For barley transformation, the transient expression of marker genes in barley cells using bombardment was demonstrated in 1989 (Kartha *et al.*, 1989; Mendel *et al.*, 1989). Subsequently, reporter genes and selectable marker genes were stably expressed in barley cells (Ritala *et al.*, 1993). The successful regeneration of fertile transgenic barley plants by bombardment was first reported using the selectable marker gene *nptII* (Ritala *et al.*, 1994). Also, the *bar* gene, *uidA* gene and BADV coat protein gene were introduced into barley by bombardment (Wan and Lemaux, 1994). Many successful barley transformations, including the two previously described reports, have shown that immature embryos or isolated scutella are suitable target tissues (Table 1). Hundreds of immature embryos can be easily isolated and bombarded in a single day.

Table 1. Methods for the production of fertile transgenic barley

Method	Explant[a]	Introduced gene (Promoter)						Reference
Protoplasts								
PEG	SC-derived protoplasts	*nptII*	(*Act1*)					Funatsuki *et al.* (1995)
	SC-derived protoplasts	*nptII*	(*Arf, Act1*)			+ *BaYMVcp*	(*Arf, Act1*)	Kuroda *et al.* (1996)
		nptII	(*Arf, Act1*)			+ *BaMMVcp*	(*Arf, Act1*)	
	SC-derived protoplasts	*nptII*	(*Act1*)			+ *TS* β-amy	(β-amy)	Kihara *et al.* (1997, 2000)
	SC-derived protoplasts	*nptII*	(*Act1*)	+ *uidA*	(β-amy)			Okada *et al.* (2000)
	PC-derived protoplasts	*nptII*	(*Act1*)					Kihara *et al.* (1998)
Electroportation	MSC-derived protoplasts	*nptII*	(*35S*)					Salmenkallio-Marttila *et al.* (1995)
Bombardment								
	IEs	*nptII*	(*35S*)					Ritala *et al.* (1994)
	IEs, IE-derived calluses	*bar*	(*Ubi1*)	+ *uidA*	(*Ubi1*)	+ *BYDVcp*	(*35S*)	Wan and Lemaux (1994)

Table 1. Continued

Method	Explant[a]	Introduced gene (Promoter)						Reference
	MS-Derived embryos	*bar*	*(35S)*	+ *uidA*	*(Adh1)*			Wan and Lemaux (1994)
		bar	*(Ubi1)*	+ *uidA*	*(Ubi1)*	+ *BYDVcp*	*(35S)*	
	MSs	*bar*	*(35S)*	+ *uidA*	*(Act1)*			Jahne *et al.* (1994)
	IEs	*pht*	*(35S)*	+ *uidA*	*(Adh1)*			Hagio *et al.* (1995)
	IE-derived scutella	*bar*	*(Ubi1)*	+ *uidA*	*(Ubi1)*			Koprek *et al.* (1996)
	IEs	*bar*	*(Ubi1)*	+ *uidA*	*(pEmu)*	+ TS β-glu	(α-amy)	Jensen *et al.* (1996)
	IEs	*bar*	*(Act1)*	+ *uidA*	*(Act1)*			Rikiishi *et al.* (1996)
	IEs	*bar*	*(Ubi1)*	+ *uidA*	*(Act1)*			Kundsen *et al.* (1996)
	IE-derived scutella	*bar*	*(35S)*	+ *uidA*	*(Adh1)*			Weir *et al.* (1996)
		uidA: *nptII*	*(Act1)*					

Table 1. Continued

Method	Explant[a]	Introduced gene (Promoter)						Reference
	IEs	*bar*	(*Ubi1*)	+ *uidA*	(*Act1*)	+ *lysC*	(35 S)	Brinch-Pedersen *et al.* (1996)
		bar	(*Ubi1*)	+ *uidA*	(*Act1*)	+ *dapA*	(35 S)	
	IEs or IE-derived scutella	*bar*	(*Ubi1*)	+ *uidA*	(*Ubi1*)	+ *TS* β-*glu*	(α-my)	Mannonen *et al.* (1997)
	MSs	*bar*	(*Ubi1*)	+ *uidA*	(*Ubi1*)			Yao *et al.* (1997)
		pht	(*35S*)			+ *BaMMVcp*	(*35S*)	Hagio *et al.* (1997)
	MSs	*pat*	(*35S*)			+ *Vst1*	(*Vst1*)	Leckband and Lorz (1998)
	IEs	*bar*	(*Ubi1*)	+ *uidA*	(*Hor2*, *Hor3*)			Cho *et al.* (1999a)
	IEs	*bar*	(*Ubi1*)			+ *wtrxh*	(*Hor2*)	Cho *et al.* (1999b)
	IEs or IE-derived scutella	*nptII*	(*35S*)			+ *TS* β-*glu*	(α–amy)	Nuutila *et al.* (1999)
	IEs	*codA*	(*Act1*)			+ *AcTPase*	(*Ubi1*, *AcTPase*)	Koprek *et al.* (1999)

Table 1. Continued

Method	Explant[a]	Introduced gene (Promoter)	Reference
		P450 (*rbcS*)	
	IEs	*lysC* (*Ubi1*)	Brinch-Pedersen *et al.* (1999)
	SMCs	*bar* (*Ubi1*) + *uidA* (*Ubi1*)	Zhang *et al.* (1999)
	IEs	*bar* (*Ubi1*) + *uidA* (*Ubi1*)	Harwood *et al.* (2000)
		bar (*Ubi1*) + *luc* (*Ubi1*)	
Agrobacterium			
	IE-derived scutella	*bar* (*Ubi1*) + *uidA* (*Act1*)	Tingay *et al.* (1997)
	IE-derived scutella	*bar* (*Ubi1*) + *xynA* (*Hor2*, *GluB-1*)	Patel *et al.* (2000)
	IEs	*bar* (*Ubi1*) + *sGFP* (*35S*) + *TS* β-*glu* (*Hor3*)	Horvath *et al.* (2000)

SC : Suspension cell; PC : Primary callus; MSC : Microspore culture; IE : Immature embryo; MS : Microspore; SMC : Shoot meristematic culture

On the other hand, the haploid tissue such as microspores was also found to be a suitable source for plant transformation. Microspores are single-celled and form a highly synchronized population. They can be regenerated in large numbers through direct embryogenesis. In addition, they have a haploid genome and can easily be regenerated to homozygous, diploid plants, which is significant for breeding purpose. Jähne *et al.* (1994) reported the recovery of homozygous transgenic barley plants following microspore bombardment. Although Leckland and Lörz (1998) also produced transgenic barley containing *pat* and *Vst1* by the same microspore bombardment system, their fertile transgenic plants were both homozygous and hemizygous. They suggested that the hemizygous nature was derived from integration of the transgene after spontaneous chromosome doubling. Yao *et al.* (1997) reported the same phenomenon, therefore, cytological studies of the in vitro development of microspores would be needed for the best procedures to achieve homozygous transgenic plants in the first generation.

Zhang *et al.* (1999) indicated that shoot meristematic cultures (SMCs) derived from germinated seedlings, which possess high regenerability and enhanced genomic stability relative to embryogenic callus, were also suitable target tissues for bombardment.

2.3. *Agrobacterium*-mediated transformation

Agrobacterium-mediated transformation is now routinely and widely used for the production of transgenic dicotyledonous plants. This method needs no cell wall removal from target cells as well as bombardment, and often only a single copy of the transgenes are integrated in transgenic plants. Therefore, many groups have attempted to establish this relatively easy and convenient system to obtain transgenic cereals. Unfortunately, these attempts were unsuccessful before the 1990s. However, this situation changed after the successful reports using maize and rice (Gould *et al.*, 1991; Chan *et al.*, 1992; Hiei *et al.*, 1994).

In barley, the first stable barley transformation with *Agrobacterium*-infection was achieved in 1997 (Table 1). Tingay *et al.* (1997) demonstrated stable barley transformation by co-cultivation of *Agrobacterium tumefaciens* carrying the binary vector with *bar* and *uidA* genes and immature embryos which were injured by bombardment. Patel *et al.* (2000) and Horvath *et al.* (2000) also reported the generation of transgenic barley using *Agrobacterium*-mediated transformation.

3. GENOTYPE SPECIFICITY

In cereals, it is well known that the abilities for callus formation and plant regeneration depend on the interaction of the genotype and culture medium. In the barley tissue culture using immature embryos as explants, Lührs and Lörz (1987) reported that the frequency of somatic embryo formation and of plant regeneration was under genotypic control. Therefore, in the case when immature embryos are used for the target tissue for bombardment or *Agrobacterium*- mediated transformation, the genotype used in the experiments is a key factor for the efficient production of transgenic barley. One variety, Golden Promise, which showed a high plant regeneration ability (Lührs and Lörz, 1987), has been widely applicable for these transformation methods (Table 2). On the other hand, plant regeneration from microspore cultures is also highly genotype dependent. Olsen (1987) reported that one variety, Igri, showed a high formation frequency of microspore-derived embryoids and plant regeneration, indicating a superior genotype for the microspore culture. Successful transformations with microspore bombardment have been limited to only this variety (Table 2).

For direct gene transfer to protoplasts, fertile transgenic barley plants have been obtained in three varieties (Igri, Kymmpi and Golden Promise), although the use of suspension cell-derived protoplasts, which is the most reliable protoplast transformation method, is applicable to only one variety, Igri (Table 2). Suspension cells from immature embryo-derived calluses of this variety yielded high regenerable protoplasts even after a long-term subculture (Kihara and Funatsuki, 1994; Funatsuki and Kihara, 1994).

4. CHARACTERISTICS OF INTRODUCED GENES

4.1. Selectable marker genes

All of the direct gene transfer to protoplasts, transformed cells are selected with the neomycin phosphotransferase II (*nptII*) gene of *Escherichia coli*, which confers resistance to the antibiotic kanamycin or G418 (Table 1). On the other hand, the *bar* gene of *Streptmyces hygroscopicus*, which encodes the enzyme phosphinothricin acetyltransferase (PAT), conferring resistance to herbicide basta (bialaphos, glufosinate), is the most widely used marker gene for bombardment or *Agrobacterium*-transformation (Table 1). Also, the *pat* gene, which was isolated from different *Streptmyces species*, is a

Table 2. Genotypes used in the successful experiments

Method	Genotype	Reference
Protoplasts	Golden Promise	Kihara *et al.* (1998)
	Igri	Funatsuki *et al.* (1995); Kuroda *et al.* (1996) Kihara *et al.* (1997, 2000); Okada *et al.* (2000)
	Kymmpi	Salmenkallio-Marttila *et al.* (1995)
Bombardment	Golden Promise	Wan and Lemaux (1994), Hagio *et al.* (1995) Koprek *et al.* (1996); Jensen *et al.* (1996) Brinch-Pedersen *et al.* (1996), Knudsen *et al.* (1996) Cho *et al.* (1999a,b); Nuutila *et al.* (1999) Koprek *et al.* (1999); Harwood *et al.* (2000)
	Igri	Wan and Lemaux (1994); Jahne *et al.* (1994) Yao *et al.* (1997); Leckband and Lorz (1998)
	Kymppi	Ritala *et al.* (1994); Mannonen *et al.* (1997) Nuutila *et al.* (1999)
	Haruna Nijo, Dissa	Hagio *et al.* (1995)
	Dera, Corniche, Salome, Femina	Koprek *et al.* (1996)
	Lenins	Rikiishi *et al.* (1996)
	SB92559, TR232, TR328	Weir *et al.* (1996)
	New Golden	Hagio *et al.* (1997)
	Harrington	Zhang *et al.* (1999)
Agrobacterium	Golden Promise	Tingay *et al.* (1997); Patel *et al.* (2000) Horvath *et al.* (2000)

useful marker gene. Wehrmann *et al.* (1996) demonstrated the structural and functional equivalence of the PAT proteins derived from both genes.

In the past few years, alternative approaches have been attempted for the selection of transgenic barley. Brinch-Pedersen *et al.* (1999) showed that the *lysC* gene encoding a lysine-threonine feedback insensitive form of asparate kinase (AK) was a useful marker. They selected transformants with MS medium containing millimolar concentrations of lysine plus threonine. Koprek *et al.* (1999) indicated that the bacterial *codA* gene encoding cytosine deaminase, which converts the non-toxic 5-fluorocytosine into the toxic 5-fluorouracil, and a bacterial cytochrome *P450* mono-oxygenase gene, the product of which catalyses the dealkylation of sulfonylurea compound, R7402, into cytotoxic metabolite, were applicable for the negative selection of transformants.

4.2. Reporter genes

The introduction of reporter genes has proved to be most useful in studies of gene regulation and function. One reporter gene, β-glucuronidase (*gus*) gene, which was encoded by the *uidA* locus of *Escherichia coli,* can be readily evaluated by histochemical assay. Cho *et al.* (1999a) and Okada *et al.* (2000) introduced this reporter gene into barley plants, and indicated the endsperm and subaleurone-specific gene expression of the barley hordein and β-amylase promoter region, respectively.

4.3. Disease resistant genes

The coat protein gene contributes to the resistance of virus diseases. Wan and Lemaux (1994) reported the production of transgenic barley containing a barley yellow dwarf virus coat protein (BYDVcp) gene. Kuroda *et al.* (1996) and Hagio *et al.* (1997) also produced transgenic barley lines containing the barley yellow mosaic virus coat protain gene (BaYMVcp) or barley mild mosaic coat protein (BaMMVcp) gene.

Leckband and Lörz (1998) reported the transformation and expression of a stilbene synthase (*Vst1*) gene of *Vitis vinifera* L. in barley for increased fungal resistance. Stilbene synthase, also termed resveratol synthase, is the enzyme which synthesizes the phenolic phytoalexin trans-resveratrol. They indicated that the expression of this enzyme into barley plants enhanced disease resistance as well as in tobacco and rice (Hain *et al.*, 1990, 1993; Stark-Lorenzen *et al.*, 1997).

4.4. Improvement of seed quality genes

In malting and brewing, the kilning and mashing processes, which employ high temperatures, reduce various enzyme activities in the barley grain and malt. Therefore, increasing the thermostability of these enzymes has been attempted by the production of transgenic barley. Jensen *et al.* (1996) and Horvath *et al.* (2000) introduced thermostable β-glucanase genes, which were isolated from *Bacillus species*, under the control of barley α-amylase and D hordein promoter, respectively. Also, Mannonen *et al.* (1997) and Nuutila *et al.* (1999) introduced thermostable β-glucanase genes, which were isolated from *Tricoderma reesei,* under the control of a barley α-amylase promoter. Their results showed that the introduced genes were expressed and thermostable β-glucanase was synthesized in the transgenic barley grains. Kihara *et al.* (1997) introduced the thermostable β-amylase gene, which was constructed by site-directed mutagenesis and random mutagenesis to achieve the substitution of seven amino acids of the original barley β-amylase (Okada *et al.*, 1995; Yoshigi *et al.*, 1995). β-Amylase is a heat labile enzyme, and its themostability has a close relation to the apparent attenuation limit which is a major malting quality parameter for the determination of the fermentability in brewing (Kihara *et al.*, 1998a). Their results showed that the introduced gene under control of a barley β-amylase promoter region (Okada *et al.*, 2000) was expressed in the maturing barley grain, and was stably transmitted to the progeny (Kihara *et al.,* 2000).

Cho *et al.* (1999) introduced the wheat thioredoxin *h* gene (*wtrxh*) under the control of the barley B1 hordein promoter. They indicated that the overexpression of thioredoxin *h* in the endosperm of the germinated grain affected up to a 4-fold increase in the activity of the starch debranching enzyme, pullulanase, the enzyme that specifically cleaves the α-1,6 linkage in starch.

In order to modify the flux through the asparate family pathway and enhance the accumulation of the corresponding amino acids, Brinch-Pedersen *et al.* (1996) generated transgenic barley plants introduced mutant *Escherichia coli* genes, *lysC* and *dapA*, encoding the lysine feed-back insensitive forms of asparate kinase (AK) and dihydrodipicolinate synthase (DHPS). They indicated that there was a 2-fold increase in the free lysine, arginine and asparagine and a 50% reduction in the free proline in the mature seeds of the DHPS transformants, while no changes were observed in the seeds of the two analyzed AK transgenic lines.

Patel *et al.* (2000) reported the production of transgenic barley expressing a modified xylanase (*xynA*) gene from the rumen fungus, *Neocallimastix patriciarum*, under control of the seed specific rice gluterin or barley hordein promoter region. Currently, xylanase from microbial sources are used to improve the efficiency of feed grain utilization in monogastric animals. They indicated that xylanase was stably maintained in the grain during maturation, desiccation and the post-harvest storage.

5. CONCLUSIONS AND FUTURE PROSPECTS

Advances in the three kinds of transformation techniques (direct gene transfer to protoplasts, microprojectile bombardment and *Agrobacterium*-infection) with efficient plant regeneration from cultured cells has led to the production of transgenic barley within the relatively short period of seven years (1994-2000). Microprojectile bombardment has been widely used for the genetic transformation in barley as well as in other cereals. This system needs no protoplast procedure, indicating no long-term culture period for the suspension establishment. *Agrobacterium*-mediated transformation has the same advantage as the bombardment method, with two additional merits, which are the low copy number of the transgene integration into a genome and no specialized equipment requirement. On the other hand, direct gene transfer with PEG also needs no specialized equipment, and transformation experiments could be repeated, once the embryogenic suspension cultures are established. However, this method demands more labor-intensive and long-term *in vitro* culture for the suspension production. Kihara *et al.* (1998c) reported that a long-term suspension culture caused a somaclonal variation in several agronomic characters. To circumvent these problems, Salmenkallio-Marttila *et al.* (1995) and Kihara *et al.* (1995, 1998b) demonstrated that three- to four-week old microspore cultures and immature embryo-derived primary calluses were good protoplast sources for the rapid production of transgenic barley. At present, it is difficult to compare the efficiencies between different transformation systems as described in this article. Therefore, It may be worthwhile to continue development of each transformation method.

At least some of the transgenic barley are expected to be commercially introduced in the 21st century. Advances using these transformation technologies, however, will have to be careful and will be depend on public acceptance of the genetic modified barley and their products.

REFERENCES

Brinch-Pedersen H, Galili G, Knudsen S and Holm PB (1996) Engineering of the asparate family biosynthetic pathway in barley (*Hordeum vulgare* L.) by transformation with heterologous genes encoding feed-back-insensitive aspartate kinase and dihydrodipicolinate synthase. *Plant Mol. Biol.*, **32** : 611-620.

Brinch-Pedersen H, Olsen O, Knudsen S and Holm PB (1999) An evaluation of feed-back insensitive asparate kinase as a selectable marker for barley (*Hordeum vulgare* L.) transformation. *Hereditas*, **131** : 239-245.

Chan MT, Lee TM and Chang HH (1992) Transformation of indica rice (*Oryza sativa* L.) mediated by *Agrobacterium tumefaciens*. *Plant Cell Physiol.*, **33** : 577-583.

Christou P, Ford TL and Kofron M (1991) Production of transgenic rice (*Oryza sativa* L.) plants from agronomically important indica and japonica varieties via electric discharge particle acceleration of exogenous DNA into immature zygotic embryos. *Bio/Technology*, **9** : 957-962.

Cho MJ, Choi HW, Buchanan BB and Lemaux PG (1999a) Inheritance of tissue-specific expression of barley hordein promoter-*uidA* fusions in transgenic barley plants. *Theor. Appl. Genet.*, **98** : 1253-1262.

Cho MJ, Wong JH, Marx C, Jiang W, Lemaux PG and Buchanan BB (1999b) Overexpression of thioredoxin *h* leads to enhanced activity of starch debranching enzyme (pullulanase) in barley grain. *Proc. Natl. Acad. Sci USA*, **96** : 14641-14646.

Fromm ME, Morrish F, Armstrong C, Williams R, Thomas J and Klein TM (1990) Inheritance and expression of chimeric genes in the progeny of transgenic maize plants. *Bio/Technology*, **8** : 833-839.

Funatsuki H, Lörz H and Lazzeri PA (1992) Use of feeder cells to improve barley protoplast culture and regeneration. *Plant Sci.*, **85** : 179-187.

Funatsuki H and Kihara M (1994) Influence of primary callus induction conditions on establishment of barley cell suspensions yielding regenerable protoplasts. *Plant Cell Rep.*, **13** : 551-555.

Funatsuki H, Kuroda H, Kihara M, Lazzeri PA, Müller E, Lörz H and Kishinami I (1995) Fertile transgenic barley generated by direct gene transfer to protoplasts. *Theor. Appl. Genet.*, **91** : 707-712.

Gould J, Devery W, Hasegawa O, Ulian EC, Peterson G and Smith RH (1991) Transformation of *Zea mays* using *Agrobacterium tumefaciens* and the shoot apex. *Plant Physiol.*, **95** : 426-434.

Hagio H, Hirabayashi T, Machii H and Tomotsune H (1995) Production of fertile transgenic barley (*Hordeum vulgare* L.) plant using the hygromycin-resistance marker. *Plant Cell Rep.*, **14** : 329-334.

Hagio H, Kashiwazaki S, Hirabayashi T and Ohsawa Y (1997) Production of transgenic barley plants resistant to Barley Mild Mosaic Virus by particle bombardment. *J. Soc. In Vitro Biol.*, **33** : 68.

Harwood WA, Ross SM, Cilento P and Snape JW (2000) The effect of DNA/gold particle preparation technique, and particle bombardment device, on the transformation of barley (*Hordeum vulgare*). *Euphytica*, **111** : 67-76.

Hain R, Bieseler B, Kindl H, Schröder G and Stöcker R (1990) Expression of a stilbene synthase in *Nicotiana tabacum* results in synthesis of phytoalexin resveratol. *Plant Mol. Biol.*, **15** : 325-335.

Hain R, Reif HJ, Krause E, Langebartels R, Kindl H, Vornam B, Wiese W, Schmelzer E, Schreier PH, Stöcker RH and Stenzel K (1993) Disease resistance results from foreign phytoalexin expression in a novel plant. *Nature*, **361** : 153-156.

Hiei Y, Ohta S, Komari T and Kumashiro T (1994) Efficient transformation of rice (*Oryza sativa* L.) mediated by *Agrobacterium* and sequence analysis of the boundaries of the T-DNA. *Plant J.*, **6** : 271-282.

Horvath H, Huang J, Wong O, Kohl E, Okita T, Kannangara CG and von Wettstein D (2000) The production of recombinant proteins in transgenic barley grains. *Proc. Natl. Acad. Sci. USA*, **97** : 1914-1919.

Jähne A, Lazzeri PA and Lörz H (1991a) Regeneration of fertile plants from protoplasts derived from embryogenic cell suspensions of barley (*Hordeum vulgare* L.). *Plant Cell Rep.*, **10** : 1-6.

Jähne A, Lazzeri PA, Jäger-Gussen M and Lörz H (1991b) Plant regeneration from embryogenic cell suspensions derived from anther cultures of barley (*Hordeum vulgare* L.). *Theor. Appl. Genet.*, **82** : 74-80.

Jähne A, Becker D, Brettschneider R and Lörz H (1994) Regeneration of transgenic, microspore-derived, fertile barley. *Theor. Appl. Genet.*, **89** : 525-533.

Jensen LG, Olsen O, Kops O, Wolf N, Thomsen KK and von Wettstein D (1996) Transgenic barley expressing a protein-engineered, thermostable (1,3-1,4) -β-glucanase during germination. *Proc. Natl. Acad. Sci. USA*, **93** : 3487-3491.

Junkker B, Zimmy J, Lührs R and Lörz H (1987) Transient expression of chimeric gene in dividing and non-dividing cereal protoplasts after PEG-induced DNA uptake. *Plant Cell Rep.*, **6** : 320-332.

Kartha KK, Chibbar RN, Georges F, Leung N, Caswell K, Kendall E and Qureshi J (1989) Transient expression of chloramphenicol acetyltransferase (CAT) gene in barley cell cultures and immature embryos through microprojectile bombardment. *Plant Cell Rep.,* **8** : 429-432.

Kihara M and Funatsuki H (1994) Fertile plant regeneration from barley (*Hordeum vulgare* L.) protoplasts isolated from long-term suspension culture. *Breeding Sci.*, **44** : 157-160.

Kihara M and Funatsuki H (1995) Fertile plant regeneration from barley (*Hordeum vulgare* L.) protoplasts isolated from primary calluses. *Plant Sci.*, **106** : 115-120.

Kihara M, Okada Y, Kuroda H, Saeki K, Yoshigi N and Ito K (1997) Generation of fertile transgenic barley synthesizing thermostable β-amylase. *Proceedings of the 26th Congress of European Brewery Convention*: 83-90

Kihara M, Kaneko T and Ito K (1998a) Genetic variation of β-amylase thermostability among varieties of barley (*Hordeum vulgare* L.) and relation to malting quality. *Plant Breed.*, **117** : 425-428.

Kihara M, Saeki K and Ito K (1998b) Rapid production of fertile transgenic barley (*Hordeum vulgare* L.) by direct gene transfer to primary callus-derived protoplasts. *Plant Cell Rep.*, **17** : 937-940.

Kihara M, Takahashi S, Funatsuki H and Ito K (1998c) Field performance of the progeny of protoplast-derived barley plants. *Breeding Sci.*, **48** : 1-4.

Kihara M, Okada Y, Kuroda H, Saeki K, Yoshigi N and Ito K (2000) Improvement of β-amylase thermostability in transgenic barley seeds and transgene stability in progeny. *Mol. Breed.* (in press).

Koprek T, Hänsch R, Nerlich A, Mendel RR and Schulze J (1996) Fertile transgenic barley of different cultivars obtained by adjustment of bombardment conditions to tissue response. *Plant Sci.*, **119** : 79-91.

Koprek T, McElroy D, Louwerse J, Williams-Carrier R and Lemaux PG (1999) Negative selection systems for transgenic barley (*Hordeum vulgare* L.): Comparison of bacterial *codA*- and cytochrome *P450* gene-mediated selection. *Plant J.*, **19** : 719-726.

Krautwig B and Lörz H (1995) Cereal protoplasts. *Plant Sci.*, **111** : 1-10.

Knudsen S, Jorgensen AB, Sandager L, Brinch-Pedersen H, Sorensen LD, McElroy D and Holm PB (1996) Segregation of the *bar* and *gus* genes in transgenic barley. *Proceedings of the V International Oat Conference & VII International Barley Genetic Symposium Vol.* **2** : 414-416.

Kuroda H, Kashiwazaki S, Kihara M, Hibino H and Ito K (1996) Production of fertile transgenic barley plants expressing barley yellow mosaic virus or barley mild mosaic virus capsid protein gene. *Proceedings of the V International Oat Conference & VII International Barley Genetic Symposium Vol.* **2** : 417-419.

Lazzeri PA, Brettschneider R, Lührs R and Lörz H (1991) Stable transformation of barley via PEG-induced direct DNA uptake into protoplasts. *Theor. Appl. Genet.*, **75** : 16-25.

Leckband G and Lörz H (1998) Transformation and expression of a stilbene synthase gene of *Vitis vinifera* L. in barley and wheat for increased fungal resistance. *Theor. Appl. Genet.*, **96** : 1004-1012.

Lee BT, Murdoch K, Topping J, Kreis M and Jones MGK (1989) Transient expression in aleurone protoplasts isolated from developing caryopses of barley and wheat. *Plant. Mol. Biol.*, **13** : 21-29.

Lürhs R and Lörz H (1987) Plant regeneration *in vitro* from embryogenic cultures of spring- and winter-type barley (*Hordeum vulgare* L.) varieties. *Theor. Appl. Genet.*, **75** : 16-25.

Mannonen L, Ritala A, Nuutila AM, Kurten U, Aspegren K, Teeri T H, Aikasalo R, Tammisola J and Kauppinen V (1997) Thermotolerant fungal glucanase in malting barley. *Proceedings of the 26th Congress of European Brewery Convention*: 91-100.

Mendel RR, Müller B, Schulze J, Kolesnikov V and Zelenin A (1989) Delivery of foreign genes to intact barley cells by high-velocity microprojectiles. *Theor. Appl. Genet.*, **78** : 31-34.

Nuutila AM, Ritala A, Skadsen RW, Mannonen L and Kauppinen V (1999) Expression of fungal thermotolerant endo-1,4-β-glucanase in transgenic barley seeds during germination. *Plant Mol. Biol.*, **41** : 777-783.

Okada Y, Yoshigi N, Sahara H and Koshino S (1995) Increase in thermostability of recombinant barley β-amylase by random mutagenesis. *Biosci. Biotech. Biochem.*, **59** : 1152-1153.

Okada Y, Kihara M, Kuroda H, Yoshigi N and Ito K (2000) Cloning and sequencing of the promoter region of the seed specific β-amylase gene from barley. *J.Plant Physiol.*, **156** : 762-767.

Olsen L (1987) Induction of microspore embryogenesis in cultured anthers of *Hordeum vulgare*. The effect of ammonium nitrate, glutamine and aspargine as nitrogen sources. *Carlsberg Res. Comm.*, **52** : 393-404.

Patel M, Johnson JS, Brettell RIS, Jacobsen J and Xue GP (2000) Transgenic barley expressing a fungal xylanase gene in the endosperm of the developing grains. *Mol. Breed.*, **6** : 113-123.

Rikiishi K, Takumi S, Shimada T, Maekawa M and Takeda K (1996) Efficient transformation in barley variety 'Lenins' by particle bombardment. *Proceedings of the V International Oat Conference & VII International Barley Genetic Symposium Vol.* **2** : 435-437.

Ritala A, Mannonen L, Aspegren K, Salmenkallio-Marttila M, Kurten U, Hannus R, Lozano JM, Teeri TH and Kauppinen V (1993) Stable transformation of barley tissue culture by particle bombardment. *Plant Cell Rep.*, **12** : 435-440.

Ritala A, Aspegren K, Kurten U, Salmenkallio-Marttila M, Mannonen L, Hannus R, Kauppinen V, Teeri TH and Enari TM (1994) Fertile transgenic barley by particle bombardment of immature embryos. *Plant Mol. Biol.*, **24** : 317-325.

Salmenkallio-Marttila M, Aspegren K, Akerman S, Kurten U, Mannonen L, Ritala A, Teeri TH and Kauppinen V (1995) Transgenic barley (*Hordeum vulgare* L.) by electroporation of protoplasts. *Plant Cell Rep.*, **15** : 301-304.

Stark-Lorenzen P, Nelke B, Hänler G, Mühlbach HP and Thomzik JE (1997) Transfer of a stilbene synthase gene to rice (*Oryza sativa* L.). *Plant Cell Rep.*, **16** : 668-673.

Teeri TH, Patel GH, Aspergen K and Kauppinen V (1989) Chloroplast targeting of neomycin phosphotransferase II with a pea transit peptide in electroporated barley mesophyll protoplasts. *Plant Cell Rep.*, **8** : 187-190.

Tingay S, McElroy D, Kalla R, Fieg S, Wang M, Thornton S and Brettell R (1997) *Agrobacterium tumefaciens*-mediated barley transformation. *Plant J.*, **11** : 1369-1376.

Vasil V, Castillo AM, Fromm ME and Vasil IK (1992) Herbicide resistant fertile transgenic wheat plants obtained microprojectile bombardment of regenerable embryogenic callus. *Bio/Technology*, **10** : 667-674.

Vasil IK (1994) Molecular improvement of cereals. *Plant Mol. Biol.*, **25** : 925-937.

Wan Y and Lemaux PG (1994) Generation of large numbers of independently transformed fertile barley plants. *Plant Physiol.*, **104** : 37-48.

Wehrmann A, Vilet AV, Opsomer C, Botterman J and Schulz A (1996) The similarities of *bar* and *pat* gene products make them equally applicable for plant engineers. *Nature Biotechnology*, **14** : 1274-1278.

Weir BJ, Lai KJ, Caswell K, Leung N, Rossnagel BG, Baga M, Kartha KK and Chibbar RN (1996) Transformation of spring barley using the enhanced regeneration system and microprojectile bombardment. *Proceedings of the V International Oat Conference & VII International Barley Genetic Symposium Vol.* **2** : 440-442.

Yao QA, Simion E, William M, Krochko J and Kasha KJ (1997) Biolistics transformation of haploid isolated microspores of barley. *Genome*, **40** : 570-581.

Yoshigi N, Okada Y, Maeba H, Sahara H and Tamaki T (1995) Construction of a plasmid used for the expression of a sevenfold-mutant barley b-amylase with increased thermostability in *Escherichia coli* and properties of the sevenfold-mutant β-amylase. *J. Biochem.*, **118** : 562-567.

Zhang S, Cho MJ, Koprek T, Yun R, Bregitzer P and Lemaux PG (1999) Genetic transformation of commercial cultivars of oat (*Avena sativa* L.) and barley (*Hordeum vulgare* L.) using *in vitro* shoot meristematic cultures derived from germinated seedlings. *Plant Cell Rep.*, **18** : 959-966.

Chapter 8

GENETIC TRANSFORMATION OF MAIZE

Kan Wang*, Bronwyn Frame and Lise Marcell

Iowa State University, Ames, Iowa 50011-1010, USA

Summary

Maize is both a model plant system for fundamental studies and an important crop species. It is grown in more countries than any other crop and is a major source of food for both humans and animals throughout the world. Current advancements in molecular biology have enabled the large scale, rapid identification and isolation of plant genes. Determination of function for thousands of genes, and application of that knowledge to crop improvement, is now one of the major challenges facing plant biologists. Plant transformation is a key technology for functional analysis of genes via complementation, overexpression, or gene silencing strategies and will be equally essential in the application of genomic technology to crop improvement. Significant progress has made on the development of maize genetic transformation technology in the past decade. In this review, we will focus on the two most efficient maize transformation systems currently in use — the Biolistic gun and Agrobacterium. We will also review three major factors critical to successful transformation, namely, target cells, delivery methods and selection systems. Their merits and limitations will be analyzed and compared.

Keywords : Maize, genetic transformation, tissue culture, *Agrobacterium tumefaciens,* biolistic gun, selectable markers

---

*Corresponding author : E-mail : kanwang@iastate.edu

1. INTRODUCTION

The first attempt at maize transformation, may be even the first attempt at plant transformation, was reported some 35 years ago (Coe and Sarkar, 1966). In the experiment, total maize seedling DNA (purple, red-anthered maize plants, homozygous *B, A, Y Pl,* Yg_2 *C Sh Wx,* R^r) was extracted and directly injected into apical meristems of developing maize seedlings (homozygous *b, A, y Pl,* yg_2 *c sh wx,* R^s). It was anticipated that certain characteristics such as green leaf sectors (*Yg*), purple sheath, husk, culm, glume sectors (*B*) or purple anthers (R^r) contributed by the donor plant would be observed in recipient plants, if the transformation were successful. Out of 242 treated plants, no phenotypic effects indicative of transformation were observed. The authors concluded that "Competence, nucleases, penetration, susceptibility of loci, numbers treated, and degradation are indicated as problems that may need to be considered and solved in order to achieve transformation in higher plants."

At the turn of this century, genetic transformation of maize – a cereal crop of major commercial importance and once considered one of the recalcitrant monocotyledonous (monocot) plants – has become a routine procedure for several genotypes in most private and a number of public research laboratories. Most of the problems pointed out by Coe and Sarkar have been addressed. It is generally recognized that three major factors have contributed to this success in maize transformation: competent target cells, an effective delivery method and a reliable selection system.

During the mid '80s, *Agrobacterium*-mediated transformation – a biological genetic transformation system – was considered as an effective delivery means for dicotyledonous (dicot) plants only (De Cleene, 1985). Among the numerous experiments using various physical delivery systems, the particle bombardment procedure (the Biolistic gun) proved to be most effective in producing fertile transgenic maize plants (Gordon-Kamm *et al.*, 1990).

Although a few laboratories were transforming maize using *Agrobacterium* with some success (Graves and Goldman, 1986; Gould *et al*, 1991), the real potential for this delivery system was not recognized until a high efficiency *Agrobacterium*-mediated maize transformation system was demonstrated (Ishida *et al.*, 1996).

Today, numerous monocots have been infected or transformed with

Agrobacterium (for review, see Komari and Kubo, 1999). For maize, the *Agrobacterium*-mediated system is reported to result in both a higher transformation efficiency and a higher percentage of single or low copy number transgenes in the maize genome when compared to the biolistic gun-mediated procedure (Ishida *et al.*, 1996; Zhao *et al.*, 1998).

In this review, we will focus on the two most efficient maize transformation systems currently in use – the Biolistic gun and *Agrobacterium.* We will also review the three major factors critical to a successful transformation, namely, target cells, delivery methods and selection systems. Their merits and limitations will be analyzed and compared.

2. TARGET TISSUE

To qualify as an appropriate target tissue/cell for transformation, a plant cell must be competent for both transformation and regeneration. Unlike tobacco plants, in which most cells possess totipotency, only certain cells in maize can be regenerated (Armstrong, 1994). In addition, explants used for tissue culture should be readily available. Competent target tissue obtained from starting sources such as mature seed, leaf or roots are generally more desirable than materials such as immature seed, flowers or pollen depending on plant species, because the latter require a large area of high quality greenhouse space and intensive care to generate an adequate supply of healthy material for transformation experiments. This requirement can be a major limitation for public laboratories. Furthermore, regenerated plants should be fertile and free of somaclonal variation. Since it is known that prolonged *in vitro* culture decreases plant fertility and induces chromosomal rearrangement (Lee and Phillips, 1987), the ideal target tissues would be those that do not require an *in vitro* culture period.

2.1. BMS and protoplast culture

The Black Mexican Sweet (BMS) maize suspension culture line played an important role in the history of maize transformation. BMS callus lines (ATCC 54022) formed finely aggregated suspension cultures and yielded protoplasts capable of reforming their cell wall and dividing (Sheridan, 1982). Although these cultures were not able to regenerate to plants, ease of maintenance and protoplast production made it particularly useful in facilitating the optimization of various transformation parameters.

Both transient gene expression and stable transformants were first obtained when BMS protoplasts were transformed using electroporation (Fromm *et al.*, 1985, 1986; Huang and Dennis, 1989) or PEG-mediated DNA uptake (Armstrong *et al.*, 1990). Intact BMS cells were used to optimize the particle delivery of DNA (Klein *et al.*, 1988a, 1988b) and silicon carbide whisker-mediated transformation (Kaeppler *et al.*, 1990, 1992).

There were several reasons why protoplasts were attractive target cells for transformation. Firstly, removal of the cell wall eliminated a major barrier to DNA delivery. Secondly, protoplast cultures were analogous to bacterial culture systems with the advantages inherent in culturing large numbers of single cells (> 10^6) in a defined medium (Negrutiu *et al.*, 1987). As such, it was possible to define precise conditions for efficient DNA uptake. Thirdly, procedures for direct gene transfer involving polyethylene glycol (PEG) treatment or electroporation were well established for many species and had been used with a range of cereals, including maize. In addition, there was speculation that the cell wall removing process in protoplast preparation might trigger a "wound response" in plant cells, thus increasing cell competence for transformation and regeneration (Potrykus, 1990). And finally, multicellular structures resulting from protoplast culture are generally of single-cell origin (Thompson *et al.*, 1986), so the generation of chimeras after transformation could be avoided.

It should be noted that while protoplasts isolated directly from maize plant tissue such as leaf or root are, as with other cereal species, extremely recalcitrant in culture, they are of value in transient expression experiments where information on promoter function or gene regulation may be obtained in a short time period (Maas *et al.*, 1988; Kovtun *et al.*, 2000).

The first transgenic maize plant generated from protoplasts (also the first transformed maize plant generated by any method) was reported in 1988 (Rhodes *et al.*). Protoplasts were isolated from embryogenic suspension cultures of the maize inbred line A188. The suspension culture was of immature embryo origin, electroporation was the delivery method, and kanamycin was the selective agent. In addition, BMS cultures were used as a feeder layer to stimulate protoplast division. Of the thirty-eight plants regenerated from ten different transformed lines, none set any seed.

Shortly after this first report, success in regeneration of fertile plants (albeit non-transgenic) from protoplast cultures of a tropical maize inbred (Prioli and Söndahl, 1989) and an elite maize inbred line (Shillito *et al.*, 1989) was demonstrated.

The only reports of fertile transgenic maize plants generated from protoplast were presented by one research group (Golovkin *et al.*, 1993; Omirulleh *et al.*, 1993). Protoplasts were isolated from a maize embryogenic suspension culture of genotype HE/89, a line developed by selection for favorable tissue culture response (Morocz, 1990). Relatively low frequencies of developmental abnormalities were observed in the plants regenerated from this line and more than 60% of transformed plants (produced using PEG-mediated transformation and phosphinothricin as a selective agent) set seed. This work suggested that young cultures from a tissue culture-adapted genotype contributed to the unprecedented high regeneration frequency and fertility.

2.2. Immature zygotic embryo (IZE) and IZE-derived embryogenic culture

The fact that DNA could be delivered into plant cells by physical means and expressed in intact cells effectively (Klein *et al.*, 1988c, 1989) greatly encouraged researchers to seek alternative target tissues to protoplasts for maize transformation. Cultures of the maize inbred line A188 were well known for their ability to regenerate to plants at any age (Green and Phillips, 1975). By 1988, the biolistic parameters optimized for BMS cells were being used to demonstrate reasonably efficient delivery and transient expression of the β-glucuronidase gene in scutellar cells of immature A188 embryos (Klein *et al.*, 1988b). However, no stable transformants were recovered from bombarded embryos in these initial experiments.

The target tissue used to produce the first fertile transgenic maize plants was a rapidly-growing embryogenic suspension culture (Gordon-Kamm *et al.*, 1990; Fromm *et al.*, 1990). Suspension cultures were initiated from independent Type II callus pieces (Armstrong and Green, 1985) derived from immature embryos excised from the hybrid A188 × B73 or B73xA188. Subsequently, transformants were also obtained from a Type II callus culture from the same hybrid line (Walters *et al.*, 1992).

Following these initial successes, the production of fertile transgenic maize plants using the biolistic gun was adapted to progressively more differentiated target tissues such as Type I callus (Wan *et al.*, 1995) and precultured immature embryos (Koziel *et al.*, 1993; Songstad *et al.*, 1996).

Meanwhile, fertile transgenic maize plants were produced by electroporating Type I callus and embryos (D'Halluin *et al.*, 1992) and by using silicon carbide whiskers to transform embryogenic suspension cultures (Frame *et al.*, 1994).

2.3. The advantage and disadvantage of targeting type I and type II callus cultures

Type II callus or suspension cultures have been considered particularly amenable to transformation due to their friability, lack of differentiation, and rapid cell growth, all of which facilitate effective *in vitro* selection and minimize the chance of escapes (Fromm, 1994). However, the major disadvantage of Type II callus culture in maize transformation is that it is not readily induced in most inbred germplasms.

Type I callus, on the other hand, can be induced to develop from immature zygotic embryos of several maize inbred genotypes (Brettschneider *et al.*, 1997, Wan *et al.*, 1995), thereby facilitating the transformation of some elite germplasms using this target tissue. The disadvantage of using Type I callus for transformation, however, is its slow growth and compact growth habit. This, along with its high degree of differentiation, could potentially lead to escapes if more intensive subculturing is not imposed to overcome these limitations (Pareddy *et al.*, 1997).

2.4. Meristemic tissues

In addition to somaclonal variation and infertility resulting from long-term *in vitro* culture, one major disadvantage in callus culture based maize transformation is the narrow range of responding genotypes. Transgenic maize plants have been reported from only a handful of amenable genotypes: A188 or A188 × B73 derivatives (Armstrong, 1994; Ishida *et al.*, 1996), H99, Pa91, and Fr61 (Brettschneider *et al.*, 1997; D'Halluin *et al.*, 1992; Wan *et al.*, 1995), and a small number of elite inbred lines (Koziel *et al.*, 1993; Aves *et al.*, 1992). This short list of genotypes has led scientists to examine strategies that avoid tissue culture.

One such approach is to target cells in an organized meristem for transformation. These cells would then produce a shoot without undergoing callus formation. If the transformed meristematic cells divide to produce lineages including reproductive tissue (germline), it would be possible to obtain transgenic gametes and hence fully transgenic progeny. As well as avoiding somaclonal variation, this approach would be less time consuming and possibly less affected by variation in culturability of different genotypes.

Attempts at maize apical meristem tissue transformation were reported as early as 1966. One of the first attempts at maize transformation was the injection of genomic DNA from dominantly-marked maize plants into the apical meristems of recessive seedlings (Coe and Sarkar, 1966). Over 26,000 progeny were screened, but no evidence of transformation was obtained.

In a later study, Graves and Goldman (1986) reported that maize seedlings could be infected by the wild type virulent strains of *Agrobacterium tumefaciens*. The infected tissues formed tumors, which were subsequently able to grow on hormone-free medium and contained opines. The technique involved the inoculation of *Agrobacterium* culture into an area of rapidly dividing cells of wounded corn seedling (Goldman and Graves, 1993). Experiments using *Agrobacterium* strains carrying selectable marker genes encoding hygromycin phosphotransferase (*hpt*) or 5-enol-pyruvylshikimate 3-phospate synthase (EPSPS) were conducted. Seedlings were assayed for respective enzymatic activities as well as for resistance to hygromycin or glyphosate by either incubating infected plant tissue in medium containing hygromycin or by spraying the herbicide glyphosate on whole plants. Although these authors demonstrated the detection of some enzymatic activity (EPSP synthase) in infected plants, other analyses such as Southern blot hybridization or progeny segregation was not presented.

Meristem tissue from mature embryos or germinating seedlings was targeted using either the biolistic-mediated (Cao *et al.*, 1990) or *Agrobacterium*-mediated transformation method (Gould *et al.*, 1991). In the work of Cao *et al.* (1990), embryos from mature seed were dissected and bombarded with a construct carrying the screenable marker gene *gus*. Plantlets were induced from the embryo sections after being cultured on MS medium containing 1 μg/ml of NAA. No selection was employed in the process.

In Gould's report (1991), maize shoot apices were isolated from germinating seedlings and infected by an *Agrobacterium* strain carrying the *gus* gene cassette and the *npt*II gene cassette as selectable marker. The inoculated shoot apices were cultured on MS-based medium containing 0.1 mg/L kinetin for a short period and then moved to hormone-free medium. Kanamycin was used for selection. Shoots rooted spontaneously on this medium within 1-3 weeks after isolation of the apex.

Both papers presented molecular evidence that the construct DNA was indeed delivered into the targeted tissue and detected in plantlets generated from the transformation. However, the reports failed to provide evidence that transgenes introduced into plant cells by these methods were transmitted to progeny in the next generation.

In the work reported by Lowe *et al.* (1995), the authors demonstrated successful transformation by bombardment of embryonic axes of immature maize embryos at the coleoptilar stage (Abbe and Stein, 1954). These embryos developed into chimeric plants with transgenic sectors containing the *npt*II and the *gus* gene at a high frequency. They could not obtain transformed progeny from these sectors if no further manipulation was applied. To enlarge transgenic sectors for enhancing the probability of germline transmission, the authors proliferated apical meristems from chimeric plants on medium containing a cytokinin (2 mg/L BA). This cytokinin-induced shoot multiplication following bombardment of immature embryos has generated transgenic progeny in a sweet corn hybrid and an elite field corn inbred (Lowe *et al.*, 1995).

In the system reported by Zhong *et al.* (1996), explants containing a shoot tip, leaf primodia, and a portion of young leaves and stem were excised from 7-d-old maize seedling and placed in a meristem multiplication medium (containing casein hydrolysate and BA, Zhong *et al.*, 1992) to produce a meristem culture. This multiple shoot culture was bombarded with DNA carrying the *bar* gene and the *gus* gene. Bombarded shoot clumps underwent further selection on shoot multiplication medium containing bialophos. Transgenic shoots from 12 genotypes were recovered.

In both cases, meristems were transformed by microprojectile bombardment to produce chimeric plants with transgenic sectors, and transmission to progeny was demonstrated from some of these sectors. Lowe *et al.* (1995) used immature embryo as starting material, first bombarded, then induced shoot proliferation, while Zhong *et al.* (1996) first induced shoot multiplication from mature embryos, then bombarded and selected. Tissue culture and regeneration using the meristem system was less genotype-dependent when compared to the embryogenesis system using immature embryos or callus cultures (Lowe *et al.*, 1995; Zhong *et al.*, 1992, 1996). However, stable transformation efficiency from the system was low, and multiple shoot proliferation could be laborious for many genotypes. To date, the fundamental limitation of the meristem transformation system has been the generation of chimeric plants and their lack of transgene transmission to subsequent generations.

2.5. Germ line tissue

2.5.1. Male gametophyte

Considerable effort has been focused on producing transgenic plants through the transformation of germ line tissues. The attraction of this approach is quite clear. Firstly, no tissue culture is involved, thus it is fast, regeneration-independent, and somaclonal variation free. Moreover, each seed produced from a successful pollination using transformed pollen will be an independent event. Therefore, it can potentially generate a large number of independent events in one experiment. Finally, in maize, pollen is abundant and readily available, making it an attractive target material for transformation.

Initial reports of corn pollen transformation used pollen mixed with a DNA solution or paste to produce seed (De Wet *et al.*, 1985; Ohta *et al.*, 1986). In these experiments, total genomic DNA extracted from a donor corn plant that displayed a phenotype distinct from the recipient plant was used. While the authors observed seed phenotypes characteristic of the donor plant, no molecular biology analyses were performed, and the nature of the phenotypic changes in the seed was not addressed.

Using a similar approach, Sanford *et al.* (1985) found no evidence for pollen-mediated transformation in maize. From over eighty pollinations in which 800 seeds were screened, no apparent transformants displayed the expected phenotypes. The authors concluded that under their conditions the frequency of transformation, if any, was under 0.02%.

Using reporter genes encoding β-glucuronidase (GUS) or chloramphenicol acetyltransferase (CAT), a number of laboratories demonstrated the delivery and expression of these reporter genes in maize pollen or microspores via DNA uptake (Hess, 1987; Arnizen *et al.*, 1991; Fennell and Hauptmann, 1992), biolistic gun (Klein *et al.*,

1988a) or electroporation (Fennell and Hauptmann, 1992; Saunders and Matthews, 1997).

In the pollen-mediated transformation method disclosed by Arnizen *et al.* (1991), a plasmid DNA carrying a selectable marker cassette (35S promoter and *npt* II gene) was introduced into maize pollen. The DNA was first suspended at a concentration of 10 μg/ml in a delivery solution consisting of 5-10% sucrose, 50 μg/ml boric acid and 10 μg/ml calcium nitrate. In some experiments, polyethylene glycol (PEG) 4000, light paraffin oil and proteinase K were included in the solution. This DNA solution was sprayed onto the silks. Mature pollen, previously collected, was immediately placed onto the stigmatic surface and allowed to germinate and grow through the DNA solution. Approximately 130,000 seeds obtained following pollen transformation were germinated and screened for kanamycin resistance. Among the 165 kanamycin resistant plants identified, 152 plants were further analyzed by slot blot and Southern hybridization. In the end, DNA samples from five of the tested plants showed hybridization to bands matching the sizes of the introduced construct. No progeny segregation data was reported.

Horikawa *et al.* (1997) described a procedure in which maize pollen was bombarded with DNA (the *bar* gene) coated magnetite particles (Fe_3O_4). The pollen selected magnetically after bombardment was used to pollinate the silks. Out of 570 seedlings screened for herbicide resistance, fourteen appeared to be resistant to herbicide spray. PCR analysis of these resistant plants indicated that three were positive for the *bar* gene. No Southern hybridization or progeny analysis data were presented.

An alternative approach, used by Saunders and Matthews (1997), involved the transformation of germinating pollen by electroporation followed by pollination. Although the authors presented convincing evidence for successful delivery of plasmid DNA into germinating pollen, the routine application of this method to stable maize transformation remains to be demonstrated.

One of the major challenges in working with maize pollen is that it is sensitive to *in vitro* manipulation. The viability of maize pollen declines drastically over time. Ohta (1986) reported that if as little as 5 minutes elapsed between the time of preparing the DNA/pollen mixture and placing it on the silks, seed set was severely reduced and no variant kernels were obtained.

It is interesting to note that despite the considerable effort dedicated to research in pollen transformation, reports of success are, to date, limited. Even more frustrating is that many of the techniques reported could not be reproduced in other laboratories. Although the general assumption is that DNA introduced into the pollen grain will automatically be integrated into the germ line upon pollination, it is important to note that mature pollen grains are made up of sperm cells and vegetative cells. Only DNA integrated into sperm cells (germ line cells) will be transmitted to the next generation (Bedinger and Russell, 1994).

2.6. Example of establishment of inbred line tissue culture

In our own laboratory, biolistic transformation of Hi II immature zygotic embryos and Type II callus is efficiently and routinely implemented (Frame *et al.*, 2000). While this hybrid genotype exhibits an excellent tissue culture response (Armstrong *et al.*, 1991), transgenic maize plants generated from it are agronomically inferior and often must undergo several generations of back crossing to eliminate the undesirable traits inherited from Hi II. In addition, the ability to assess the expression of genes in different genetic backgrounds provides a more robust and reliable assessment of gene constructs and their interactions with genetic elements known to vary in the maize genome. As such, we have developed protocols for transforming several inbred maize genotypes. Although these protocols are not as efficient as our Hi II system, they are reproducible.

In the following examples, we outline tissue culture and biolistic transformation protocols for the three maize inbred lines B73, H99 and Oh43 [seed obtained from P. Schnable and A. Hallauer (B73), M. Lee (H99) and P. Scott (Oh43) at Iowa State University]. These examples also illustrate bombardment procedures for three common target tissues, namely, immature zygotic embryos and Type I (H99 and Oh43) or Type II (B73) callus derived from these inbred lines.

In all experiments, the constructs used in bombardments were either plasmid pAHC25 (kindly provided by P. Quail, Christensen and Quail, 1996) containing a selectable marker cassettee (the *bar* gene driven by the ubiquitin promoter) and a visual marker gene (the *gus* gene driven by the ubiquitin promoter), or the pAHC25 derivative containing only the ubiquitin-*bar* selectable cassette (Frame *et al.*, 2000). All bombardments were carried out using 0.6 micron gold particles. Osmotic medium used to pre- and post- treat target tissue at bombardment contained 0.2M mannitol and 0.2M sorbitol after Vain *et al* (1993b). Tissue cultures were maintained at 28°C in the dark and wrapped with parafilm unless stated otherwise. Regeneration of putative transgenic events for all three inbred lines was carried out at 25°C on MS based medium (Murashige and Skoog, 1962) as described by Armstrong and Green (1985) and McCain *et al.* (1988). Finally, putative transgenic events of H99 and Oh43 plants were deemed transgenic only if R_0 plants regenerated from them survived a glufosinate spray test in the green house (Brettschneider *et al.*, 1997).

2.6.1. B73

2.6.1.1. Tissue culture : B73 plants were grown to maturity in the greenhouse and ears harvested 10 - 12 days post-pollination. Immature zygotic embryos (~1.5 mm) were aseptically dissected (axis side down) to N6-based medium (Chu *et al.*, 1975) containing 10 mM silver nitrate, 1 mg/l 2,4-D, 100 mg/L casein hydrolysate and 25 mM proline as described by Songstad *et al.* (1991). Plates were wrapped with vent tape. Two weeks later, the number of zygotic embryos producing Type II callus (stocked somatic embryos emerging from scutellum) was assessed. On average, 12% of B73 embryos produced Type II callus, with response by ear ranging between 5 and 38%. On this medium, zygotic embryos either produced Type II callus or didn't respond at all. No Type I callus formation was observed. While sustained callus development on zygotic embryos was not always observed, this did not hinder our efforts to establish several Type II, fast-growing and highly regenerable B73 callus lines for bombardment.

In separate experiments with H99 (described below), we had observed a favorable effect of mannitol on Type I callus formation from immature zygotic embryos. We therefore amended the medium described above to contain 0, 18 or 37 g/L mannitol, with the aim of increasing the callus response rate of cultured B73 embryos. Field-derived B73 immature zygotic embryos were dissected to the three media and the percent response (number of embryos forming Type II callus over the number of embryos plated) averaged over two field seasons. On the non-mannitol standard medium, average response was 7.0%. Embryos responded at 22% when cultured on medium containing 18 g/L mannitol, and at 19% when the medium was supplemented with 37 g/L mannitol. While the response trend was similar in the two field seasons described, the overall response rates were three-fold higher in 1998 than 1997.

When dicamba (15 μM) was compared with 2,4-D (4.5 μM) as the auxin source in the 18 g/L mannitol medium, we observed a reduction in the number of responding B73 embryos on dicamba-containing medium and sustained Type II callus development only on the 2,4-D based medium.

2.6.1.2. ***Transformation :*** In spite of the improved Type II callus response from embryos plated on medium containing mannitol, the average rate of B73 immature zygotic embryo response remained too low to warrant targeting this explant for biolistic transformation. Instead, we bombarded B73 Type II callus lines, each derived from a single immature zygotic embryo, as follows. Callus lines were subcultured weekly on N6-based medium without mannitol as described (Songstad *et al.*, 1991) and were bombarded 5-7 days post subculture. Target plates were prepared 4 hours prior to bombardment by plating 30, 2-3 mm callus pieces to the center of a plate of N6 osmotic medium. Bombardment parameters were 650 psi, ¼″ gap, and a 6 cm target distance. A 150 μ mesh (Gordon-Kamm *et al.*, 1990; Wilson *et al.*, 1995) was used in all bombardments. Three milligrams of DNA-coated gold were diluted to 160 μl with 100% ethanol before loading 10 μl per macro-carrier. Callus was transferred to maintenance medium 1 hour after bombardment and selection was initiated 10 days later on 1 mg/L bialaphos.

In our initial experiments, selection was carried out on non-proline containing medium following our standard practice for Hi II, Type II callus bombardments. B73 Type II callus on this medium underwent considerable browning and grew very poorly after bombardment. The rate of clone recovery, based on the number of bialaphos resistant clones recovered over the number of pieces of callus bombarded, was 0.7% (3 putative events from 440 bombarded callus pieces). Inclusion of 25mM proline in selection medium favored B73 callus growth but also necessitated an increase from 1 to 2 mg/L bialaphos at first selection to control excessive growth of the bombarded callus pieces. Percent transformation efficiency from these later experiments averaged 7.8% (16 events from 204 bombarded pieces), and has ranged between 2.2% to 43% based on individual callus line. Average seed set on green house grown transgenic B73 plants was 32 kernels, ranging from 0 to 88 kernels per plant.

2.6.2. H99

2.6.2.1. ***Tissue culture :*** Immature zygotic embryos were dissected from three field grown ears of H99 and plated to two media after

D'Halluin *et al.* (1992) containing either 37.2 g/L mannitol (medium C) or not (medium C-1). In this study, the rate of callus formation for embryos on medium C was 67% and on medium C-1 was 0%. The callus produced on C medium was a characteristically vigorous, regenerable mixture of Type I and late Type II callus. Response rates of H99 green house grown immature zygotic embryos on C medium were close to 100%, and unlike B73, the variability in % response from ear to ear was not pronounced. While we have also observed 85–100% Type I callus response rates of H99 embryos dissected to MS based medium after Green and Phillips (1975) and dicamba-based N6 media after Carvalho *et al.* (1997), we have routinely used medium C for H99 tissue culture/transformation experiments.

2.6.2.2. Transformation : Because the frequency of Type I callus response from greenhouse derived H99 embryos was consistently high (~100%), we targeted immature zygotic embryos for biolistic transformation. Greenhouse grown immature zygotic embryos (1.2-1.8 mm) were aseptically dissected in a 3.5 cm grid to Whatman No. 5 filter paper (axis side down) on C medium. Plates were wrapped with vent tape and incubated for 3-5 days prior to bombardment. Embryos on filter paper were transferred to osmotic medium for 4 hours prior to and 16 hours after bombardment. Intact embryo explants were transferred to selection medium containing 3 mg/L bialaphos 10-13 days after bombardment. Two to three weeks later, selection was enhanced to 5 mg/L bialaphos, at which time callus forming from the scutellum of single embryos was broken into small pieces and spread out on the selection plate. Pieces from the same embryo were circled on the plate-bottom so their single explant lineage could be followed. After two weeks on 5 mg/L bialaphos, bright white lobes of Type I somatic embryos were visible and only these sectors were taken to a third selection on 5 mg/L bialaphos.

Our highest transformation efficiency was achieved using 1,100 psi rupture disks, a 6 cm target distance and a large gap distance (1.0 inch) as described by Brettschneider *et al.* (1997). In agreement with these authors, we observed a higher rate of callus formation on embryos bombarded with diluted gold in two replicated experiments. Using our standard protocol (diluting 3 mg of DNA-coated gold in 120 μl of 100% ethanol before loading macros), the average % embryo response was 62%, while diluting the gold 3-fold (to 360 μl) prior to loading the macros resulted in a 90% response rate for bombarded embryos. The average number of blue foci (transient GUS assay) per embryo for these two treatments was 57 and 32, respectively. The average response rate of non-bombarded embryos in these experiments was 97%. In spite of the improved response rate observed using diluted gold, the stable clone recovery rate in these experiments was slightly higher using non-diluted gold (5.7%) than using diluted gold (3.3%).

In 11 consecutive experiments in which H99 green house immature zygotic embryos were bombarded, we recovered putative clones at efficiencies ranging from 1.5 to 13% for an average of 4.0% (44 events from 1108 bombarded embryos). Seed set on transgenic plants averaged 39 kernels per plant and ranged from 0 to 120 kernels.

2.6.3. Oh43

2.6.3.1. Tissue culture : Oh43 tissue culture and transformation work in our laboratory is based on N6 medium containing the auxin dicamba, as described by Carvalho *et al.* (1997). In preliminary field trials we compared the effect of 15 and 30 μM dicamba on the number of immature zygotic embryos forming Type I callus. On 30 μM dicamba medium (D30), 100% of plated embryos produced Type I callus, compared to 67% on 15 μM dicamba medium (D15). On both media, rapidly growing, tightly lobed, highly regenerable Type I callus was formed. Callus initiated on D30 was easily maintained on D15.

The Type I callus response of Oh43 greenhouse embryos was compared on D15 medium in which dicamba (15 μM) was substituted with 2,4-D (4.5 μM). While the initial rate of embryo response was 100% on both media, the response on 2,4-D was far less vigorous and eventually unsustainable compared to that on D15.

2.6.3.2. Transformation : While we have successfully targeted both immature zygotic embryos and Type I callus of Oh43 for biolistic transformation in our laboratory, here we describe only the Type I callus protocol. Immature zygotic embryos from green house grown Oh43 plants (~1.5 mm) were aseptically dissected (axis side down) to D15 medium. Plates were wrapped with vent tape and incubated in the dark for 2 weeks. Type I callus produced from the scutellar base of these embryos was serially subcultured every three weeks for 2 months prior to bombardment. Four hours prior to bombardment, target plates were prepared by subculturing ~ 30 healthy Type I callus pieces (2-3 mm) to osmotic medium in a 3.5 cm grid. Bombardment parameters were: 650 psi, ¼ inch gap, and a 6 cm target distance. A 150 μ mesh was placed between the launch assembly and the target tissue in all bombardments and 3 mg of DNA-coated gold was diluted to 160 µl with 100% ethanol prior to loading the macros (10 µl/macro). Callus pieces were transferred off osmotic medium to D15 one hour after bombardment, and selection was initiated on 2 mg/l bialaphos 10-13 days later. Two weeks later, selection was enhanced to 5 mg/L bialaphos at which time the original bombarded callus pieces were broken into small pieces on the surface of the selection medium. As with H99, bright white lobes of Type I somatic embryos visible on second selection were transferred to a third selection on 5 mg/L bialaphos. The average transformation efficiency achieved using this Type I callus transformation protocol was 3.2% (20 events/632 pieces bb). Average seed set per transgenic plant was 35 kernels and ranged from 0 to 87 kernels.

2.6.4. General observations

In implementing these tissue culture protocols, we have observed a particularly strong ear effect in B73. While the tissue culture response of H99 and Oh43 embryos was relatively consistent in that 100% of embryos from all greenhouse ears would produce callus on designated medium, the embryo response in the B73 system was more ear dependent. Several non-responding ears of B73 have been observed. Even in cases where all embryos do respond to produce callus before or after bombardment, as is the case for Hi II immature zygotic embryos, this so-called "ear" effect can influence transformation efficiency. For example, when 11 Hi II ears were bombarded using the same biolistic gun parameters and selection protocol in our laboratory, efficiencies of transformation achieved by ear ranged from 0 – 23% (B. Frame, unpublished).

While proline was not included during the selection scheme described here for H99 and Oh43, or, for that matter, in our standard selection scheme for bombarded Type II callus or immature zygotic embryo of Hi II, its inclusion during B73 Type II callus selection greatly improved the transformation efficiency for this genotype.

All three inbreds were regenerated as described by Armstrong and Green (1985) and McCain *et al.* (1988). Somatic embryos (Type I or Type II) were matured on MS-based medium containing 6% sucrose, no hormones, and 3 or 5 mg/L bialaphos, and germinated in the light on MS-based medium containing 3% sucrose. While ease of regeneration from these inbred lines was comparable, R_0 plant phenotype was not. Oh43 R_0 transgenic plants were the least vigorous of the three inbreds described here, exhibiting a very thin stock at the plant base, causing plants to lodge in their pots. Excessively husked ears were common on Oh43 transgenic plants such that donor pollen had to be poured into the ear to reach the non-emerged silks. Fertility of transgenic plants generated from the three inbred lines was nevertheless comparable.

The choice of target tissue used for bombardment of these three genotypes was decided to some extent by the nature of their response on any given tissue culture medium. In our experiments, the low response rate of B73 immature zygotic embryos precluded them being used as a primary target tissue in bombardment. Because we rarely saw any good quality Type I callus induction on B73 embryos, we targeted the Type II callus lines that developed at low frequency, but vigorously, from the culture system described. In our experience, Type II callus selection is less tedious than Type I callus selection in which bombarded explants had to be broken down on subsequent selections to ensure that effective selection pressure was imposed.

Finally, because immature zygotic embryos of both H99 and Oh43 produced an excellent Type I callus response on their designated medium, we could choose to bombard either the embryo or Type I callus explant derived from it. While the embryo is preferred because it requires less pre-transformation tissue culture manipulation thereby minimizing somaclonal variation and labor, transformation can be achieved using both of these target tissues.

3. DELIVERY SYSTEMS

One of the principle problems identified by Coe and Sarkar (1966) in their pioneering attempt at maize transformation was that "The cell wall, a massive barrier to large structures, may have to be disrupted mechanically or chemically, or otherwise circumvented". During the past two decades, numerous methods have been developed to deliver genetic material into plant cells. Here we will review four delivery systems that have successfully generated fertile transgenic maize plants beginning with the biolistic gun in 1990 and ending with the most recent development using the *Agrobacterium*-mediated method.

3.1. The biolistic gun

The first fertile transgenic maize plants were reported in 1990 (Gordon-Kamm *et al.*; Fromm *et al.*). Both research groups used high-velocity microprojectile bombardment as the gene delivery system. In the mid-to-late 1980's, there was no efficient way to deliver biological material into intact plant cells. Existing methods included DNA uptake, PEG fusion, electroporation, microinjection or *Agrobacterium*-mediated transformation. Uptake, fusion, or electroporation mechanisms allowed treatment of cells en masse, however, they required removing the plant cell wall. The major obstacle in these methods was to obtain a regenerable maize culture from transformed protoplasts (Fromm *et al.*, 1986; Rhodes *et al.*, 1988). Even when the transformed protoplasts were successfully regenerated, the plants were sterile (Rhodes *et al.*, 1988). Microinjection, on the other hand, introduced DNA into intact cells, but required that one target cells individually or in small aggregates and was therefore relatively inefficient (Crossway *et al.*, 1986).

At the time, the *Agrobacterium*-mediated transformation technique that was working efficiently for dicotyledonous plants was believed to be inapplicable to monocots. As a result, a novel mechanism for delivering biological substances into "walled" living cells received considerable attention. The greatest single advantage of this "so-called" particle bombardment method was that it could introduce biological material directly into any cell type. In addition, large numbers of cells could be bombarded simultaneously, thereby avoiding the single cell manipulation inherent in microinjection techniques (Sanford *et al.*, 1987).

It is no exaggeration to say that direct DNA transfer methods based on particle bombardment have revolutionized plant genetic engineering. Ten years after the first fertile maize plants were reported, more than 20% of the corn acreage in the United States is now planted to transgenic maize hybrids with important agronomical traits such as insect resistance, disease tolerance or herbicide tolerance [(National Corn Growers Association (http://www.ncga.com/biotechnology/main/index.html)]. Globally, Bt corn was the second most dominant crop to be grown commercially in 2000. It occupied 6.8 million hectares, equivalent to 15% of the global GM crop area [International Service for the Acquisition of Agri-biotech Applications (http://www.isaaa.org/)]. In addition, maize transformation technology is becoming one of the important enabling tools for academic researchers to better understand the genetics and biology of this important crop plant.

3.1.1. Type of particle gun

At least three different types of particle bombardment devices have been reported to date. They are classified based on the driving forces and the types of macrocarriers they employ. The BioRad PDS-1000/He biolistic gun (Bio-Rad Laboratories, Hercules, CA) is a gas (helium) pressure-driven device using plastic film as a macrocarrier (Klein *et al.*, 1987). This is the most widely used acceleration system for many plants. The first fertile transgenic corn was produced using this device (Gordon-Kamm *et al.*, 1990; Fromm *et al.*, 1990).

ACCELL™ is an electrical discharge particle bombardment device using a metalized sheet as a macrocarrier (McCabe and Christou, 1993). Since the device can accelerate DNA coated gold beads to any desired velocity by varying the input voltage, it can control the particle penetration into cells more precisely when compared to PDS-1000/He system. This device has allowed transformation of deeper cell layers thereby reaching tissues competent for both transformation and regeneration that were not accessible using alternative bombardment procedures (McCabe and Christou, 1993).

The Particle Inflow Gun (PIG) is a gas-stream driven device using no macrocarriers (Finer *et al.*, 1992). The PIG accelerates microprojectiles with a gentle burst of gas. As a result, this gene gun may produce less of a shock wave than that produced when the macrocarrier impacts the stopping screen in the PDS-1000/He. Bombardment with a PIG (Vain *et al.*, 1993a) was less injurious to some genotypes than the PDS-1000/He allowing regeneration and transformation of the more recalcitrant varieties. However, it is not clear if the PIG has comparable velocities to the PDS-1000/He. It therefore may not be appropriate for stable transformation in systems where dividing and regenerable cells reside below the surface of the targeted tissue (McCabe and Christou, 1993).

All three biolistic guns can be used to deliver DNA into plant cells effectively. In this review, we focus on BioRad PDS1000/He biolistic gun that has been widely used in maize transformation.

3.1.2. Improvement of transformation efficiency using the biolistic gun

Efforts to improve the efficiency of maize transformation using the biolistic gun have focused, in part, on altering biolistic parameters (Brettschneider *et al.*, 1997; Kausch *et al.*, 1995; Klein *et al.*, 1988; Sanford *et al.*, 1993) osmotic treatment of target tissue (Vain *et al.*, 1993b) and gold particle size (Frame *et al.*, 2000).

3.1.2.1. Optimization of bombardment parameters : Brettschneider *et al.* (1997) observed similar levels of transient *gus* expression when helium pressures of 900 or 1200 psi were used to bombard H99 immature zygotic embryos, but achieved a higher rate of stable events from bombardments using the 1,200 psi treatment. These latter results support observations by Dunder *et al.* (1995) who recommended accelerating microprojectiles into maize embryos that give rise to Type I callus at higher velocity than embryos that produce Type II callus. This is probably because Type I callus originates from internal meristems (Vasil *et al.*, 1985). Experiments in our laboratory, support this argument. The Hi II system produces a vigorous Type II callus response from scutellar tissues of immature zygotic embryos, and we have achieved higher transformation efficiencies from 650 psi (11%) than 1,100 psi (3.3%) bombardments in experiments where this comparison has been made (B. Frame, unpublished results).

3.1.2.2. Effect of gold particle size on transformation efficiency : The effect of reducing gold particle size from 1.0 μ to 0.6 μ on the transformation efficiency of Type II callus cultures of Hi II was reported by Frame *et al*, 2000. Transformation efficiency was reported as the number of independent, bialaphos resistant calluses recovered after 10 weeks of selection on 2 mg/L bialaphos per 100 callus pieces bombarded. Results from 5 experiments in which the same callus lines were compared across particle size showed a 14% average transformation efficiency for the 0.6 μ particle bombardment treatment compared with 1.0% for the 1 μ gold particle treatment. These results agreed with those for Type I callus reported by Randolph-Anderson *et al.* (1997). The beneficial effect of the smaller particle size on transformation efficiency was attributed to less sustained damage being incurred by callus cells bombarded with 0.6 μ as compared to 1.0 μ gold particles.

3.1.2.3. Effect of osmotic treatment on transformation efficiency : Sorbitol and mannitol osmotic treatments were shown to increase stable transformation rates in maize cell suspensions (Vain *et al.*, 1993b). As well, 12% sucrose was used to favor the recovery of stable events from bombarded immature zygotic embryos of maize (Dunder *et al.*, 1995). The osmotic pre-treatment was believed to plasmolyze the cells prior to impact, thereby reducing the likelihood of cells bursting and death. Post bombardment osmotic treatments were thought to favor a gradual recovery from bombardment induced cell injury thereby increasing the chances of recovering stable events from targeted cells. In Type II callus bombardments, 0.2 M sorbitol/0.2 M mannitol and 12% sucrose osmotic pretreatments resulted in similar, increased rates of stable clone recovery compared to the non-treated control (Frame *et al.*, 2000).

3.2. Other physical delivery systems

A number of free DNA delivery systems for generating fertile maize plants were reported during the mid-1990s (D'Halluin *et al.*, 1992; Golovkin *et al.*, 1993; Omirulleh *et al.*, 1993; Frame *et al.*, 1994; Laursen *et al.*, 1994; Pescitelli and Sukhapinda, 1995). Most of this research was conducted in private laboratories. The development of alternative transformation technologies other than the biolistic delivery system was largely motivated by the need to by-pass the legal hurdles in the gene gun patent for commercial production of transgenic corn and to circumvent the difficulties in protoplast regeneration. Two systems that generated fertile transgenic plants are discussed here.

3.2.1. Electroporation

The first stably transformed maize cells produced by electroporation were reported as early as 1986 (Fromm *et al.*), in which protoplast cultures from the non-regenerable maize cell line, BMS, were used. Subsequently, using this delivery system, Rhodes *et al.* (1988) reported the regeneration of transgenic maize plants. Protoplast cultures were isolated from an 18-month old embryogenic cell suspension culture of maize inbred A188. The authors were the first to report convincing data of stably transformed maize plants, although the R_0 transgenic plants were not fertile.

To avoid the use of protoplasts, D'Halluin *et al* (1992) developed an electroporation based transformation method for DNA delivery into maize immature zygotic embryos and into Type I callus cultures of the maize inbred lines H99 or Pa91. In this method, embryos or Type I callus cultures were treated for 1-3 minutes with a 0.3% solution of macerozyme, an enzyme used to degrade pectic substances, prior to electroporation. Without the enzyme treatment, no transient gene expression (*npt* II gene) could be detected. However, NPT II activity could be detected reproducibly 4-6 days after electroporation if target tissues were pretreated with macerozyme. The authors reasoned that DNA delivery via electroporation required wounding of the tissue. Because this procedure targeted immature embryos and Type I callus, some maize genotypes previously considered inaccessible for electroporation-mediated transformation (due to their lack of Type II callus response), could now be transformed.

Production of fertile transgenic maize was also reported using electroporation of suspension culture cells (Laursen *et al.*, 1994) and Type II callus cultures (Pescitelli and Sukhapinda, 1995). Laursen *et al* (1994) pretreated target tissues (suspension cultures from a hybrid line A188 × B73 and a B73-related inbred line) with pectin-degrading enzyme (0.5% Pectolyase Y-23) while Pescitelli and Sukhapinda (1995) combined mechanical wounding and plasmolysis of Type II callus cultures of the hybrid lines B73 × A188 and High II prior to electroporation.

Although transgenic lines generated via electroporation have transgene integrity and copy numbers comparable to the biolistic method (Gordon-Kamm *et al.*, 1990; D'Halluin *et al.*, 1992), the method's relatively low efficiency may account for the fact that few current publications refer to its use.

3.2.2. Whiskers

Whisker transformation is technically very simple and does not require expensive equipment or consumables. Whiskers, cells and plasmid DNA are combined in, for example, an Eppendorf tube and mixed on a vortex. Silicon carbide whiskers have great intrinsic hardness and fracture readily to give sharp cutting edges (Greenwood and Earnshaw, 1984). They are obtained by the thermal reduction of silica in a reducing atmosphere, one source of silica being rice husks (Mutsuddy, 1990). Industrially, silicon carbide whiskers are used as abrasives, in the manufacture of cutting tools, and in the production of composite materials.

When whiskers (typically with an average diameter of less than 1 μm) are mixed with plant cells and plasmid DNA, cell penetration appears to occur (Kaeppler *et al.*, 1990). While the exact mechanism for whiskers transformation is unclear, it seems likely that the whiskers function as numerous needles, facilitating DNA entry into cells during the mixing process (Coffee and Dunwell, 1994). In their study, a 250 μl packed cell volume of a BMS maize suspension culture was added to a premixed DNA/whisker suspension and vortexed for 60 seconds. Transient *gus* activity was reproducibly observed in all samples of BMS culture two days after transformation. The same group later reported the recovery of stably transformed callus clones using this technique (Kaeppler *et al.*, 1992).

The first fertile, transgenic maize generated by the whiskers method was reported in 1994 (Frame *et al.*). Transformation was carried out on regenerable suspension cultures of the hybrid A188 × B73. Success in producing fertile transgenic maize plants using this approach was attributed, in part, to the use of sorbitol:mannotol osmotic treatments pre- and post-whisker transformation that enhanced DNA delivery and the subsequent recovery of transformed tissue. A similar beneficial effect of such treatments had been observed using the biolistic method (Russell *et al.*, 1992a, 1992b; Vain *et al.*, 1993b).

The use of silicon carbide whiskers to transform maize is now a routine and high throughput method in some industry research laboratories (Petolino *et al.*, 2000; Bullock *et al.*, 2001). Recent whisker clone production involves a scaled-up collision system using a commercial paint shaker that creates collisions within very large batches of cultured maize suspension cells and silicon carbide whiskers (Bullock *et al.*, 2001). About 47% of transformation events were reported to have single or low copy number insertions of the transgene.

One of the undesirable aspects of the whiskers-mediated transformation method has been its dependence on embryogenic suspension cultures as a source of target cells. Genotype limitation and fertility loss are two commonly cited technical problems associated with suspension cultures (Phillips *et al.*, 1988). Recently, it was reported that fertile transgenic maize plants have been produced using non-suspension cultures as a target tissue for the whisker method (Petolino *et al.*, 2000). Type II callus cultures were established from a B73 × A188 hybrid cross. Stable transformants and fertile plants were generated from the friable embryogenic callus culture mixed with whiskers and plasmid DNA carrying the *gus* and *bar* gene cassettes. Although the efficiency reported was low (3 Basta-resistant clones recovered from 100 whiskers-treated samples), this technique represents a reproducible method for the generation of fertile transgenic maize.

3.3. *Agrobacterium*-mediated transformation of maize

3.3.1. Early work

Routine, efficient *Agrobacterium*-mediated transformation of dicotyledonous plants was first reported in 1985 (Horsch *et al.*). Because monocotyledonous plants are not natural hosts for *Agrobacterium tumefaciens*, the development of transformation systems using this vector for monocots lagged behind that of dicots. To date, the most widely used transformation method of the important cereal monocots, including maize, has been microprojectile bombardment (for review, see Armstrong, 1999). Despite the development of routine direct DNA delivery techniques for maize, several disadvantages are associated with their use. One of the main drawbacks of using direct DNA delivery techniques is that transgenes are inserted as multiple and/or rearranged copies in a high proportion of the transformants, leading to unstable transgene expression or transgene silencing (for review, see Fagard and Vaucheret, 2000).

In 1986, Grimsley *et al* were the first to demonstrate that *Agrobacterium tumefaciens* was capable of infecting maize. cDNA of maize streak virus was delivered to maize plants by *Agrobacterium* and the plants became systemically infected. Subsequent attempts in *Agrobacterium*-mediated monocot transformation resulted in a number of encouraging publications. These include observations of *Agrobacterium* attachment to the surfaces of maize cells (Graves *et al.*, 1988), identification of *vir*-inducing substances by monocotyledons including maize (Primich-Zachwieja and Minocha, 1991), and evidence of expression of the *gus* gene in maize tissues infected with *Agrobacterium* (Gould *et al.*, 1991; Ritchie *et al.*, 1993; Schläppi and Horn, 1992; Shen *et al.*, 1993).

3.3.2. The super binary vector system

The first well-documented report of fertile transgenic maize plants produced using *Agrobacterium* was published by a research group at Japan Tobacco Inc (Ishida *et al.*, 1996). Large numbers of transgenic plants were obtained from immature embryos of the maize inbred line A188 infected with an *Agrobacterium* strain carrying a "super-binary" vector system. The *bar* gene, in conjunction with phosphinothricin (PPT), were used as a selection system. Transformation efficiency ranged from 5% to 30% with 70% fertility of the regenerants. The researchers provided the following evidence for transformation. Firstly, the intron-*gus* gene, which is not expressed in *Agrobacterium* cells, was expressed strongly in infected maize immature embryos. Secondly, more than 30 lines were analyzed by Southern hybridization. The size of the DNA fragment containing junctions between the T-DNA and plant DNA varied in different transformation events, indicating the random integration of the T-DNA into the maize genome. Analysis of the T-DNA boundaries by DNA sequencing revealed that the junctions were located in or near the 25-bp border repeats similar to those in transgenic dicotyledons and rice. Furthermore, inheritance of the transgenes was followed to the R_2 generation with demonstration of the Mendelian segregation of transgenes.

Success in these experiments was attributed to use of the super binary vector system. The *Agrobacterium* strain (LBA4404) used carries additional copies of *vir*B, *vir*C and *vir*G (Komari and Kubo, 1999). The two most notable outcomes resulting from the use of this vector system were the high transformation efficiency achieved (up to 30%) and the high percentage of independent events carrying a single or low copy number of the transgenes. Little rearrangement was observed in these events. The quality and the quantity of transgenic maize generated by this method have not been reported for any other delivery systems for monocotyledons.

Using the super-binary vector system, Zhao *et al.* (1998) reported similar success transforming immature embryos of the hybrid line, Hi II. Under their conditions, the embryo-based transformation frequency could be as high as 32% to 50%. In addition, when comparing transformation events derived from *Agrobacterium* with those from bombardment, the authors found that out of 133 bombardment events, 8% were low copy number with simple integration. In contrast, out of 107 *Agrobacterium* events analyzed, as high as 58% were low copy number with simple integration. The rate of "desirable" events (low copy/simple insertion) recovered from the *Agrobacterium* transformation method was 62%, compared to 1% from the biolistic method.

The most recent report of *Agrobacterium*-mediated maize transformation used phosphomannose-isomerase (*pmi*) gene and mannose as a selection system (Negrotto *et al.*, 2000). Using the super binary vector system and immature embryos of A188, the authors successfully obtained transgenic maize using a non-antibiotic or non-herbicide resistant gene as a selectable marker.

3.3.3. Factors involved in *Agrobacterium*-mediated maize transformation

Although high frequency *Agrobacterium*-mediated transformation has been reported in the studies described above, these frequencies have not been reproduced in public maize transformation laboratories to date. Factors contributing to the lack of reproducibility in the public sector could include: 1) lack of critical details in protocol and media descriptions in published reports, 2) lack of access to specialized binary vectors by public researchers, 3) intellectual property issues barring information or protocol transfer to the public sector by private industry and 4) lack of resources and critical mass in transformation research.

The significant advantages of using an *Agrobacterium*-based transformation system for maize (high frequency transformation, low copy numbers, simple transgene insertion, increased stability of transgene expression, low cost relative to biolistics, and potential for introducing large DNA fragments into the plant genome), make it imperative that optimized protocols be developed, published and implemented by maize researchers in the public sector.

In their recent review paper, Komari and Kubo (1999) summarized a number of key factors involved in *Agrobacterium*-mediated cereal transformation. They are 1) target tissue for infection, 2) vector system and *Agrobacterium* strains, 3) infection and co-cultivation conditions, 4) selection scheme, and 5) plant genotype.

3.3.4. Example of *Agrobacterium*-mediated maize transformation with the super binary vector

Following the protocol described by Zhao *et al.* (1999), we have achieved routine *Agrobacterium*-mediated maize transformation of Hi II immature zygotic embryos using the super binary vector system in strain LBA4404, as described by Ishida *et al.* (1996). In the experiments described below, we used the basic mother vector/strain system that was kindly provided by Dr. Komari of Japan Tobacco Inc. However, in our construct pTF3, the selectable marker was the ubiquitin-*bar* cassette and the screenable marker was the ubiquitin-*gus* cassette from pAHC25 (Christensen and Quail, 1996). A brief outline of the protocol (after Zhao *et al.*, 1999) follows.

F_1 plants of the hybrid line Hi II were grown to maturity in the green house and selfed or crossed to their sister plants. Nine to 12 days after pollination, ears were harvested and stored for up to 2 days at 4°C. Ears were surfaced sterilized in 50% bleach with Tween 20 (1 drop/liter) and washed three times with sterile water after which embryos (~1.5 mm) were dissected to an eppendorf tube (2 ml) containing plant infection medium (Inf) supplemented with 100 μM acetosyringone (AS). Embryos were washed twice in this medium, the final wash removed, and 1 ml of *Agrobacterium* suspension was added to the washed embryos for infection.

To prepare the bacterial suspension, *Agrobacterium* was grown for

2 days at 28°C on solid *Agrobacterium* medium (YEP) amended with approporiate antibiotics before being suspended to an OD_{550} of 0.72 in plant infection medium (Inf) supplemented with 100 μM AS. This Agrobacterial suspension was diluted 1:1 with Inf+AS and 1 ml added to the immature zygotic embryos as described above.

At infection, the contents of the tube were gently rocked several times, then incubated on the bench top for 5 minutes. After infection, embryos were transferred to the surface of freshly prepared co-cultivation medium. Excess *Agrobacterium* suspension was pipetted off the medium surface and away from the embryos that were then oriented axis side down for co-cultivation. Plates were wrapped with vent tape and embryos were co-cultivated at 23°C for 3 days after which they were transferred to resting medium (28°C) containing 250 mg/L cefotaxime.

Selection on 1.5 mg/L bialaphos was initiated 3-5 days later, and increased to 3 mg/L after two weeks. Counter selection of 250 mg/L cefotaxime was maintained throughout. Putative clones could be identified as early as 4 weeks after infection, and were regenerated on MS medium (Armstrong and Green, 1985; McCain *et al.*, 1988) containing 3 mg/L bialaphos and 250 mg/L cefotaxime. From 13 independent experiments using this protocol we have recovered 75 bialaphos resistant clones, for an overall efficiency of 4.3%.

Of the 15 events crossed to date in the greenhouse, all have set seed (transgenic as female), for an average per plant seed set of 73 kernels. R_1 progeny from 5 of these events were screened for segregation of *bar* gene expression by spraying 9-12 days old germinating plantlets with a 250 mg/L glufosinate solution (Brettschneider *et al.*, 1997). In four of these events, the ratio of glufosinate resistant vs glufosinate susceptible plants was 1:1, suggesting that the *bar* gene was successfully transmitted through the female zygote in a Mendelian fashion. Copy numbers of the *bar* gene in nine events analyzed using Southern blotting ranged from 1 to 5, in agreement with Zhao *et al.* (1998).

4. MARKERS

Marker genes, including selectable markers and screenable markers, are critical tools for the genetic manipulation of microorganisms, mammalian cultures, and plant tissues. Typically, a marker gene is co-transformed with a gene of interest into a recipient cell, followed by expression of the delivered genes. Transformed events can either be assessed by assays for expression of a reporter gene or identified by expression of a selectable gene that confers cell viability to transformed cells in the presence of a selective agent.

A recent review (Båga *et al.*, 1999) comprehensively outlined most selectable and screenable marker genes used for cereal crops. In maize, successful transformation has been achieved using reporter genes such as *cat* (chloramphenicol acetyl transferase gene, Fromm *et al.*, 1985), *gus* (β-glucuronidase gene, Klein *et al.*, 1988a; Gordon-Kamm *et al.*, 1990; Frame *et al.*, 1994; Ishida *et al.*, 1996), *C1* and *R* (the anthocyamin genes, Ludwig *et al.*, 1990; Lusardi *et al.*, 1994), *luc* (the firefly luciferase gene, Callis *et al.*, 1987; Fromm *et al.*, 1990), and *gfp* (the green fluorescent protein gene, Sheen *et al.*, 1995; Reichel *et al.*, 1996). The ideal reporter genes should code for products that can be detected directly or catalyze specific reactions whose products are detectable.

Selectable marker genes play an important role in genetic transformation. Since the incorporation of genes into plant chromosomes during the transformation process occurs at an extremely low frequency, it is essential that the gene of interest is introduced into a plant cell together with a selectable marker gene so that only the plant cells that contain and express the selectable marker genes will survive and grow on culture media containing the selective agent. Many selectable marker genes used for dicotyledonous plants have been tested for maize transformation. These include antibiotic resistant markers such as *npt II* (neomycin phosphotransferase for kanamycin resistance, D'Halluin *et al.*, 1992; Lowe *et al.*, 1995; Rhode *et al.*, 1988), *hpt* (hygromycin phosphotransferase for hygromycin B resistance, Walters *et al.*, 1992), *dhfr* (dihydrofolate reductase for methotrexate resistance, Golovkin *et al.*, 1993), and *aadA* (streptomycin resistance, Gordon-Kamm *et al.*, 1999; Lowe *et al.*, 1997). In general, antibiotic resistant marker genes that have been used successfully in dicot plants are not effective for the transformation of monocot plants. They either have an inherently high level of natural resistance in monocots, or exhibit phytotoxicity during plant regeneration (Dekeyser *et al.*, 1989; Zhang *et al.*, 1988).

Another important category of selectable marker genes is that of herbicide resistant genes. In fact, the use of the herbicide resistant gene *bar* (phosphinothricin acetyltransferase for bialaphos resistance) was one of the critical factors that contributed to the successful production of the fertile, transgenic maize (Gordon-Kamm *et al.*, 1990). Other herbicide resistant genes such as EPSPS (5-enolpyruvylshikimate-3-

phosphate synthase for glyphostate resistance, for Roundup Ready Corn, see http://www.biotechbasics.com/product_information/roundup_corn_-general_info.html) and ALS (acetolactate synthase for sulfonylureas and imidazolinones resistance, Fromm *et al.*, 1990) have also been used in maize.

From among all selectable marker genes, the *bar* and *pat* genes that confer resistance to bialaphos have been most widely used in maize transformation. In the following section, we will discuss our observations in utilizing this selection system for the maize transformation.

Recently, a positive selection system has been successfully used in maize transformation (Evans *et al.*, 1996; Negrotto *et al.*, 2000). In this system, mannose, a chemical of non-antibiotic and non-herbicidal nature, was used as a selectable agent and the phosphomannose isomerase gene (*pmi* from *E. coli*) as the selectable marker. PMI converts mannose-6-phosphate to fructose-6-phosphate. The transformed plant cells that are capable of utilizing mannose as a carbon source acquire a growth advantage (positive selection) on mannose-containing media.

4.1. The *bar* and *pat* genes

Phosphinothrincin or PPT is an analogue of glutamate and competitively inhibits glutamine synthetase (GS), an enzyme that plays a central role in the assimilation of ammonia and in the regulation of nitrogen metabolism in plants (Skokut *et al.*, 1978). It is the only enzyme in plants that can detoxify ammonia released by nitrate reduction, amino acid degradation and photorespiration. The inhibition of glutamine synthetase by PPT results in the accumulation of ammonia in cells. It is this accumulation, rather than a lack of glutamine, which causes the death of plant cells (Wild *et al.*, 1987; Sauer *et al.*, 1987; Givan *et al.*, 1988).

$$\text{Glutamate} + NH_4^+ + \text{ATP} \xrightarrow[\text{Glutamine Synthetase}]{} \text{Glutamine} + \text{ADP} + \text{Pi} + H^+$$

Two PPT derivatives have been successfully used as selective agents for maize transformation. Bialaphos, a natural tri-peptide compound consisting of PPT and two alanine residues (Murakami *et al.*, 1986; Thomspon *et al.*, 1987), is an antibiotic produced by *Streptomyces hygroscopicus* as an extracellular product (for review, see Thompson and Seto, 1995). This compound, the active ingredient of herbicide Herbiace® (Meiji Seika Ltd, Japan), is produced by fermentation of *S.*

hygroscopicus. Glufosinate, on the other hand, a synthetic commercial formulation that is the ammonium salt of PPT, is an active ingredient of the herbicide BASTA® (Aventis, France).

The *bar* (bialaphos resistance) gene was isolated from the bialaphos biosynthetic cluster of *Streptomyces hygroscopicus* and was so-named because it confers resistance to the herbicide BASTA® (Murakami *et al.*, 1986; Thompson *et al.*, 1987). The *pat* (phosphinothricin acetyltransferase) gene was isolated from *Streptomyces viridichromogenes* and was so-named because the resistance mechanism was found to be acetylation of PPT (Strauch *et al.*, 1988).

The sequence of the bacterial *bar* and *pat* genes has been compared (Wohlleben *et al.*, 1988). Both genes code for proteins of 183 amino acids that show an overall identity of 85%. From the 28 differing amino acids, 15 were identified as non-conservative replacements. Differences in the amino acid sequences are more frequent in the C- and N- terminal regions, indicating that the central part of the protein may carry the functional domain. The translation start codon for both bacterial *bar* and *pat* genes was GTG instead of ATG (Thompson *et al.*, 1987; Wohlleben *et al.*, 1988). The overall identity of the two genes at the nucleotide level was 87%. The G+C content of the two bacterial genes is similar; around 93% of codons have a G or C in the third position (Wohlleben *et al.*, 1988). The products of these two genes were purified and compared for structural and functional differences (Wehrmann *et al.*, 1996). The similarity of the enzymes suggested that both genes could be used equally for plant transformation.

Most successful reports of maize transformation to date have used the *bar* gene as the selectable marker (Fromm *et al.*, 1990; Gordon-Kamm *et al.*, 1990; Spencer *et al.*, 1990; Koziel *et al.*, 1993; Frame *et al.*, 1994). In these reports, the GTG start codon of the bacterial *bar* gene was converted to ATG. The fact that the *bar* gene has been much more widely used (and reported) in the transformation of both cereal and dicot species may be attributed to the greater accessibility of the *bar* gene and its widespread dissemination among academic groups. It is assumed that there are no significant differences in the versions of the *bar* gene which are in use.

4.2. Comparison of bialaphos and glufosinate as selective agents for maize transformation

Both bialaphos and glufosinate have demonstrated their effectiveness in selecting transformants, although bialaphos is somewhat better at controlling non-transgenic callus growth than glufosinate, especially when L-proline is added to the selection medium (Dennehey *et al.*, 1994). Our standard protocol has included bialaphos (Shinyo Sangyo Co., Ltd., 19-11, Hiroo 5 –Chome, Shibuya-ku, Tokyo 150, Japan) as the selection agent in our maize transgenic production pipeline (Frame *et al.*, 2000). Using bialaphos, we have avoided producing escapes. However, bialaphos is not readily available in North America and is more expensive than glufosinate. To investigate the feasibility of glufosinate [Sigma, glufosinate ammonia (pestanal), 45520R] as the selection agent for a large production pipeline, we conducted a study comparing glufosinate with bialaphos. The selection medium in these experiments did not contain proline.

Immature zygotic embryos were bombarded 3 days post-dissection as described (Frame *et al.*, 2000). Embryos were then taken to selection 10-14 days after bombardment. At selection, one half of the callus pieces were transferred intact to callus-inducing growth medium supplemented with 2 mg/l glufosinate (G2) and the other half were transferred to medium containing 2 mg/l bialaphos (B2). Putative clones were picked to 2 mg/l bialaphos.

The results of this comparison are shown in Table 1. Although the total number of herbicide resistant calli recovered from both bialaphos and glufosinate media did not differ significantly ($P = 0.28$), the number of false positive events identified was significantly higher in the calli from the glufosinate selection ($P = 3.32 \times 10^{-6}$). A false positive event is one that appeared to be real (based on rate of callus growth) after two selection subcultures but, when transferred to its own plate (2 mg/L bialaphos) did not continue to proliferate and eventually died. Those putative events were deemed "false" and discarded.

We repeated the experiment using 4 mg/l glufosinate (G4) compared with 2 mg/L bialaphos (B2) and found equivalent numbers of putative transgenic events from both treatments ($P = 0.98$). Similarly, we observed more false positive events from G4 than B2 plates (Table 2). However, the increased concentration appeared to reduce the incidence of "false picks" ($P = 0.02$).

In our experiments, glufosinate used at 2 mg/L or 4 mg/L proved less than ideal for selection because growth continued much longer than on B2 medium. Callus pieces were also drier and more friable on glufosinate medium than on bialaphos medium and would often break apart at the

Table 1. A comparison of glufosinate (2 mg/L) and bialaphos (2 mg/L) on the recovery of herbicide resistant events from bombarded embryos.

Exp#	No. embryos bombarded	On glufosinate (2 mg/L)		No. embryos bombarded	On bialaphos (2 mg/L)	
		% positive events	% false positive events		% positive events events	% false positive events
1	105	10	10.5	105	14	3.8
2	101	14	18.8	105	9	1.0
3	150	13	10.0	150	9	0.7
4	130	10	8.5	131	9	0.0
5	105	8	15.2	105	9	0.0
6	100	16	19.0	100	14	0.0
7	217	4	3.7	202	6	1.5
8	223	11	4.0	238	8	2.1
Total	1131			1136		
Average		10.8	11.2		9.8	1.1

P - value for difference = 0.28 (percentage of positive events)
P - value for difference = 3.32×10^{-6} (percentage of false positive events)

slightest jarring of the plates. Many of the glufosinate plates had to be sub-divided a second time to avoid cross contamination of events. Finally, we had to transfer the callus pieces from glufosinate to bialaphos (our standard selection agent) to avoid losing individual events. The continued callus growth on glufosinate produced ambiguous results when picking putative transgenic events. Many more "false picks" were reported from the glufosinate than the bialaphos-containing medium, although ultimately, similar numbers of resistant callus pieces were produced from both selective agents (Tables 1 and 2).

Our observations agree with those reported by Dennehey *et al* (1994). In their studies bialaphos was a more effective selective agent for corn than glufosinate. In the absence of proline, about three-fold higher levels of glufosinate than bialaphos were needed to inhibit the growth of embryogenic cells. In addition, L-proline interfered significantly with glufosinate selection, but not with bialaphos selection.

These studies showed that both bialaphos and glufosinate could be used to successfully select transgenic events from bombarded embryos. Because glufosinate selection is not as "tight" as bialaphos, it should either be used at no less than 4 mg/L in selection medium and may not, therefore, be as cost effective as previously thought. Finally, it may be more effective to initiate selection sooner than 10-14 days after bombardment when using glufosinate as a selection agent.

5. CONCLUSIONS AND PERSPECTIVES

Maize is not only a model plant system for fundamental studies but also one of the most important crop species (Freeling and Walbot, 1994). It is grown in more countries than any other crop and is a major source of food for both humans and animals throughout the world.

The discovery of the unique ability of cross-kingdom DNA transfer of *Agrobacterium tumefaciens* and its application in plant genetic engineering, has forever changed our ways of conducting scientific research as well as agricultural practice. The development of maize transformation technology during the past decade has made the plant more valuable as a genetic model system as well as a crop.

Since the first report of successful transformation of maize in 1990 (Gordon-Kamm *et al.*), the biolistic transformation of corn has become a routine procedure in most industry laboratories. For most public research groups, however, this technology remains non-straightforward

Table 2. A comparison of glufosinate (4 mg/L) and bialaphos (2 mg/L) on the recovery of herbicide resistant events from bombarded embryos.

Exp#	No. embryos bombarded	On glufosinate (4 mg/L)		No. embryos bombarded	On bialaphos (2 mg/L)	
		% positive events	% false positive events		% positive events events	% false positive events
9	150	5	11.3	150	7	0.0
10	150	5	2.0	150	8	0.7
11	165	7	4.8	165	2	0.0
12	300	5	3.3	300	6	0.7
Total	765			765		
Average		5.5	5.4		5.8	0.3

P - value for difference = 0.98 (percentage of positive events)
P - value for difference = 0.02 (percentage of false positive events)

because they do not have access to the physical or human resources necessary to establish an efficient transformation effort. In addition, serious limitations exist with the technology, including low throughput, difficulty in inbred line transformation, unpredictable transgene copy numbers, gene rearrangements and undesirable levels of transgene silencing through generations. In fact, the limitations associated with current public maize transformation technology were identified by the maize research community as one of the most serious bottlenecks in the advancement of maize research (43rd Annual Maize Genetic Conference, March 16-18, 2001).

To this end, in 1995, Iowa State University established one of the first public Plant Transformation Facilities (PTF) in the United States. Supported by the university, local farmers' associations, as well as private endowment funds, the facility provides expertise and services for genetic transformation of corn and soybeans for academic researchers. To date, PTF remains the only public laboratory that provides such services for maize.

In addition to providing genetic transformation for the public community, PTF conducts rigorous research programs to further improve and develop high throughput transformation technologies that will advance functional genomic research. PTF also teaches and trains graduate students and other scientific professionals in the area of transgenic research. This practice, *i.e.* having a plant transformation facility in the public sector, is proving to greatly enhance the research capability of university plant scientists. Moreover, the research effort carried out at the transformation facility is promoting a healthy and open interaction and scientific atmosphere among university researchers. The facility and its research program can be viewed in http://www.agron.iastate.edu/ptf/web/mainframe.htm.

Current advancements in molecular biology have enabled the large scale, rapid identification and isolation of plant genes. Determination of function for thousands of genes, and application of that knowledge to crop improvement, is now one of the major challenges facing plant biologists. Plant transformation is a key technology for functional analysis of genes via complementation, over-expression, or gene silencing strategies and will be equally essential in the application of genomic technology to crop improvement. Clearly, challenges and opportunities exist and the race in maize technology development is still intense.

In a recent review, Armstrong (1999) summarized a number of active research areas in the improvement of transformation technology. Breakthroughs or incremental developments can be expected from these areas in the next decade. Most of these areas, such as controlled transgene insertion and expression, plastid transformation, large DNA fragment transformation and gene stacking, genotype independent transformation and high throughput methods are issues of common interest to maize and other plant researchers and are currently being addressed. For example, what determines the tissue culture responsiveness of different maize genotypes to plant media? Can we better control transgene expression in desired organs, or tissues and at different developmental stages? How is plant gene expression regulated globally after transgene insertion? Just as plant transformation technology is an enabling tool for genomics research, so will progress in genomics greatly influence the technology development of transformation.

ACKNOWLEDGEMENTS

This work is supported by the Iowa Corn Promotion Board, the Agricultural Experiment Station, the Office of Biotechnology, the Plant Science Institute and the Baker Endowment Advisory Council for Excellence in Agronomy at Iowa State University.

REFERENCES

Abbe EC and Stein OL (1954). The growth of the shoot apex in maize : embryogeny. *Am. J. Bot.,* **41** : 285-293.

Armstrong CL (1994). Regeneration of plants from somatic cell cultures : Applications for in vitro genetic manipulation. *In* : (Eds. Freeling M and Walbot V) *The Maize Handbook,* Springer-Verlag New York, Inc. pp 663-670.

Armstrong CL (1999). The first decade of maize transformation: a review and future perspective. *Maydica,* **44** : 101-109.

Armstong CL and Green CE (1985). Establishment of friable, embryogenic maize callus and the involvement of L-proline. *Planta,* **164** : 207-214.

Armstrong CL, Green CE and Phillips RL (1991). Development and availability of germplasm with high Type II culture formation response. *Maize Genet. Coop. Newsletter,* **65** : 92-93.

Armstrong CL, Petersen WI, Buchholz WG, Bowen BA and Sulc SL (1990). Factors affecting PEG-mediated stable transformation of maize protoplasts. *Plant Cell Rep.,* **9** : 335-339.

Arnizen CL, DeBonte Jr LR, Evans DA, Loh WH and O'Dell JT (1991). Pollen-mediated gene transformation in plants. US patent # 5,049,500.

Aves K, Genovesi D, Willetts N, Zachwieja S, Mann M, Spencer T, Flick C and Gordon-Kamm W (1992). Transformation of an elite maize inbred through microprojectile bombardment of regenerable embryogenic callus. *In Vitro Cell Dev. Biol. Plant,* **28** : 74a.

Båga M, Chibbar R and Kartha K (1999). Expression and regulation of transgenes for selection of transformants and modification of traits in cereals. *In* : *Molecular Improvement of Cereal Crops* (Ed. Vasil I) Kluwer Academic Publishers, The Neherlands, pp 83-132.

Bedinger P and Russell SD (1994). Gametogenesis in maize. *In* : *The Maize Handbook* (Eds. Freeling M and Walbot V) Springer-Verlag, New York pp 48-60.

Brettschneider R, Becker D and Lorz H (1997). Efficient transformation of scutellar tissue of immature maize embryos. *Theor. Appl. Genet.,* **94** : 737-748.

Bullock P, Dias D, Bagnall S, Cook K, TeRonde S, Ritland J, Spielbauer D, Abbaraju R, Christensen J and Heideman N (2001). A high efficiency maize "whisker" transformation system. Abstracts of 43rd Annual Maize Genetic Conference: pp. 78.

Callis J, Fromm M and Walbot V (1987). Introns increase gene expression in cultured maize cells. *Genes Dev.,* **1** : 1183-1200.

Cao J, Wang YC, Klein TM, Sanford J and Wu R (1990). Transformation of rice and maize using the biolistic process. *In*: *Plant Gene Transfer* (Eds. Lamb, CJ and Beachy RN) UCLA Symp. Mol. Cell Biol., Vol. 129 Wiley-Liss Inc, New York pp. 21-33.

Carlvalho CHS, Bohorova N, Bordallo PN, Abreu LL, Valicente FH, Bressen W and Paiva E (1997). Type II callus production and plant regeneration in tropical maize genotypes. *Plant Cell Rep.,* **17** : 73-76.

Christensen AH and Quail PH (1996). Ubiquitin promoter-based vectors for high-level expression of selectable and/or screenable marker genes in monocotyledonous plants. *Transgenic Res.,* **5** : 213-218.

Chu CC, Wang CC, Sun CS, Hsu C, Yin KC, Chu CY and Bi FY (1975). Establishment of an efficient medium for anther culture of rice through comparative experiments on the nitrogen source. *Sci. Sin.,* **18** : 659-668.

Coe EH and Sarkar KR (1966). Preparation of nucleic acids and a genetic transformation attempt in maize. *Crop Sci.,* **6** : 432-435.

Coffee R and Dunwell J (1994) Transformation of plant cells. United State Patent #5, 302, 523.

Crossway A, Hauptli H, Houck CM, Irvine JM, Oakes JV and Perani LA (1986). Micromanipulation techniques in plant biotechnology. *BioTechniques,* **4** : 320-334.

De Cleene M (1985). The susceptibility of monocotyledons to *Agrobacterium tumefaciens*. *Phytopathologie Zeitschrift,* **113** : 81-89.

Dekeyser R, Claes B, Marichal M, Van Montagu M and Caplan A (1989). Evaluation of selectable markers for rice transformation. *Plant Physiol.,* **90** : 217-223.

Dennehey BK, Petersen WL, Ford-Santino C, Pajeau M and Armstrong C (1994). Comparison of selective agents for use with the selectable marker gene *bar* in maize transformation. *Plant Cell, Tiss. Organ Cult.,* **36** : 1-7.

De Wet JMJ, Bergquist RR, Harlan JF, Brink DE, Cohen CE, Newell CA and De Wet AE (1985). Expgenous gene transfer in maize (*Zea mays*) using DNA-treated pollen. *In*: *Experimental manipulation of ovule tissue* (Eds. Chapma, G.P., Mantell, S.H. and Daniels, R.W.) Longman, London pp. 197-209.

D'Halluin K, Bonne E, Bossut M, De Beuckeleer M and Leemans J (1992). Transgenic maize plants by tissue electroporation. *Plant Cell,* **4** : 1495-1505.

Dunder E, Dawson J, Suttie J and Page G (1995). Maize transformation by microprojectile bombardment of immature embryos. In: *Gene Transfer to Plants* (Eds. Potrykus I and Spangenberg G) Springer, Berlin pp. 127-138.

Evans R, Wang A, Hanten J, Alterdorf R and Metter I (1996). A positive selection system for maize transformation. *In Vitro Cell Dev. Biol. Plant,* **32** : 72a.

Fagard M and Vaucheret H (2000). (Trans)Gene silencing in plants: How many mechanisms? *Annu. Rev. Plant Physiol. Plant Mol. Biol.,* **51** : 167-194.

Fennell A and Hauptmann R (1992). Electroporation and PEG delivery of DNA into maize microspores. *Plant Cell Rep.,* **11** : 567-570.

Finer JJ, Vain P, Jones MW and McMullen MD (1992). Development of the particle inflow gun for DNA delivery to plant cells. *Plant Cell Rep.,* **11** : 323-328.

Frame BR, Drayton PR, Bagnall SV, Lewnau CJ, Bullock WP, Wilson HM, Dunwell JM, Thompson JA and Wang K (1994). Production of fertile transgenic maize plants by silicon carbide whisker-mediated transformation. *The Plant J.,* **6** : 941-948.

Frame B, Zhang H, Cocciolone S, Sidorenko L, Dietrich C, Pegg S, Zhen S, Schnable P and Wang K (2000). Production of transgenic maize from bombarded Type II callus: effect of gold particle size and callus morphology on transformation efficiency. *In Vitro Cell. Dev. Biol.- Plant,* **36** : 21-29.

Freeling M and Walbot V (1994). *The Maize Handbook.* Springer-Verlag New York, Inc.

Fromm ME (1994). Production of transgenic maize plants by microprojectile-mediated gene transfer. In: *The Maize Handbook* (Eds. Freeling, M. and Walbot, V.) Springer-Verlag, New York pp. 677-684.

Fromm ME, Taylor LP and Walbot V (1985). Expression of genes transferred into monocot and dicot plant cells by electroporation. *Proc. Natl. Acad. Sci. USA,* **82** : 5824-5828.

Fromm ME, Taylor LP and Walbot V (1986). Stable transformation of maize after gene transfer by electroporation. *Nature,* **319** : 791-793.

Fromm ME, Morrish F, Armstrong C, Williams R, Thomas J and Klein TM (1990). Inheritance and expression of chimeric genes in the progeny of transgenic maize plants. *Bio/Technology,* **8** : 833-839.

Givan CV, Joy KW and Kleczkowski LA (1988). A decade of photorespiratory nitrogen cycling. *TIBS,* **13** : 433-437.

Golovkin VM, Abraham M, Morocz S, Bottka S, Fefer A and Dudits D (1993). Production of transgenic maize plants by direct DNA uptake into embryogenic protoplasts. *Plant Sci.,* **90** : 41-52.

Goldman SL and Graves AC (1993). Process for transformating corn and the products thereof. US patent #5, 177, 010.

Gordon-Kamm WJ, Spencer TM, Mangano ML, Adams TR, Daines RJ, Start WG, O'Brien JV, Chambers SA, Whitney J, Adams R, Willetts NG, Rice TB, Mackey CJ, Krueger RW, Kausch AP, Lemaux PG (1990). Transformation of maize cells and regeneration of fertile transgenic plants. *Plant Cell,* **2** : 603-618.

Gordon-Kamm WJ, Baszczynski CL, Bruce WB and Tomes DT (1999). Transgenic cereals – *Zea mays* (maize). *In*: *Molecular Improvement of Cereal Crops* (Ed Vasil, I) Kluwer Academic Publishers, Dordrecht/Boston/London pp 43-82.

Gould J, Devey M, Hasegawa O, Ulian EC, Peterson G and Smith RH (1991). Transformation of *Zea mays* L. using *Agrobacterium tumefaciens* and the shoot apex. *Plant Physiol.,* **95** : 426-434.

Graves ACF and Goldman SL (1986). The transformation of *Zea mays* seedlings with *Agrobacterium tumefaciens*. *Plant Mol. Biol.,* **7** : 43-50.

Graves AE, Goldman SL, Banks SW and Graves AC (1988). Scanning electron microscope studies of *Agrobacterium tumefaciens* attachment to *Zea mays, Gladiolus* sp., and *Triticum aestivum*. *J. Bacteriol,* **170** : 2395-2400.

Green CE and Phillips RL (1975). Plant regeneration from tissue cultures of maize. *Crop Sci.,* **15** : 417-427.

Greenwood NN and Earnshaw A (1984). Silicon carbide. *In*: *Chemistry of the elements,* Pergamon Press, Oxford pp. 386-394.

Grimsley NH, Hohn B, Hohn T and Walden R (1986). "Agroinfection", an alternative route for viral infection of plants by using the Ti plasmid. *Proc. Natl. Acad. Sci. USA,* **83** : 3282-3286.

Hess D (1987). Pollen-based techniques in genetic manipulation. *Int. Review of Cytology,* **107** : 367-395.

Horikawa Y, Yoshizumi T and Kakuta H (1997). Transformation through pollination of mature maize (*Zea mays* L.) pollen delivered bar gene by particle gun. *Grassland Science,* **43** : 117-123.

Horsch RB, Fry JE, Hoffmann N, Eichholtz D, Rogers SG and Fraley RT (1985). A simple and general method for transferring genes into plants. *Science,* **227** : 1229-1231.

Huang Y and Dennis ES (1989). Factors influencing stable transformation of maize protoplasts by electroporation. *Plant Cell Tiss. Org. Cult.,* **18** : 281-296.

Ishida Y, Saito H, Ohta S, Hiei Y, Komari T and Kumashiro T (1996). High efficiency transformation of maize (*Zea mays* L.) mediated by *Agrobacterium tumefaciens*. *Nat. Biotechnol.,* **14** : 745-750.

Kaeppler HF, Gu W, Somers DA, Rines HW and Cockburn AF (1990). Silicon carbide fiber-mediated DNA delivery into plant cells. *Plant Cell Rep.,* **9** : 415-418.

Kaeppler HF, Somers DA, Rines HW and Cockburn AF (1992). Silicon carbide fiber-mediated stable transformation of plant cells. *Theor. Appl. Genet.,* **84** : 560-566.

Kausch AP, Adams TR, Mangano M, Zachwieja SJ, Gordon-Kamm W, Daines R, Willets NG, Chambers SA, Adams Jr W and Anderson A (1995). Effects of microprojectile bombardment on embryogenic suspension cell cultures pf maize (*Zea mays* L.) used for genetic transformation. *Planta,* **196** : 501-509.

Klein TM, Fromm M, Weissinger A, Tomes D, Schaaf S, Slettern M, and Sanford JC (1988a). Transfer of foreign genes into intact maize cells using high velocity microprojectiles. *Proc. Natl. Acad. Sci. USA,* **85** : 4305-4309.

Klein TM, Gradziel T, Fromm ME and Sanford JC (1988b). Factors influencing gene delivery into *Zea mays* cells by high-velocity microprojectiles. *Bio/Technology,* **6** : 559-563.

Klein TM, Harper EC, Svab Z, Sanford JC, Fromm ME and Maliga P (1988c). Stable genetic transformation of intact *Nicotiana* cells by particle bombardment process. *Proc. Natl. Acad. Sci. USA,* **85** : 8502-8505.

Klein TM, Kornstein L, Sanford JC and Fromm ME (1989a). Genetic transformation of maize cells by particle bombardment. *Plant Physiol.,* **91** : 440-444.

Komari T and Kubo T (1999). Methods of genetic transformation: *Agrobacterium tumefaciens*. *In*: *Molecular Improvement of Cereal Crops* (Ed. Vasil I) Kluwer Academic Publishers, Dordrecht/Boston/London pp 43-82.

Kovtun Y, Chiu WL, Tena G and Sheen J (2000). Functional analysis of oxidative stress-activated MAPK cascade in plants. *Proc. Natl. Acad. Sci. USA,* **97** : 2940-2945.

Koziel MG, Beland GL, Bowman C, Carozzi NB, Crenshaw R, Crossland L, Dawson J, Desai N, Hill M and Kadwell S, Launis K, Lewis K, Maddox D, McPherson K, Meghji MR, Merline E, Rhodes R, Warren GW, Wright M and Evola SV (1993). Field performance of elite transgenic maize plants expressing an insecticidal protein derived from Bacillus thuringiensis. *Bio/Technol,* **11** : 194-200.

Laursen CM, Krzyzek RA, Flick CE, Anderson PC and Spencer TM (1994). Production of fertile transgenic maize by electroporation of suspension culture cells. *Plant Mol. Biol.,* **24** : 51-61.

Lee M and Phillips RL (1987). Genomic rearrangements in maize induced by tissue culture. *Genome,* **29** : 122-128.

Lowe K, Bowen B, Hoerster G, Ross M, Bond D, Pierce D, Gordon-Kamm B (1995). Germline transformation of maize following manipulation of chimeric shoot meristems. *Bio/Technology,* **13** : 677-682.

Lowe K, Ross M, Sandahl G, Miller M, Hoerster G, Church L, Tagliani L, Bond D and Gordon-Kamm W (1997). Transformation of the maize apical meristem: Transgenic sector reorganization and germline transmission. *In* : *Genetics, Biotechnology and Breeding of Maize and Sorghum* (Ed Tsaftaris AS), Royal Society of Chemistry, Cambridge pp. 94-97.

Ludwig S, Bowen B, Beach L and Wessler S (1990). A regulatory gene as a novel visible marker for maize transformation. *Science,* **247** : 449-450.

Lusardi M, Neuhaus-Url G, Potrykus I and Neuhaus G (1994). An approach towards genetically engineered cell fate mapping in maize using the *Lc* gene as a visible marker: transactivation capacity of *Lc* vectors in differentiated maize cells and microinjection of *Lc* vectors into somatic embryos and shoot apical meristems. *Plant J.,* **5** : 571-582.

Maas C, Werr W and Starlinger P (1988). Analysis of the Zea mays sucrose synthase promoter by transient gene expression in maize protoplasts. *Curr. Plant Sci. Biotechnol. Agric.,* **7** : 377-378.

McCabe D and Christou P (1993). Direct DNA transfer using electric discharge particle acceleration (ACCELLÔ technology). *Plant Cell Tiss. Org. Cult.,* **33** : 227-236.

McCain JW, Kamo KK and Hodges TK (1988). Characterization of somatic embryo development and plant regeneration from friable maize callus cultures. *Bot. Gaz.,* **149** : 16-20.

Morocz S, Donn G, Nemeth J and Dudits D (1990). An improved system to obtain fertile regenerants via maize protoplasts isolated from a highly embryogenic suspension culture. *Theor. Appl. Genet.,* **80** : 721-726.

Murakami T, Anzai H, Satoh A, Nagaoka D and Thompson CJ (1986). The bioalaphos biosynthetic genes of *Streptomyces hygroscopicus*: molecular cloning and characterization of the gene cluster. *Mol. Gene. Genet.,* **205** : 42-50.

Murashige T and Skoog F (1962). A revised medium for rapid growth and bioassays with tobacco tissue cultures. *Physiol. Plant.,* **15** : 473-497.

Mutsuddy BC (1990). Electrokinetic behavior of aqueous silicon carbide whisker suspensions. *J. Am. Ceram. Soc.,* **9** : 2747-2749.

Negrutiu I, Shillito R, Potrykus I, Biasini G and Sala F (1987). Hybrid genes in the analysis of transformation conditions. *Plant Mol. Biol.,* **8** : 363-373.

Negrotto D, Jolley M, Beer S, Wenck AR and Hansen G (2000). The use of phosphomannose-isomerase as a selectable marker to recover transgenic maize plants (*Zea mays* L.) via *Agrobacterium* transformation. *Plant Cell Rep.,* **19** : 798-803.

Ohta Y (1986). High-efficiency genetic transformation of maize by a mixture of pollen and exogenous DNA. *Proc. Natl. Acad. Sci. USA,* **83** : 715-719.

Omirulleh S, Abraham M, Golovkin M, Stefanov I, Karabaev M, Mustardy L, Morocz S and Dudits D (1993). Activity of a chimeric promoter with the doubled CaMV 35S enhancer element in protoplast-derived cells and transgenic plants in maize. *Plant Mol. Biol.,* **21** : 415-428.

Pareddy D, Petolino J, Skokut T, Hopkins N, Miller M, Welter M, Smith K, Clayton D, Pescitelli S and Gould A (1997). Maize transformation via helium blasting. *Maydica,* **42** : 143-154.

Pescitelli SM and Sukhapinda K (1995). Stable transformation via electroporation into maize Type II callus and regeneration of fertile transgenic plants. *Plant Cell Rep.,* **14** : 712-716.

Petolino JF, Hopkins NL, Kosegi BD and Skokut M (2000). Whisker-mediated transformation of embryogenic callus of maize. *Mol. Gen. Genet.,* **207** : 245-250.

Potrykus I (1990). Gene transfer to plants: assessment and perspectives. *Physiol. Plant.,* **79** : 125-134.

Primich-Zachwieja S and Minocha SC (1991). Induction of virulence response in *Agrobacterium tumefaciens* by tissue explants of various plant species. *Plant Cell Rep.,* **10** : 545-549.

Prioli LM and Söndahl MR (1989). Plant regeneration and recovery of fertile plants from protocplasts of maize (*Zea mays* L.). *Bio/Technology,* **7** : 589-594.

Randolph-Anderson B, Boynton JE, Dawson J, Dunder E, Eskes R, Gillham NW, Johnson A, Perlman PS, Suttie J and Heiser WC (1997). Sub-micron gold particles are superior to larger particles for efficient BiolisticÒ transformation of organelles and some cell types. *Bio-Rad. Tech. Bull.,* 2015 (95-0699).

Reichel C, Mathur J, Eckes P, Langenkemper K, Koncz C, Schell J, Reiss B and Maas C (1996). Enhanced green fluorescence by the expression of an *Aequorea Victoria* green fluorescent protein mutant in mono- and dicotyledonous plants cells. *Proc. Nat. Acad. Sci. USA,* **93** : 5888-5893.

Rhodes CA, Pierce DA, Mettler IJ, Mascarenhas D and Detmer JJ (1988). Genetically transformed maize plants from protoplasts. *Science,* **240** : 204-207.

Ritchie SW, Lui CN, Sellmer JC, Kononowicz TK, Hodges TK and Gelvin SB (1993). *Agrobacterium tumefaciens*-mediated expression of *gus*A in maize tissue. *Transgenic Res.,* **2** : 252-265.

Russell JA, Roy MK and Sanford JC (1992a). Physical trauma and tungsten toxicity reduce the efficiency of biolistic transformation. *Plant Physiol.,* **98** : 1050-1056.

Russell JA, Roy MK and Sanford JC (1992b). Major improvements in biolistic transformation of suspension-cultured tobacco cells. *In Vitro Cell Dev. Biol. Plant,* **28** : 97-105.

Sanford JC, Skubik KA and Reisch BI (1985). Attempted pollen-mediated plant transformation employing genomic donor DNA. *Theor. Appl. Genet.,* **69** : 571-574.

Sanford JC, Klein TM, Wolf ED and Allen N (1987). Delivery of substrances into cells tissues using a particle bombardment process. *Particle Sci. Tech.,* **5** : 27-37.

Sanford JC, Smith FD and Rusell JA (1993). Optimizing the biolistic process for different biological applications. *Meth. Enzymol.,* **217** : 483-509.

Saunders J and Matthews B (1997). Plant transformation by gene transfer into pollen. US patent #5, 629, 183.

Sauer H, Wild A and Rühle W (1987). The effect of phosphinothricin (glufosinate) on photosynthesis. II. The causes of inhibition of photosynthesis. *Z. Naturforsch.,* **42c** : 270-278.

Schläppi M and Horn B (1992). Competence of immature maize embryos for *Agrobacterium*-mediated gene transfer. *Plant Cell,* **4** : 7-16.

Sheen J, Hwang S, Niwa Y, Kobayashi H and Galbraith DW (1995). Green fluorescent protwin as a new vital marker in plant cells. *Plant J.,* **8** : 777-784.

Shen WH, Escudero J, Schläppi M, Ramos C, Hohn B and Koukolíková-Nicola Z (1993). T-DNA transfer to maize cells: Histochemical investigation of b-glucuronidase activity in maize tissues. *Proc. Natl. Acad. Sci. USA,* **90** : 1488-1492.

Sheridan WF (1982). Black Mexican Sweet Corn: Its use for tissue cultures. *In: Maize for biological Research.* (Ed. Sheridan, W.F.) Charlottesville, Virginia: Plant Molecular Biology Association. Pp. 385-388.

Shillito RD, Carswell GK, Johnson CM, DiMaio JJ and Harms CT (1989). Regeneration of fertile plants from protoplasts of elite inbred maize. *Bio/ Technology,* **7** : 581-587.

Skokut TA, Wolk CP, Thomas J, Meeks JC, Shaffer PW and Chien WS (1978). Initial organic products of assimilation of ammonium and nitrate by tobacco cells cultured on different sources of nitrogen. *Plant Physiol.,* **62** : 299-304.

Songstad DD, Armstrong CL and Petersen WL (1991). $AgN0_3$ increases type II callus production from immature embryos of maize inbred B73 and its derivatives. *Plant Cell Rep.,* **6** : 699-702.

Songstad DD, Armstrong CL, Petersen WL, Hairston B and Hinchee MAW (1996). Production of transgenic maize plants and progeny by bombardment of Hi-II immature embryos. *In Vitro Cell. Dev. Biol. Plant,* **32** : 179-183.

Strauch E, Wohlleben W and Pühler A (1988). Cloning of a phosphinothricin N-acetyltransferase gene from *Streptomyces viridichromogenes* Tu494 and its expression in *Streptomyces lividans* and *Escherichia coil. Gene,* **63** : 65-74.

Thompson CJ and Seto H (1995). Bialaphos. *Biotechnology,* **28** : 197-222.

Thompson CJ, Rao Movva N, Tizard R, Crameri R, Davies JE, Lauwereys M and Botterman J (1987). Characterization of the herbicide resistant bar gene from *Streptomyces hygroscopicus. EMBO J.,* **6** : 2519-2523.

Thompson JA, Abdullah R and Cocking EC (1986). Protoplast culture of rice using media solidified with agarose. *Plant Sci.,* **47** : 123-133.

Vain P, Keen N, Murillo J, Rathus C, Nemes C and Finer J (1993a) Development of the particle inflow gun. *Plant Cell Tiss. Org. Cult.,* **33** : 237-246.

Vain P, McMullen MD and Finer JJ (1993b) Osmotic treatment enhances particle bombardment-mediated transient and stable transformation of maize. *Plant Cell Rep.,* **12** : 84-88.

Vasil V, Lu C and Vasil IK (1985). Histology of somatic embryogenesis in cultured immature embryos of maize (*Zea mays* L.). *Protoplasma,* **127** : 1-8.

Wan Y, Widholm JM and Lemaux PG (1995). Type I callus as a bombardment target for generating fertile transgenic maize (*Zea mays* L). *Planta,* **196** : 7-14.

Walters DA, Vetsch CS, Potts DE and Lundquist RC (1992). Transformation and inheritance of a hygromycin phosphotransferase gene in maize plants. *Plant Mol. Biol.,* **18** : 189-200.

Wehrmann A, Van Vliet A, Opsomer C, Botterman J and Schulz A (1996). The similarities of *bar* and *pat* gene products make them equally applicable for plant engineers. *Nat. Biotechnol.,* **14** : 1274-1278.

Wild A, Sauer H and Rühle W (1987). The effect of phosphinothricin (glufosinate) on photosynthesis. I. Inhibition of photosynthesis and accumulation of ammonia. *Z. Naturforsch.,* **42c :** 2519-2523.

Wilson HM, Bullock WP, Dunwell JM, Ellis JR, Frame B, Register III J and Thompson JA (1995). Maize, In : *Transformation of Plants and Soil Microorganisms* (Eds. Wang, K., Herrera-Estrella and Van Montagu, M) pp65-80; Cambridge University Press, Cambridge.

Wohlleben W, Arnold W, Broer I, Hilleman D, Strauch E and Pühler A (1988). Nucleotide sequence of the phosphinothricin-N-acetyl transferase gene from *Streptomyces viridichromogenes* Tu494 and its expression in *Nicotiana tabacum. Gene,* **70** : 25-37.

Zhang H, Yang H, Rech E, Golds T, Davis A, Mulligan B, Cocking E and Daavey M (1988). Transgenic rice plants produced by electroporation-mediated plasmid uptake into protoplasts. *Plant Cell Rep.,* **7** : 379-384.

Zhao ZY, Gu W, Cai T, Tagliani LA, Hondred D, Bond D, Krell S, Rudert ML, Bruce WB and Pierce DA (1998). Molecular analysis of T0 plants transformed by *Agrobacterium* and comparison of *Agrobacterium*-mediated transformation with bombardment transformation in maize. *Maize Genetic Newsletter,* **72** : 34-37.

Zhao ZY, Gu W, Cai T, Pierce DA (1999). Methods for *Agrobacterium*-mediated transformation. United States Patent #5, 981, 840.

Zhong, Srinivasan C and Sticklen MB (1992). In-vitro morphogenesis of corn (Zea mays L.). *Planta,* **187** : 483-489.

Zhong H, Sun B, Warkentin D, Zhang S, Wu R, Wu T and Sticklen B (1996). The competence of maize shoot meristems for integrative transformation and inherited expression of transgenes. *Plant Physiol.,* **110** : 1097-1107.

Chapter 9

BIOTECHNOLOGY OF PEARL MILLET

MV Subba Rao★ and V Manga

Cytogenetics and Biotechnology Laboratory, Botany Department, Andhra University, Visakhapatnam - 530 003, India

Summary

Pearl millet with its rich genetic diversity, with considerable background information on conventional genetic and breeding aspects and with well established basic biotechnological protocols offers a fertile arena for a fruitful combination of the conventional and novel technologies. The available information on characterization of genetic diversity in protein/enzyme, molecular and in vitro characters has been reviewed. Somatic hybridization and in vitro selection methods have helped to generate and screen for new variability. Attempts for gene tagging have been made using biochemical, molecular and in vitro characters. Limited information is also available on cryopreservation, genetic transformation and fluorescence in situ hybridization. The future lines of crop improvement programs are likely to involve the biotechnological methods with Marker Assisted Selection playing the directive role.

Keywords : Biotechnology, gene tagging, isoenzymes, pearl millet, *Pennisetum*, RFLP, tissue culture

1. INTRODUCTION

Pearl millet is now regarded as an "indispensable cereal crop of the dry and semi - arid regions" (Rai, 2000). It is a dual purpose crop being

★Corresponding author

valued for both its grain and fodder. The improvement achieved in pearl millet production so far (from 3.24 mt in 1950 - 55 to 6.61 mt in 1994-98 amounting to 100% increase (Rai, 2000) owes entirely to the classical breeding methods only. Further success in crop improvement programs can be achieved in relatively short periods and probably in more economical way by a judicious combination of conventional breeding programs and the novel biotechnological methodologies. This article reviews the present status of information available on the two major aspects of pearl millet biotechnology *viz.* cell tissue and organ culture and DNA manipulation.

2. CHARACTERIZATION OF GENETIC DIVERSITY IN PENNISETUMS

Characterization and efficient use of genetic diversity are prerequisites for any productive crop improvement program. Owing to its natural cross pollinated nature, pearl millet shows extensive variation at intraspecific level. There is also considerable variation at interspecific level. Use of morphometric characterization of the genetic diversity has been amply documented by the genetic resource unit of ICRISAT, and other such centers (see Appa Rao, 1999).

However, being subjected to environmental influence, these morphometric markers have limited use in tracing the successful transfer of genes from one generation to the other during breeding programs. In this context, biochemical markers in the form of proteins and enzymes and more so, the molecular markers based on DNA characters (such as RFLP, RAPD, AFLP) offer distinct advantage over the conventional morphometric markers.

2.1. Biochemical characterization

According to the genepool system of classification of genetic diversity related to a domesticated plant species (Harlan and de Wet, 1971), the wild species of *Pennisetum* fall into three gene pools. All those taxa that undergo natural cross-pollination with the cultivated species (*P. glaucum*) are included in the primary genepool. *P. purpureum* which hybridizes with the cultivated species rather infrequently due to reproductive barriers and produces often weak or partly sterile hybrids is included in the secondary gene pool. The tertiary gene pool comprises several taxa which either do not cross with pearl millet, or hybridize with it only rarely, producing completely sterile hybrids. The inter relationships reflected by the gene pool concept, which was based on crossability relationships and morphological diversity also finds support from the studies on isoenzyme variation by Lagudah and Hanna (1989, 1990), Sujatha (1984), Saideswara Rao *et al.* (1986), Subba Rao *et al.* (1988) and Hussain *et al.* (1990).

Isozyme polymorphism noticed in the various geographical races of pearl millet is well correlated with their photoperiod sensitivities, but at the same time seems to be not much influenced by the environment. The early and late maturing cultivars among the west African collections could be distinguished based on the zymogram patterns (Tostain *et al.*, 1987; Tostain and Marchais, 1989). The enzyme diversity was more in the early maturing types as compared to the late maturing ones. Founder effect and genetic drift from a unique gene source were suggested to be responsible for this isozyme diversity in relation to geographic distribution.

Enzymes that have been used to characterize genetic variability at intra and interspecific levels included Acid phosphatase, Alcohol dehydrogenase, Amino peptidase, Amylase, Catalase, Esterase, Glutamic-oxalo transaminase, Malate dyhydrogenase, Peroxidase, 6-Phosphogluconate dehydrogenase, Phosphogluco-mutase, and Shikimate dehydrogenase (Hussani *et al.*, 1990; Lagudah and Hanna, 1987; To stain *et al.*, 1987; Tostain and Riandey, 1984,85; Sujatha, 1984; Saideswara Rao *et al.*, 1986; Subba Rao *et al.*, 1988). For a more detailed account of the contributions of isozyme and protein studies to our knowledge about the genus *Pennisetum*, a reference can be made to Subba Rao *et al.* (1999).

2.2. Molecular characterization

The advances in molecular biology, especially during the last two decades, have made available such remarkable new techniques as gene cloning, *in situ* hybridization and polymerase chain reaction (PCR) etc. The degree of polymorphism generated at the DNA level through the use of restriction enzymes and PCR is much more than that possible either with morphometric or enzyme/protein characters.

The restriction fragment length polymorphism (RFLP) of mitochondrial DNA has provided useful information for characterizing the diversity existing in the CMS lines of pearl millet from various sources. Two of the CMS lines designated as 81A4 (= 81 Am) and ICMA 88001 (=Av) which received their cytoplasms from different accessions of *P. glaucum* subsp. *monodii* (= *P. violacecan*) of Senegal were found to be different from each other as well as from CMS lines derived from other sources (Rajeshwari *et al.*, 1994). The results were further confirmed by Rai *et al.* (1996) from the fertility restoration patterns in the hybrids. The use of RFLP analysis also lead to the identification of a new source for cytoplasmic male sterility (designated as *Aegp*) derived from an early genepool (Sujatha *et al.*, 1994). Several of the male sterile plants isolated from a large seeded gene pool were inferred to have been derived from at least five groups of cytoplasms (Chhabra, 1995).

Variability in the genes conferring restistance to downy mildew was analysed by DNA finger printing (Sastry *et al.*, 1995) and todate markers have been indentified for atleast 16 different putative downy mildew resistance genes (Hash *et al.*, 1999).

2.3. Diversity for *in vitro* response

Information on tissue culture (including protoplast, cell, tissue and organ culture) in pearl millet has originated from the use of various accessions by different laboratories. As is common in such studies, the variation in the *in vitro* response (quantity and nature of callus and frequency of regeneration) is attributable to the type of explants, hormones and other chemicals used in the medium as well as to the accessions. The diversity in the *in vitro* response that could be attributed clearly to the genotype was reported by Subba Rao and Nitzsche (1984), Mythili *et al.* (1997), Nagarathna *et al.* (1991) and Satyavathi (1998). The nature and developmental stage of the explant, the medium and phytohormonal concentrations remaining constant, variation between genotypes was observed for total callus quantity, E-callus quantity, growth rate and number of regenerated plantlets per explant. In spite of the several studies reported, the number of genotypes characterized for their *in vitro* response forms only a fraction of the total germplasm available in this crop (Bui-Dang-Ha and Pernes, 1985; Bui-Dang-Ha *et al.*, 1986; Devi *et al.*, 2000; Ketchum *et al.*, 1987; Kothari *et al.*, 1994; Nabors *et al.*, 1983; Nitsch *et al.*, 1982; Pius *et al.*, 1993; Rangan, 1976; Vasil and Vasil, 1981a, b).

3. GENERATION OF NEW GENETIC DIVERSITY THROUGH BIOTECHNOLOGICAL METHODS

Though the genetic diversity already known in pearl millet and in the genus *Pennisetum* as a whole is considerably large, the spectrum of such diversity, at least theoretically, could be endless with the changing agro-climatic conditions and with the possibility of development of new physico-pathological races in nature. As such, there is a continuous need for new genes/gene combinations. Conventional mutagenesis and hybridization can help to increase the genetic diversity but relatively at a very slow pace. This disadvantage can be overcome by the novel techniques of *in vitro* mutagenesis and somatic hybridization. Site specific mutagenesis, though theoretically has a much more potentiality, is still in its infancy and is not available for regular use in all the systems.

3.1. *In vitro* selection

In vitro selection was demonstrated for the first time in pearl millet by Bajaj *et al.* (1980) through screening for differential tolerance to ergot. The regenerants from the selected calli also showed ergot resistance, thus adding utility value to this technique. Subsequently, this technique was successfully applied for selection of 5 (2-amino ethyl)-2-cystine (AEC) tolerance (Boyes and Vasil, 1987), downy mildew resistance (Prasad *et al.*, 1984; Nagarathna *et al.*, 1993) and ergot resistance (Sharma and Chahal, 1990). Dual culture technique by co-culturing pearl millet and *Sclerospora graminicola* was adopted for the selection of downy mildew resistance (Upadhyaya *et al.*, 1992).

Cell lines tolerant to salt (NaCl) were selected using inflorescence derived calli by Rangan and Vasil (1983). These tolerant lines retained the capacity even in the absence of the selection pressure. More recently Arjun Rao (1999) established callus cultures from immature inflorescence segments of two salt tolerant and two sensitive accessions. The calli from the tolerant could withstand the same level of salinity stress as was used for the selection at plant level i.e. $17dSm^{-1}$ of saline solution (165.3 mM NaCl + 39.2 mM $MgSO_4$ in half strength Hoagland solution which is equivalent to 200 mM NaCl or 35.4% of sea water. Cell lines with NaCl tolerance have also been isolated in *P. purpureum* Bajaj and Gupta, 1987).

Callus clultures derived from etiolated 4-day old seedling shoot apices were also used to evaluate freezing response using electrolyte leakage method (Stair *et al.*, 1998). Callus tissue of *P. flaccidum* cv Carostan, a winter - hardy species had greater freezing tolerance at -3.6°C than at -1.1°C. Calli from other winter-hardy species *P. orientale* or the moderate winter-hardy species *P. mezianum* or from a poor winter-hardy species *P. ciliare* exhibited no detectable freezing tolerance at either temperatures.

3.2. Somatic hybridization

Few instances of protoplast fusion between species of *Pennisetum* and those of other poaceous genera have been reported as a means to transfer genes and produce new gene combinations. The fusion products of pearl millet and sugarcane were selected using 6PGD/SKDH (6-phosphogluconate dehydrogenase/ shikimate dehydrogenase) aryl esterase and rRNA polymorphic patterns as markers (Tabaeizadeh *et al.*, 1986). Similarly, 6PGD and mtDNA patterns were used for the selection of pearl millet + *Panicum maximum* hybrids (Ozias - Akins *et al.*, 1986, 87), while 6PGD and RFLP patterns formed the selection criteria for pearl millet + *Triticum monococcum* hybrids (Vasil *et al.*, 1988). In all these three cases, AEC resistant embryogenic cell cultures were used as the source of pearl millet protoplasts. In the case of hybridisation with *Triticum monococcum,* inactivation of pearl millet protoplasts by iodo-acetate treatment was also used as an additional marker.

4. GENE TAGGING

Understanding the linkage relationships of the marker genes will facilitate identification of the required genotypes during breeding programs for effecting an efficient gene transfer. When compared to the extent of genetic diversity known, relatively few genes effecting plant phenotype have been mapped so far (ref. Anand Kumar and Andrews, 1993). Since the morphometric characters are often subject to the influence of environmental factors or complicated gene action mechanisms and/or involvement of several loci, gene tagging among such markers will be of little use to breeders. On the other hand, biochemical and DNA based markers, which do not suffer these disadvantages, serve better as gene tags.

4.1. Using biochemical markers

As early as 1980, association between peroxidase and esterase isozyme patterns with downy mildew resistance was recorded by Gupta *et al.*, (1980). A more definite analysis of the relationship was established by Kumar *et al.* (1987) between resistance and a cathodal band (C_2) and an modal band (A_4). Subsequently, three more isoperoxidases (C_5, C_6 and C_9) were also found to be closely associated with the resistance mechanism (Chahal *et al.*, 1986, 88).

The dwarfing gene (D_2) was tagged to *Skdh A* (Shikimate dehydrogenase) at 9 ± 5 cM and to *Adh A* (Alcohol dehydrogenase) at

20 cM (Tostain, 1985). It was suggested that the linkage between the d_2 and the codominant *Skdh A* alleles will help in the creation of isogenic semidwarf lines of cultivars. Though the genetics of some other enzymes have been analysed (Banuett-Bourillon, 1982a,b; Banuett-Bourillon and Hague, 1979; Gepts and Clegg, 1989; Sandmeier *et al*., 1981; Subba Rao *et al*., 1989; Tostain, 1985; Trigui *et al*., 1986), no information is available regarding their association with other genetic markers. The genetic diversity in the biochemical characters seems to be not fully exploited as far their potential to serve as gene tags is concerned.

4.2. Using molecular markers

Though recently started, considerable progress has been made in the usage of molecular markers for the purpose of tagging with agronomically important genes. The relationship between mitochondrial genes (mt DNA), nuclear genes and cytoplasmic male sterility in pearl millet was analysed by Smith *et al*. (1987). These authors demonstrated that a 4.7 kb fragment revealed by the Pst I digests of mt DNA was present in the male sterile lines, but was missing in the corresponding male fertile lines as well as the revertants. A 10 kb fragment was common to all the male sterile lines examined. Another fragment of 13.6 kb was additionally present in only one of the revertants as well as in all A and B lines.

Later Smith and Chowdhary (1991) identified two sets of repeated sequences in the 4.7, 10.9 and 13.6 kb fragments which accommodated the rrn_{18}, rrn_5 and cox_1 genes in them. The mt DNAs of CMS lines possessed three copies each of rrn_5 and rrn_{18} genes, but the revertants had only a single dose of these genes. The CMS characters in A_1, Av, A_{egg} and A_5 lines were suggested to be due to rearrangements in cox_1 gene, where as the male sterility in A_4 line might be involving rearrangement in the atp_6 /cox_3 cluster (Delorme *et al*., 1997).

A significant contribution made by Liu *et al*. (1994) was the development of RFLP map of nuclear DNA using 200 genomic probes. 181 loci were mapped in 7 linkage groups with a total map length of 303 cM (the hypothetical maximum being 400 cM) and an average of 2 cM distance between the loci. Availability of the RFLP maps has facilitated tagging of four major QTLS responsible for downy mildew resistance by Jones *et al*. (1995). RFLP analysis of the segregating populations derived from resistant susceptible crosses had resulted in assigning two loci responsible for resistance against the pathogen races from Niger and Nigeria to the fourth linkage group LG4, one locus offering resistance for the pathogen from India to LG 1 and the fourth locus covering resistance to the pathogen race from Senegal to LG2.

RAPD and RFLP markers corresponding to the primer 5' CTGCAGA(C)C(A)C(T)T(G)C(T)C(G)AAA(C)CAG 3' were used to follow the transfer of apomictic trait from *P. squamlatum* to pearl millet (Ozias-Akins *et al.*, 1993). These markers were found to be tightly linked with apomixis and were always present in the obligate apomicts.

Fluorescent *in situ* hybridization (FISH) technique has also been used to provide chromosome markers that have potential in modifying and characterizing alien chromosome segments introduced from the wild relatives (King *et al.*, 1993). Liu *et al.* (1997) located two pairs of rDNA sites with significantly different signal strength which were localized at the distal ends of the short arms of two chromosome pairs of pearl millet and P. *violaceum*. Using RFLP analysis, one of these two pairs of rDNA sites (Nor - PI) was mapped on the fifth linkage group of pearl millet. The 18s-5.8s-26s rDNA probe was found to bind to the telomeric regions of chromosomes 6 and 7 in the three primary gene pool species (*P. glaucum*, *P. violaceum* and *P. mollissimum*) and to chromosomes I and 4 in the tertiary gene pool species, *P. schweinfurthii* (Mortel *et al.*, 1996). The 5s rDNA probe was localized on the short arm of chromosome 4 of the three primary gene pool species and to short arm of chromosome 2 in *P. schweinfurthii*.

4.3. Using *in vitro* characters

Satyavathi (1998) analysed the joint segregation patterns of four *in vitro* characters (total callus quantity, E-callus quantity growth rate and regeneration frequency) and four phenotypic markers (dwarf nature, purple pigmentation of internode, midrib, margin and sheath, bristled ear head and basal branching of the earhead). The dwarfing gene 'd_2' was found to be co-inherited with E-callus quantity and regeneration frequency. QTL analysis for determining the linkage values using RFLP markers is in progress.

5. BIOTECHNOLOGY AND CROP IMPROVEMENT

In addition to being useful for generating new variability, characterizing genetic diversity and for gene tagging, biotechnology has some other applications also which include:

5.1. Germplasm conservation

Difficulties in the availability of suitable explants at the desired time, and the necessity of periodic subculturing can be overcome by the technique of cryopreservation which is now preferred for long term storage and maintenance of germplasm. Gnanapragasam and Vasil (1992) have worked out a protocol for the cryopreservation of immature embryos, embryogenic callus and cell suspension cultures of gramineous species including pearl millet. A 3-year old cell suspension culture derived from immature embryos of pearl millet was used to study and optimize each step of the cryopreservation procedure. The cell cultures were pregrown in media supplemented with 0.33 M mannitol for 3 days and cryoprotected by the gradual addition of 0.5 M sorbitol and 0.7 M dimethyl sulfoxide. No reduction in viability was observed in cells stored in liquid nitrogen for upto 3 years. Plant regeneration from cryopreserved calli was observed to be similar to that from unfrozen controls. An added advantage observed during this study was that cryopreservation also acted as a selection process because the non-embryogenic cells which usually have vacuolated cells become lethally damaged due to freezing while the embryogenic cells with dense cytoplasm could survive the low temperature effects.

5.2. Genetic transformation

So far, there is a single instance of gene transfer in pearl millet. Using micro projectile bombardment technique involving rDNA constructs - pBARGUS and $pAHC_{25}$, the expression of GUS was observed in transformed cell suspension cultures of *P. glaucum* and *P. purpureum* (Taylor *et al.*, 1993).

5.3. Marker assisted selection (MAS)

Most of the characters that are of agronomic importance are quantitative in nature, being subjected to polygenic control, gene interaction and involvement of modifiers. Therefore, it becomes very difficult, sometimes impossible, to delineate the individual effects of one component from the other by classical approaches. Suitable manipulation of these quantitative trait loci (QTL) often becomes complicated and not amenable for conventional breeding procedures. These inherent features in handling of QTLS could be overcome if these loci are tagged tightly (due to close linkage) to reliably detectable markers such as RFLP etc. If such a marker is analysed instead of the QTL directly, the complexity of QTLs becomes reduced to a simple level equaling a

Mendelian character. This would also help to eliminate linkage drag and confounding of environmental effects. By using such suitable markers each linked closely to an individual QTL component of a character, it becomes possible to analyse the individual effects of such QTLs and also each one of them could be followed /traced over generations in a breeding program. In the case of breeding for stress or disease resistance, the marker assisted selection (MAS) will enable the breeder to recognize and select a particular level or intensity of resistance even in the absence of pathogen or the factor/agent responsible in the stress thus simplifying considerably the selection process. This (MAS) approach will also enable the production of a range of desired genotypes by regulating the combinations of QTLs in pyramiding genes (Pederson and Leath, 1988).

The availability of RFLP maps in pearl millet has opened up the possibilities for a successful application of MAS especially in the context of resistance breeding (Jones *et al.*, 1995). The effectiveness of MAS for downy mildew resistance is currently under evaluation at ICRISAT Asia center (Hash *et al.*, 1999).

6. FUTURE OUTLOOK

Pearl millet, with its rich genetic diversity, with considerable background information on conventional genetic and breeding aspects and with well established basic biotechnological protocols offers a fertile arena for a fruitful combination of the conventional and novel technologies. Besides breeding for new varieties or developing better cultivation practices, pearl millet breeders now look for the more extended and alternative usages of pearl millet (Harinarayana *et al.*, 1999). Development of new varieties with ability to grow on marginal and saline soils, varieties with specific constituents of nutritional value such as lysine, threonine, tryptophan, carotene (Vit A), lines with increased biomass production, pyramiding of genes conferring resistance to pathogen and abiotic stress are some of the future breeding strategies. These have a potential for the use of *in vitro* selection, protoplast fusion and gene transfer technologies with MAS playing the directive role.

REFERENCES

Ananda Kumar K, and Andrew DJ (1993) Genetics of qualitative traits in pearl millet : A review. *Crop Sci.,* **33** : 1-20.

Appa Rao S (1999) Genetic Resources. In : *Pearl Millet Breeding.* (Eds. Khairwal, IS, Rai KN, Andrew DJ and Harinarayana G.) Oxford & IBH publishibg Co, N. Delhi. pp 49-82.

Arjuna Rao PV (1999) Charaterisation and genetic analysis of salinity tolerance in same inbreds of pearlmillet, *Pennisetum glaucum*. Ph. D thesis. Andhra University, Visakhapatnam.

Bajaj YPS and Gupta R (1987) Plants from salt tolerant cell lines of napier grass, *Pennisetum purpureum* Schum. *Indian J. Experiment Biol.,* **25** : 58-60.

Bajaj YPS, Phul PS and Sharma SK (1980) Differential tolerance of tissue cultures of pearl millet to ergot extract. *Indian J. Exp. Biol.,* **18** : 429-432.

Banuett-Bourillon F (1982a) Linkage of alcohol dehydrogenase structural genes in pearl millet (*Pennisetum typhoides*). *Biochem. Genet.,* **20** : 359-367.

Banuett-Bourillon F (1982b) Natural variants of pearl millet (*Pennisetum typhoides*) with altered levels of Set II alcohol dehydrogenase activity. *Biochem. Genet.,* **20** : 369-383.

Banuett-Bourillon F and Hague DR (1979) Genetic analysis of alcohol dehydrogenase isozymes in pearl millet (*Pennisetum typhoides*). *Biochem. Genet.,* **17** : 537-551.

Boyes CJ and Vasil IK (1987) *In vitro* selection for tolerance to S-(2-aminoethyl)-L-cysteine and overproduction of lysine by embryogenic calli and regenerated plants of *Pennisetum americanum* (L.). *K. Schum. Plant Sci.,* **50** : 195-203.

Bui - Dang - Ha D, and Pernes J (1985) Variability observed in millet (*P. typhoides*) after anther culture. *Bulletin de la societe Botanique de France, Actualites Botaniques.,* **132** : 150.

Bui - Dang - Ha D, Me quinion M J and Pernes J (1986) Genetic changes after and rogenesis in pearl millet, *Pennisetum americanum*, study of progenies from pollen culture of an F_1 hybrid (Massue Ligni). *International Atomic Energy Agency,* 195-205.

Chahal SS, Kumar R, Sidhu JS and Minocha JL (1986) Peroxidase isozyme pattern in pearl millet lines resistant and susceptible to ergot. *Indian Phytopath.,* **39** : 65-69.

Chahal SS, Kumar R, Sidhu JS and Minocha JL (1988) Peroxidase isozyme pattern in pearl millet lines resistant and susceptible to downy mildew. *Plant Breed.,* **101** : 256-259.

Chhabra AK (1995) Molecular characterization of cytoplasmic-nuclear male sterility (CMS) sources and tall/dwarf near-isogenic lines in pearl millet. Ph.D. Thesis. CSS Haryana Agricultural University, Hisar, India.

Delorme V, Keen CL, Rai KN and Leaver CJ (1997) Cytoplasmic-nuclear male sterility in pearl millet : Comparative RFLP and transcript analysis of isonuclear male-sterile lines. *Theor. Appl. Genet.,* **95** : 961-968.

Devi P, Zhong H, Stickler HB (2000) *In vitro* morphogenesis of pearl millet *Pennisetum glaucum* : efficient production of multiple shoots and inflorescences from shoot apices. *Plant Cell Rep.,* **19** : 546-550.

Gepts P and Clegg MT (1989) Genetic diversity in pearl millet (*Pennisetum glaucum* (L.) R. Br.) at the DNA sequence level. *J. Heredity,* **80** : 203-208.

Gnanapragasam S and Vasil IK (1992) Cryopreservation of immature embryos, embryogenic callus and cell suspension cultures of graminaeceous species. *Plant Sci.,* **83** : 205-215.

Gupta DN, Roy MK, Mehta SL, Singhal GL and Murthy BR (1990) Peroxidase and esterase isoenzymes in downy mildew susceptible and resistant cultivars of pearl millet. In: *Trends in Genetical Research on Pennisetums.* (Eds. Gupta VP and Minocha JL). Punjab Agricultural University, Ludhiana, India. Pp 173-78.

Harinarayan G, Anand Kumar K, Andrews DJ (1999) Pearl millet in global Agriculture. In : *Pearl Millet Breeding (*Eds. Khairwal IS, Rai KN, Andrews DJ and Harinarayan G.) Oxford - IBH publishing Co., New Delhi. pp 479-506.

Harlan JR and dewet JMJ (1971) Towards a rational taxonomy of cultivated plants, *Taxonomy,* **20** : 509-517.

Hash CT, Single SD, Thankur RP and Talukolar BS (1999). Breeding for disease resistance. In : *Pearl Millet Breeding (*Eds. Khairwal IS, Rai KN, Andrews DJ and Harinarayana G.) Oxford and IBH publishing Co New Delhi. pp 337-380.

Hussain SN, Saideswara Rao Y, Manga V and Subba Rao MV (1990) Chaemotaxonomic studies on the genus *Pennisetum* L. Rich. by isozyme analysis. *Beitrage zur Biologie der Pflanzenzuchtung.,* **65** : 211-222.

Jones ES, Liu CJ, Gale MD, Hash CT and Witcombe JR (1995) Mapping quntitative trait loci for downy mildew resistance in pearl millet. *Theor. Applied Genet.,* **91** : 448-456.

Ketchum JLF, Gamborg OL, Hanning GE and Nabors MW (1987) Millet tissue culture for crops - project progress report. Department of Botany, Colarado State University, Fort Collins, Collarado. 38-39.

King IP, Purdie KH, Oxford SE, Reader SM and Miller TE (1993) Detection of homoeologous chiasma formation in *Triticum durum* × *Thinopyrum bessaribicum* hybrids using genomic *in situ* hybridization. *Heredity,* **71** : 369-372.

Kothari SL, Sivadas P and Chandra N (1994) Differentiation and development of somatic embryos in suspension cultures and plantlet formation in *Pennisetum americanum. J. Indian Bot. Soc.,* **73** : 233-236.

Kumar R, Chahal SS, Sidhu JS and Minocha JL (1987) Peroxidase isozyme studies in pearl millet related to downy mildew resistance. *Plant Sci. Research,* **2** : 113-115.

Lagudah ES and Hanna WW (1989) Species relationship in the *Pennisetum gene pool* : Enzyme polymorphism. *Theor. Appl. Genet.,* **78** : 801-808.

Lagudah ES and Hanna WW (1990) Patterns of variation for seed proteins inthe *Pennisetum* gene pool, *J. Heredity,* **81** : 25-29.

Lui CJ, King IP, Pittaway TS, Abbos S, Reader SM, Miller TE and Gale MD (1997) Physical and genetical mapping of rDNA siters in *Pennisitum* (pearl millet). *Heredity,* **78** : 529-531.

Liu CJ, Witcombe JR, Pittaway TS, Nash M, Hash CT, Busso CS and Gale MD (1994) An RFLP-based genetic map of pearl millet (*Pennisetum glaucum*). *Theor. Appl. Genet.,* **89** : 481-487.

Martel E, Ricroch A and Sarr A(1996). Assessment of genane organisations among diploid species (24 = 2X = 14) belonging to primary and tertiary gene pools of pearl millet using fluorescent *in situ* hybridization with rDNAproloes. *Genome,* **39** : 680-687.

Mythili PK, Satyavathi VV, Pavan Kumar G, Rao MVS and Manger V (1997) Genetic analysis of short term callus culture and morphogenesis in pearl millet, *Pennisetum glaucum. Plant Cell Tiss. Org. Cult.,* **50** : 171-178.

Nabors MW, Heyser JW, Dukes TA and Demott KJ (1983) Long duration, high frequency plant regeneration from cereal tissue cultures. *Planta,* **157** : 139-142.

Nagarathna KC, Prakash HS, and shetty HS (1991) Genotypic effects on the callus formation from different explants of pearl millet B lines. *Adv. Pl. Sci.,* **4** : 82-86.

Nagarathna KC, Shetty SA, Harinarayana G and Shetty HS (1993) Selection for downy mildew resistane from the regenerants of pearl millet. *Plant Sci.,* **90** : 53-61.

Nitsch C, Andersen S, Godard M, Neuffer G, and Sheridan WF (1982) Production of haploid plants of *Zea mays* and *Pennisetum* through androgenesis. In : *Variability in plants regenerated from tissue culture* (Eds. Elizabeth D. Earle, Yves Demarley) pp 69-91.

Ozias-Akins P, Ferl RJ and Vasil IK (1986) Somatic hybridization in Gramineae: *Pennisetum americanum* (L) K. Schum (pearl millet) + *Panicum maximum* Jacq (Guinea grass). *Mol. Gen. Genet.,* **203** : 365-370.

Ozias-Akins P, Lubbers EL, Hanna WW and McNay JW (1993) Transmission of the apomictic mode of reproduction in *Pennisetum*: Co-inheritance of the trait and molecular markers. *Theor. Appl. Genet.,* **85** : 632-638.

Ozias-Akins P, Pring DR and Vasil IK (1987) Rearrangements in the mitochondrial genome of somatic hybrid cell lines of *Pennisetum americanim* (L.) K. Schum + *Panicum maximum* Jacq. *Theor. Appl. Genet.,* **74** : 15-20.

Pederson WI, Leath S (1988) Pyramiding major genes for resistance to maintain residual effects. *Annu. Rev. Phytopathol.,* **26** : 369-378.

Pius J, George L, Eapen S and Rao PS (1993) Enhanced plant regeneration in pearl millet (*Pennisetum americanum*) by ethylene inhibitors and cefotaxime. *Plant Cell Tiss. Org. Cult.,* **32** : 91-96.

Prasad B, Prabhu MS and Shanthamma C (1984) Regeneration of downy mildew resistant plants from infected tissues of pearl millet (*Pennisetum americanum*) cultured *in vitro. Curr. Sci.,* **53** : 816-817.

Rai Mangala (2000) Productivity gains focus. *The Hindu Survey of Indian Agriculture* 2000.

Rai KN, Virk DS, Harinarayana G and Rao AS (1996) Stability of male-sterile sources and fertility restoration of their hybrids in pearl millet. *Plant Breed.,* **115** : 494-500.

Rajeshwari R, Sivaramakrishnan S, Smith RL and Subrahmanyam NC (1994) RFLP analysis of mitochondrial DNA from cytoplasmic male-sterile lines of pearl millet. *Theor. Appl. Genet.,* **88** : 441-448.

Rangan TS and Vasil IK (1983) Sodium chloride tolerant embryogenic cell lines of *Pennisetum americanum* (L.) K. Schum. *Annals of Botany* **52** : 59-64.

Saideswara Rao Y, Sujatha DM, Manga V and Subba Rao MV (1986) Esterase variability in five species of the genus *Pennisetum* L. Rich (Gramineae). *Geobios,* **13** : 108-111.

Sandmeier M, Beninga M and Pernes J (1981) Genetic analysis of the relationships between spontaneous and cultivated forms of pearl millet. III. Inheritance of esterases and anodic peroxidase isozymes. *Agronomie,* **1** : 487-494.

Sastry JG, Ramakrishna W, Sivaramakrishnam S, Thakur RP, Gupta V and Ranjekar PK (1995). DNA finger printing detects genetic variability in the pearl millet downy mildew pathogen (*Sclerospora graminicola*). *Theor. Appl. Genet.,* **91** : 856-861.

Sharma SB, and Ghahal SS (1990) Development of regenerants from pearl millet cell lines resistant to culture filtrate of *Claviceps furiformis*. *Plant Dis. Res.,* **5** : 175-180.

Satyavathi VV (1998) Genetics of some callus characters and *in vitro* morphogenesis in short - term cultures of pearl millet; *Pennisetum glaucum*. Ph.D Thesis. Andhra University, Visakhapatnam, India.

Smith RL and Chowdhury MKU (1991) Characterization of pearl millet mitochondrial DNA fragments rearranged by reversion from cytoplasmic male sterility to fertility. *Theor. Appl. Genet.,* **81** : 793-799.

Smith RL, Chowdhury MKU and Pring DR (1987) Mitochondrial DNA rearrangements in *Pennisetum* associated with reversion from cytoplasmic male sterility to fertility. *Plant Mol. Biol.,* **9** : 277-286.

Stair DW, Dahmer ML, Bashaw Ec and Hussey Ma (1998) Freezing tolerance of selected *Pennisetum* species. *Int. J. Plant Sci.,* **159** : 599-605.

Subba Rao M V and Nitzsche W (1984) Genotypic differences in callus growth and organogenesis of eight pearl millet lines. *Euphytica,* **33** : 923-928.

Subba Rao MV, Mythili PK, Sree Rama Murty J, Saideswara Rao Y, Marga V (1999) Biotechnology. In : *Pearl Millet Breeding.* (Eds. Khairwal IS, Rai KN, Andrews DJ, Harinarayana G.) Oxford & IBH publishing Co., New Delhi. pp 417-444.

Subba Rao MV, Saideswara Rao Y and Manga V (1989) Genetics of five seed esterase isozymes in pearl millet. *Plant Breed.,* **102** : 133-139.

Sujatha, DM (1984) Studies on some biosystematic aspects of twelve species of the genus *Pennuisetum* L. Rich. (*Poaceae*). Ph. D. Thesis. Andhra University, Visakhapatnam, India.

Sujata, V, Sivaramakrishnan S, Rai KN and Seetha K (1994) A new source of cytoplasmic male sterility in pearl millet: RFLP analysis of mitochondrial DNA. *Genome,* **37** : 482-486.

Tabaeizadeh Z, Ferl RJ and Vasil IK (1986) Somatic hybridization in the Gramineae: *Sacharum officinarum* L. (sugarcane) and *Pennisetum americanum* (L.) K. Schum (pearl millet.). In: *Proc. Natl. Acad. Sci. USA,* **83** : 5616-5619.

Taylor MG, Vasil V and Vasil IK (1993) Enhanced GUS gene expression in cereal/grass cell suspensions and immature embryos using the maize ubiquitin-based plasmic PAHC 25. *Plant Cell Rep.,* **12** : 491-495.

Tostain S (1985) Mise en evidence d'une lieassongenetique entre ungene de nanisme et des marquers enzymatique chezle mil penicillaire (*Pennisetum glaucum* L.). *Canad. Genet. Cytol.,* **27** : 751-758.

Tostain S and Marchais L (1989) Enzyme diversity in pearl millet (*Pennisetum glaucum*). 2. Africa and India. *Theor. Appl. Genet.,* **77** : 634-640.

Tostain S and Riandey MR (1984) Polymorphisme et determinisme genetique des enzymes du mil pencillarie (*Pennisetum glaucum* L.): Etude des alcool deshydrogenases, catalases, endopeptidases et esterases. L' *Agronomie Tropicale,* **39** : 335-345.

Tostain S and Riandey MF (1985) Polymorphisme et determinisme genetique des enzymes du mil pencillarie (*Pennisetum glaucum*). Etude des malates deshydrogenases. *Agronomie,* **5** : 227-238.

Tostain S, Riandey MF and Marchais L (1987) Enzyme diversity in pearl millet (*Pennisetum glaucum*). 1. West Africa. *Theor. Appl. Genet.,* **74** : 188-193.

Trigui N, Sandmeier M, Salanoubat M and Pernes J (1986) Utilization of enzymatic and morphological data for the study of the populations and the domestication of plants 1. Separation and genetic identification of isozymes of pearl millet (*Pennisetum typhoides*). *Agronomie,* **6** : 779-788.

Upadhyaya G,Shivanna MB, Prakash HS and Shetty HS (1992) A novel approach to the establishment of dual cultures of pearl millet and *Sclerospora graminicola. Plant Cell Tiss. Org. Cult.,* **31** : 203-206.

Vasil V and Vasil IK (1979) Isolation and culture of cereal protoplasts. 1. Callus formation from pearl millet (*Pennisetum americanum*). *Theor. Appl. Genet.*, **56** : 97-99.

Vasil V and Vasil IK (1981a) Somatic embryo genesis and plant regenaration from tissus cultures of *Pennisetum americanum and P. americanum X P. purpureum* hybrid. *Amer. J. Bot.*, **68** : 864-872.

Vasil V and Vasil IK (1981b) Somatic embryo genesis and plant regenaration from suspension cultures of pearl millet. *Ann. Bot.*, **47** : 679-686.

Chapter 10

TISSUE CULTURE AND ALIEN GENE TRANSFER IN SORGHUM

N Seetharama★, PK Mythili, TS Rani, D Harshavardhan, A Ranjani and HC Sharma

ICRISAT, PO Patancheru 520 324, Andhra Pradesh, India

Summary

Genetic transformation of crops with specific genes conferring disease and pest resistance is an efficient tool to complement traditional breeding. Though transgenics of many major crops have been already produced and deployed, genetic transformation of sorghum is not yet routine and easy. A major constraint is lack of efficient and genotype independent tissue culture and regenerating system which till recently has been considered as recalcitrant. However, with the systematic work at ICRISAT and elsewhere it is now possible to regenerate sorghum from a variety of explants such as immature inflorescence or embryos and shoot tips from germinating seedlings. New techniques such as shoot meristem cultures are very efficient for production of transgenics. Similarly, advances in cell suspension and protoplast cultures, and tissue culture of wild species have opened up the possibilities for gene transfer by other methods such as somatic hybridization. To date, there have been nearly a dozen attempts to genetically transform sorghum, but plant regeneration and testing of transgenics was carried out only in a few cases. Use of commercial gene gun is most common. In this article, we demonstrate principles and use of a simpler particle-inflow gun built in-house for efficient transformation. However, use of Agrobacterium-mediated methods is also becoming popular. Most of the efforts to genetically transform sorghum are aimed at

★Corresponding author : E-mail : N.Seetharama@cgiar.org

incorporation of exotic genes for plant protection, but we can deploy many other types of genes such as those to enhance quality or adaptation to abiotic stress. Progress made in developing genotype-independent transformation and regeneration system are being matched with other technical advances such as use of novel selection system by avoiding use of antibiotic or herbicide markers, and gene deployment strategies to enhance the effectiveness and durability of transgenes. We have also examined the scope for deploying transgenics and means of integrating transgenic research with conventional crop improvement, particularly to serve the subsistence farming systems of the tropics and to increase wider public acceptance of transgenic technology. We believe that there is enormous scope to genetically alter sorghums to suit our needs, especially in the context of revolutionary advances in plant genome research, especially in functional genomics.

Keywords : Sorghum, tissue culture, genetic transformation

Abbreviations : LS - Linsmaier and Skoog. MS - Murashige and Skoog. PIG – particle- inflow gun. GUS - β-glucuronidase. PPT - phosphinothricin. PAT - phosphinothricin acetyl transferase. *npt II* - neomycin phosphotransferase. *bar* - bailophos resistance. GFP - green-fluorescence protein. *hpt* - hygromycin phosphotransferase. RAPD - random amplified polymorphic DNA. BI-BAC – binary bacterial artificial chromosome.

1. INTRODUCTION

Sorghum (*Sorghum bicolor* (L.) Moench) is cultivated on approximately 44 million hectares worldwide, and is the fifth major cereal crop after wheat, rice, maize and barley. The annual grain production is about 58 million tonnes (FAO, 1991). It is an important staple food crop, especially of the poor in Africa, South Asia and Central America. Sorghum is also an important source of animal feed and fodder. Sorghum includes at least four groups of cultivated plants: grain sorghum; sweet sorghum as forage; Sudan grass for pasture, hay and silage; and broomcorn for making brooms. The cultivated sorghum originated in Africa somewhere in the region of present-day Sudan and Ethiopia. It includes following five basic races: *bicolor*, *guinea*, *caudatum*, *kafir* and *dura* (Smith and Fredricksen, 2000). Most of the global grain sorghum is produced in arid or semi-arid regions, and it is grown mostly as a rainfed crop. Unlike in the USA or Australia, in the tropics sorghum is generally grown by the marginal farmers on small farms with low or no purchased inputs. The low crop management and the several complex stress factors are responsible for less than 1 tonne ha^{-1} yield in the tropical countries while the potential yield of this crop is at least 10-fold more.

The primary objective of most of the national and international sorghum improvement programs is to improve the yield and food quality, and to stabilize the sorghum production under dryland conditions. Prerequisites for the application of *in vitro* techniques for crop improvement are efficient and reproducible methods of regeneration from callus, cell or protoplast cultures, long-term regeneration potential, genetic stability of callus cultures and genotype independent protocols for genetic transformation. Though significant progress has been achieved at few places in the USA, Australia, Europe and India for establishing totipotent or morphogenic callus and cell cultures in cultivated and wild sorghums, successful transfer of alien genes into cultivated sorghum has met with only limited progress. Therefore, to date there is no single transgenic sorghum under commercial cultivation. Therefore, we will review the progress in sorghum tissue culture and genetic transformation and examine prospects for the near future.

2. TISSUE CULTURE AND REGENERATION STUDIES

2.1. Callus induction and regeneration from different explants

A system for rapid and uniform regeneration system with high regeneration efficiency is a prerequisite for successful genetic transformation. However, the frequency of plant regeneration reported until recently is not high enough for genetic transformation on a routine basis. *In vitro* response of different explants – shoot tip, immature inflorescence and immature embryos - were studied at ICRISAT with both wild and cultivated species of sorghum (Fig. 1) as done elsewhere on a range of media. Among the different explants, callus induction and regeneration ranged from 10 - 100 %. Linsmaier and Skoog (LS) medium supplemented with 2,4-dichloro-phenoxy acetic acid and kinetin was used for induction of friable embryogenic calli and somatic embryos from the above explants. For regeneration, invariably MS medium supplemented with BAP and NAA was employed. Shoot tip and inflorescence are

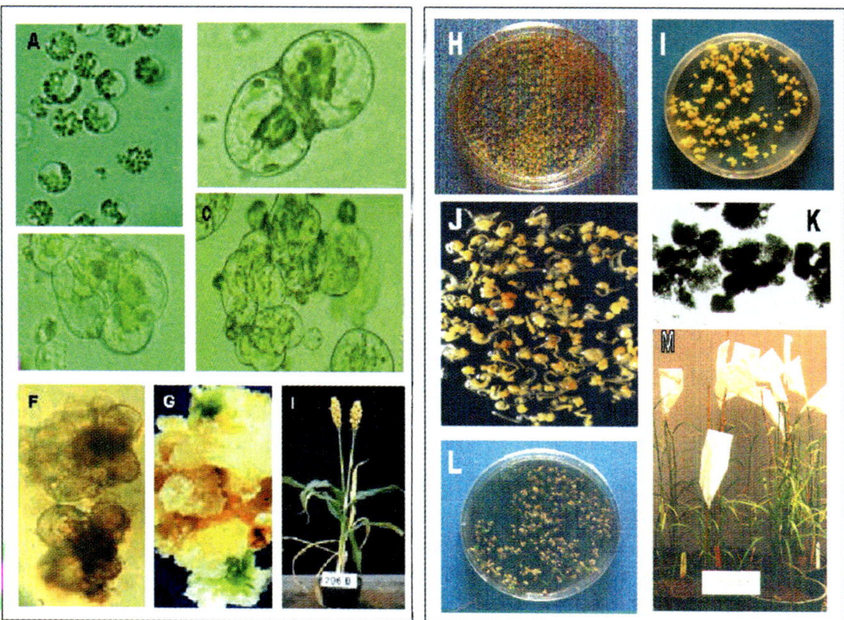

Figure 1 : Plant regeneration in cultivated (A-G) and wild sorghums (H-M). A-G : Plant regeneration from *in vitro* cultured mesophyll protoplasts of cultivated species of *S. bicolor* (line 296 B).

(A) Freshly isolated mesophyll protoplasts; (B) two-celled protoplast; (C) a protoplast producing multicellular structure with more than four divisions; (D) colony formation; (E) protoplast-derived callus; (F) differentiation of callus into shoots and (G) mature plant derived from mesophyll protoplast.

H-M: Plant regeneration from *in vitro* plated suspension cultures of wild species *S. dimidiatum.*

(H) Freshly plated suspension cells; (I) embryogenic suspension cultures; (J) germinating embryos; (L) microscopic section of embryogenic suspension cultures differentiating embryos and, (M) mature tillering plants in greenhouse.

more totipotent than the immature embryo. Regeneration pathway in all the three explants is via somatic embryogenesis. Apart from above, a rapid regeneration protocol from shoot meristem explant has been standardized recently by striking an optimal balance between a weak auxin like NAA and strong cytokinin like thidiazuron (Harshavardhan *et al.*, 2001). The isolated shoot meristems were manipulated to follow either organogenic or embryogenic pathway. There was no intermediate callus formation in both cases. Induction and multiplication was achieved on MS medium (Harshavardhan *et al.*, 2001).

2.1.1. Immature embryo

Following the publication of Strogonov *et al.* (1968), rapid advances in callus induction and regeneration of sorghum from *in vitro* cultured tissues has been achieved and the protocols are constantly improved. Thomas *et al.* (1977) reported the formation of a large number of embryos and shoot like structures from immature embryos excised 10-30 days after pollination. Callus induction and regeneration of sorghum plants from scutella cultured *in vitro* has been studied using 14 different genotypes (Sairam *et al.*, 1999). The anatomical studies on embryogenic calli derived from scutellum revealed 4 to 6 proembryos from each scutellum, with secondary embryos initiating from the primary ones after about a month under culture. Origin of somatic embryos is mainly from the sub-epidermal cells of scutellar ridge. Callus formation ranged from 14 -100% for the 14 genotypes studied, and regenerating ability varied from 10-100%. Popular seed parents 296B and BT×623, which are extensively used in sorghum improvement, have shown up to 100% callus induction and regeneration. The performance of the plants derived from seeds of the *in vitro*-grown plants, when tested in field was found to be similar to the normal control plants.

2.1.2. Shoot tip culture

Use of shoot tip for *in vitro* culture facilitates year-round availability of the explant material for tissue culture. Somatic embryogenesis was observed after incubation of such explants in dark for 6 to 7 weeks through friable embryogenic callus (FEC) phase. Multifold increase in somatic embryo induction, secondary embryoid formation and high conversion frequency of embryoids to plantlets may be helpful for increasing the transformation frequency. Linsmaier and Skoog (LS) medium supplemented with 2,4-dichloro-phenoxy acetic acid (2 mg L^{-1}) and kinetin (0.1 mg L^{-1}) was used for induction of friable embryogenic calli (FEC) and somatic embryos. Seeds from *in vitro* regenerated plants produced normal crop in the field comparable to the crop grown with the seeds of mother plant used to initiate tissue culture (Seetharama *et al.*, 2000). The protocol is suitable for both wild and cultivated species of sorghum.

2.1.3. Immature inflorescence culture

Brettell *et al.* (1980) observed embryo like structures in immature inflorescence segments. Cai and Butler (1990) reported successful plant regeneration from high tannin sorghum varieties using immature inflorescences as explant. We used immature inflorescences, while still enclosed in the boot and flag leaf and measuring 3-5 cm (M35, 296 B, BT×623 and SPV 104) showed best morphogenetic response. Among the four genotypes tested, M35-1 showed best performance. Young inflorescence ranging in length from 1.0 - 1.5 cm were cut into 4-5 pieces and placed on LS callusing medium with 2.5 mg L^{-1} 2,4-D and 0.5 mg L^{-1} kinetin and incubated in dark at 26^0C. Three weeks after transfer onto LS medium, considerable amount of callus growth was observed. This was then subcultured on to the same basal medium supplemented with 1.5 mg L^{-1} 2, 4-D and 0.5 mg L^{-1} kinetin. Callus differentiated into embryogenic and non-embryogenic masses after 2 to 3 weeks (Fig. 2A). For formation of shoots, embryogenic callus was transferred on to MS medium with 1 mg L^{-1} BAP and 0.2 mg L^{-1} NAA. Shoot initials were observed 1 week after transfer and then elongated to 2 - 3 cm in 4 weeks. Next, shoots were transferred on to half-strength MS rooting medium with 2.0 mg L^{-1} IBA and 0.2 mg L^{-1} IAA. Fully rooted plants cultured in the above medium set normal seeds when in greenhouse.

2.1.4. Shoot meristem

Cultured immature embryos and shoot tips of cereals are widely used for genetic transformation (Vasil and Philipes, 2000). Recently, we have refined the conventional shoot tip culture protocol reported earlier (Seetharama *et al.*, 2000), which is somewhat similar to the technique for shoot tip culture in corn (Zhong *et al.*, 1992). The technique involves isolation of only meristematic tissue and a pair of primordial leaves from the shoot tip, which is reprogrammed *in vitro* to produce multiple shoots or a large number of somatic embryos (Harshavardhan *et al.*, 2001). Such responding meristems are targeted for genetic transformation at appropriate developmental stages. We could successfully choose the pathway of regeneration - organogenesis or direct somatic embryogenesis

Figure 2 : Transient GUS expression in different explants after bombardment and various stages of confirmation of putative transgenics.
A : Immature inflorescence - derived embryogenic calli.
B-D : Transient *GUS* expression in different explants: (B) inflorescence - derived calli; (C) Immature embryos, and (D) isolated shoot meristems.
E-G : Regeneration from different explants after bombardment during selection: (E) shoot meristem derived somatic embryos; (F): immature embryo-derived calli; (G): immature inflorescence-derived calli.
H-K : GUS gene expression parts of a transgenic plant (T_0) (in each controls are on the left): (H) in microspore; (I) in anther; (J) in stigma and, (K) in spikelets.
L-M : Southern analysis: (L) Hybridization pattern of putative transformants # 1-6. Genomic DNA was restriction digested with the enzyme *Xba I* and hybridized with *GUS* fragment of plasmid DNA (lane P in figure L). *GUS* regions were detected in transformed plants (lanes 1, 2, 3, 5 and 6) whereas they were absent (blank lane) in non - transformed plant (lane 4) as in the case of control plant (lane C in the figure L); (M) Southern hybridization pattern of T_1 transgenics (derived from T_0 plant # 3) restriction digested with the enzyme *Xba I* and hybridized with *GUS* fragment. Lanes 1-20 are genomic DNAs from T_1 transgenics (Plant #s 1-20); Lane 21 labeled as C: control plant (genotype 296B); Lane 22: digested plasmid (*PATGUS*) DNA; Lane 23: undigested plasmid DNA; Lane 24 labeled as M: 1 Kb DNA marker (M).
N-P : Herbicide resistance tests with T1 progeny. (N) Response in resistance with different levels to *Basta* spray in control, non - transgenic plants; (O) T1 progeny, painted with 0.5 % *Basta*. Plant on the left is resistant, with no scorching symptoms, and the one on the right susceptible and, (P) Control *Basta* leaf paint experiment

- by striking an optimal balance between BAP and NAA concentrations. There was no intermediate callus formation in both cases. Multiple buds were induced on enlarged meristems using MS medium with 1.1 mg L^{-1} of TDZ, 4.0 mg L^{-1} BAP and 0.1 mg L^{-1} NAA. To maximize the number of bud initials per explant, three parameters - seed size, germination technique, and age of explant - were studied. Five to seven day old shoot meristems responded most with ≥ 80% induction of bud initials. The size of the seed used as starting material did not influence the multiple bud induction potential.

Six weeks after *in vitro* culture, each meristem produced 35-40 shoot buds. Among the three sorghum genotypes (BT×623, M35-1 and 296B) tested for their induction and regeneration response, induction percentage was 80.0% in BT×623; 72.5% in M35-1; and 54.5% in 296B. The corresponding regeneration frequencies were 91.7, 89.6 and 84.6%, respectively. Direct somatic embryogenesis (≥ 80% induction) was accomplished following a two-step culture procedure consisting of induction of multiple buds and formation of somatic embryos. A high frequency (700 - 1000) of somatic embryos per explant was obtained on MS medium with 4.0 mg L^{-1} BAP and 0.1 mg L^{-1} NAA from the translucent outgrowths or tissue stratum. Plantlets from both organogenic and embryogenic pathways rooted on half-strength MS medium with 2.0 mg L^{-1} IBA and 0.2 mg L^{-1} IAA. After one month on rooting medium, plants with well-developed roots were transferred to jiffy cups. Such plants were subsequently acclimatized in the glasshouse and were grown until maturity with normal seed set. Histological studies have confirmed that meristems proliferate through direct organogenesis and somatic embryogenesis. RAPD analysis of regenerants did not detect any somaclonal variation. We believe this system offers excellent target material for genetic transformation.

2.2. Cell and protoplast cultures

Cell suspension cultures are amenable to study cellular and developmental processes and offer a convenient system for introducing novel genes. Similarly, protoplasts can be used for gene transfer. When the target gene is available, they can be used for transformation by direct DNA uptake. If the target genes are not yet available, somatic hybridization can be used to produce novel recombinants.

2.2.1. Cell suspension cultures

A number of studies have been carried out on tissue culture of cultivated sorghum (Gamborg *et al.*, 1977; Brettel *et al.*, 1980; Wernicke and Brettell, 1982; Mythili *et al.*, 1997). Cell suspension cultures were developed in both cultivated and wild species using callus derived from shoot tip and immature inflorescence. In a wild sorghum species *S. dimidiatum*, high frequency (>80%) somatic embryogenesis from small cell clusters (300- 400 μm) was observed when the cultures were initially maintained in liquid medium with reduced levels of 2,4-D (0.25 mg L^{-1}; Fig. 1H), followed by transfer to regeneration medium (Fig 1J). Direct plating of these small clusters on regeneration medium (Fig. 1I) or transfer to liquid regeneration medium containing kinetin and BAP resulted in the development of mature somatic embryos (Fig. 1K) and plantlets (Figs. 1L and 1M). Though establishment and maintenance of cell suspensions is more difficult in cultivated than in wild species, with the improved protocol, it has been possible to obtain fine cell clusters and embryoid formation in both types (Mythili *et al.*, 1997). Cultures maintained their regenerative potential for up to 12-15 months by continuous selection of embryogenic clumps at each subculture. When perfected, cell suspensions will be useful for routine protoplast isolation, cryopreservation and genetic transformation studies.

2.2.2. Protoplast culture

In contrast to several reports on protoplast culture in cereals such as rice and wheat, only five publications are there in sorghum as mentioned by Sairam *et al.* (1999). Apart from the above five reports, there is only one report on successful plant regeneration from young mesophyll-derived protoplasts of sorghum (Sairam *et al.*, 1999). Factors important for regeneration include growth conditions of the donor plant, genotype, and protoplast isolation methodology. As with other explants, genotypic differences were significant in protoplast culture as well. Genotype IS 32266 gave maximum yield (10^8 million protoplasts g^{-1} leaf tissue). Wall regeneration started within 24 -48 hours and further divisions were seen after 6-10 days (Figs. 1A-D). Maximum division frequency was 4.8% in 296B. Microcolonies were seen 15-20 days after culture, which were sub-cultured for further proliferation and subsequently transferred to regeneration media and plants were recovered (Figs. 1E-G).

2.3. Tissue culture of wild relatives of sorghum

Wild relatives of sorghum are often the only source for high levels of insect resistance. Since they are not readily crossable with the cultivated types, a technique like somatic hybridization needs to be developed. This is of critical importance when the resistance genes in wild sorghums have not been identified or resistance is distributed across the whole genome.

Plant regeneration from wild species, *viz., S. arundinaceaum* (Boyes and Vasil, 1984), *S. almum* (George and Eapen, 1988), *S. versicolor* (Eapen and George, 1990), *S. nilolium, S. miliaceum, S. virgatum, S. plumosum, S. sudanences* and *S. aethiopicum* has been reported. However, there are very few reports on cell suspensions of cultivated and wild sorghums (Chourey and Sharpe, 1985). Embryogenic cell suspension cultures of both cultivated and wild sorghum species have been established from shoot tip and immature inflorescence derived calli. We established cell suspension cultures with high frequency somatic embryogenesis for two species of section *Parasorghum* with 2n =10 (*S. dimidiatum*) and 2n = 20 (*S. australiense*). In *S. dimidiatum,* callus is mostly friable and resembles type-II callus. This explant was found suitable for culture of both cultivated and wild species. Callus induction was 100% in both cultivated and wild species, and 80% regeneration was observed up to 9 months after culture. Regenerated plants were transferred to glasshouse and set seed. In *S. australiense*, regenerants from 6-9 months old cultures were found to be chimeric, and several cytological variations were found.

The above protocol of regeneration of small cell clusters directly in liquid medium offers a system for fundamental studies on differentiation, regeneration and gene isolation. In addition, these suspension systems will expedite the transfer of genes from wild to cultivated varieties. The suspension cultures, apart from serving as the best source for isolation of protoplasts could be directly used for genetic transformation.

3. GENETIC TRANSFORMATION STUDIES

Genetic transformation is now widely accepted as method of choice for alien gene transfer into selected cultivars for improving complex traits for which natural variation in the existing gene pool of crops is inadequate. Compared to all other major world crops, sorghum transformation is still in its infancy, and a much technical progress is yet to be achieved. Attempts to genetically transform sorghum to date have been summarized in Table 1.

3.1. Major steps in genetic transformation

Besides an amenable tissue culture and regeneration system, following

Table 1. Summary of attempts on genetic transformation of sorghum to date

S. No	Method of transformation	Explant/ culture system	Objective, gene of interest	Selection agent/ conc. used.	Observations and remarks	Reference
1	Electroporation	Cell suspension/ protoplasts	*nptII*	Kanamycin, 100 mg L^{-1}	Stable transformation. Non-morphogenic callus regenerated	Battraw and Hall (1991).
2	PDS- 1000/He (Bio-Rad)	Cell suspension culture	*nptII/hpt*	kanamycin/ hygromycin	Stable transformation. Non-morphogenic callus regenerated	Haigo *et al.* (1991)
3	PDS-1000/He	Immature embryos	*bar*	Bialophos, 3 mg L^{-1}	Plants regenerated at low frequency	Casas *et al.* (1993)
4	PDS-1000/He	Immature embryos/ shoot tips	*nptII/dhdps -raec 1*	kanamycin/ For more lysine content.	Obtained > 10 plants. Showed southern positive.	Tadesse and Jacobs (1995)
5	PDS-1000/He	Immature embryos/ inflorescence derived callus	*bar*	bialophos	Plants regenerated at low frequency	Kononowicz *et al.* (1995)
6	PDS-1000/He	Immature embryos	*bar/chitinase*	Basta/ 1-2 mg L^{-1}	Showed southern positives	Zhu *et al.* (1998)

Table 1. Continued

S. No	Method of transformation	Explant/ culture system	Objective, gene of interest	Selection agent/ conc. used.	Observations and remarks	Reference
7	PDS-1000/He	Shoot meristems isolated from germinating seedlings	*bar/HVA I*	glufosinate/ 10 mg L^{-1}	For drought tolerance. Southern conformed. Further analysis on.	Devi *et al.* (1999)
8	PIG (Particle Inflow Gun)	Immature embryos/ inflorescence derived callus	*bar*	bialophos/ 2 mg L^{-1}	Single plant reported.	Rathus *et al.* (1996)
9	PIG	Immature embryo	*bar*	Basta/ 1-2 mg L^{-1}	Casein hydrolysate was used for increasing regeneration frequency.	Rathus and Godwin (2000)
10	PIG	Shoot meristems isolated from germinating seedlings	*bar/ cry1A(b) & cry1 A(c)*	Basta/ 2 mg L^{-1}	For insect resistance. Southerns confirmed one plant.	Gray *et al.* (1999)
11	PIG	Immature inflorescence derived callus	*bar*	Basta/ 2 mg L^{-1}	Low frequency of regeneration. Abnormal shoots reported.	Rani *et al.* (2000)

Table 1. Continued

S. No	Method of transformation	Explant/ culture system	Objective, gene of interest	Selection agent/ conc. used.	Observations and remarks	Reference
12	*Agrobacterium* (LBA4404)	Immature embryos derived callus.	*gfp/bar/ chitinases (G11, TLP & RC7)*	bialophos/ 3 mg L^{-1}	Southerns not reported	Jeoung *et al.* (1999)
13	*Agrobacterium* (LBA4404)	Immature embryos derived callus.	*bar*	PPT/5 mg L^{-1}	2.1 % transformation frequency reported.	Zhao *et al.* (2000)

Abbreviations : *nptII* - neomycin phosphotransferase; *bar* - bailophos resistance; gfp - green fluorescence protein gene; *hpt* - hygromycin phosphotransferase.

factors are also critical for successful transformation: (i) suitable target gene in a convenient vector with reporter (to confirm alien gene introduction into target cell or tissue) and selectable (to eliminate those cells into which foreign DNA has not been incorporated) marker genes, (ii) method of DNA delivery into cell, and (iii) efficient testing method to confirm transformation events (stable integration). It is equally important to ensure consistent inheritance and expression of the transgene in the progeny. Lack of pleotropic effects, and consideration of biosafety issues are important before useful transgenic can be commercialized.

3.1.1. Sorghum genotypes and explants used

Ideally, genotype-independent protocols are preferred, but this is difficult to achieve. In sorghum, embryogenic calli derived from immature embryo or immature inflorescence is commonly used. In our experience, isolated shoot meristems can be very useful as an explant especially for transformation by particle bombardment (see below). To date, experience of genetic transformation with cell suspension culture is limited, and the single report on the use of protoplast is only on transgene expression, without further regeneration of sorghum plant (Battraw and Hall, 1991).

3.1.2. Genes and plasmids used for transformation studies

The plasmids used to carry the gene(s) of interest are the common commercial ones like pBI series (more in the references mentioned in Table I). At ICRISAT, we used pJS108, which contains a *bar* and a GUS gene (plasmid pPATGUS) kindly provided by Dr. Ray Wu of Cornell University. The selectable *bar* gene is under the control of 35S promoter and it encodes phosphinothricin acetyl transferase (PAT), which inactivates by acetylation, phosphinothricin (PPT - the active ingredient of herbicide Basta). The GUS reporter gene is under the control of rice actin promoter and codes for the enzyme β-glucuronidase. To study the effect of co-bombardment, a construct of synthetic *cryIA(b)* or *cryIA(c)*gene with the ubiquitin promoter (obtained from Dr I. Altosaar of Montreal University, Canada) was also used along with PATGUS construct.

3.1.3. Choice of selectable marker

Earlier attempts to genetically transform sorghum used neomycin phosphotransferase II (*npt*II) gene isolated from *E. coli*, which confers resistance to the antibiotic kanamycin [Battraw and Hall (1991); 100 mg L^{-1}; Hagio *et al.* (1991). 500 mg L^{-1}]. Selection was not satisfactory because sorghum cell cultures show natural resistance to kannamycin.

Therefore, attempts were made to look for alternate selection agents. Hagio *et al.* (1991) used the *E. coli* hygromycin phosphotransferase (*hph*) gene (1-2 mg L^{-1}), which confers resistance to hygromycin. In our experience with calli from common Indian genotypes, a higher concentration of hygromycin is required for selection (LD_{100} = 8 mg L^{-1}, and LD_{50} = 4 mg L^{-1} hygromycin). The obvious disadvantage with the use of these antibiotic marker genes is the lack of wide public acceptance, apart from environmental concerns. Further, hygromycin is highly photosensitive and cannot be easily used as the selection agent during selection of bombarded cultures such as meristems, for which adequate exposure to light is important. To date, the most successful and the most popular selection marker has been the *bar* gene of *Streptomyces hygroscopicus*, which encodes the enzyme phosphinothricin acetyltransferase (PAT) conferring resistance to the herbicide Basta (or glufosinate, biolophos and phosphinothricin) (Casas *et al.*, 1993; Rathus *et al.*, 1996; Zhu *et al.*, 1998; Zhao *et al.*, 2000 and Rani *et al.*, 2001). The concentration of selection agent added to the medium when using *bar* gene varies from 8-10 mg L^{-1} glufosinate, 2-3 mg L^{-1} bialophos, and 2-5 mg L^{-1} phosphinothricin. The biggest barrier for extensive use of the *bar* gene is the possible gene escape to pollinate wild sorghums, which may result in the development of herbicide resistant weeds. Recently, Weeks *et al.* (2000) used a new selectable marker *cah* for wheat transformation. This gene *cah* isolated from *Myrothecium verrucaria* encodes for the enzyme cyanamide hydratase, which converts cyanamide into urea. For selection of the transformed calli, they used 37.5 mg L^{-1} of cyanamide, in the culture medium. This marker may overcome the disadvantages posed by earlier markers and thus, may stimulate transgenic research, just as the commercially available selectable markers such as mannose phosphate isomerase system of *Novartis.* Until such system are readily available in the public domain, it is likely that co-bombardment followed by elimination of marker genes by employing marker-assisted selection is the most appropriate strategy.

3.2. Regeneration, rooting and transfer of putative transgenics to greenhouse

Regeneration following selection with antibiotic or herbicide resistance markers is commonly carried out with half-strength MS medium containing small amount of auxin for root initiation (e.g., 1.0 mg L^{-1} of IAA). In case of explants that showed weak rooting, an additional treatment of IBA at a concentration of 1.0 mg L^{-1} can be applied to the explant. For this, the explants are dipped in the solution of IBA 1.0 mg L^{-1} for about

30-min. before sub-culturing them onto the rooting media. Plants with well-established roots are obtained within 3 weeks after transfer onto the rooting medium. They can be transferred into the jiffy cups containing autoclaved *Soilrite*™ in a culture room for acclimatization before transfer to greenhouse for further hardening. Hardened seedlings are planted in pots in a containment facility and grown to maturity.

3.3. Methods of genetic transformation of sorghum

Among the three common methods of genetic transformation of sorghum (Table I), the biolistics method followed by the one based on agroinfection are common (Table 1). There is a single report of use of electroporation, but in this case, the plants were not regenerated (Battraw and Hall, 1991). However, with better cell suspensions and protoplast system available now, electroporation cannot be written-off. Direct uptake of exogenous DNA by immature seeds, or injection of *Agrobacterim* directly into shoot apiece (Chickwamba and Godwin, 1992) can produce transgenics without the need for tissue culture, but these methods have not been adequately tested in sorghum.

3.4. Microprojectile bombardment

Particle bombardment is an efficient method of genetic transformation of cereals when agro-infection is difficult. Particle bombardment also facilitates introduction of multiple genes in a simple fashion. To date PDS-1000/He gene delivery device of *Bio-rad L*aboratories (Richmond, California) is the most popular device used for direct gene transfer (Table I). However, this device is expensive, has high operational costs, and the time interval between two bombardments is long. Further, there are complications related to intellectual property rights, as the device is only licensed and not sold. As an alternative, we at ICRISAT have constructed a particle inflow gun (PIG) with the help of scientists of University of Queensland in Australia and *Endeavour Enterprises* at Hyderabad following the design of Finer *et al.* (1992) (Fig. 3). It is cheap and easy to construct, needs fewer supplies, and simple to operate. Therefore, we briefly describe its use here.

3.4.1. Working principle of the particle-inflow gun

Particle bombardment device is designed to pump tungsten or gold particles into target tissue under partial vacuum (using a timed burst of helium gas). The system consists of a thick-walled vacuum chamber having a door fitted with latches at two places (Fig. 3). The interior of

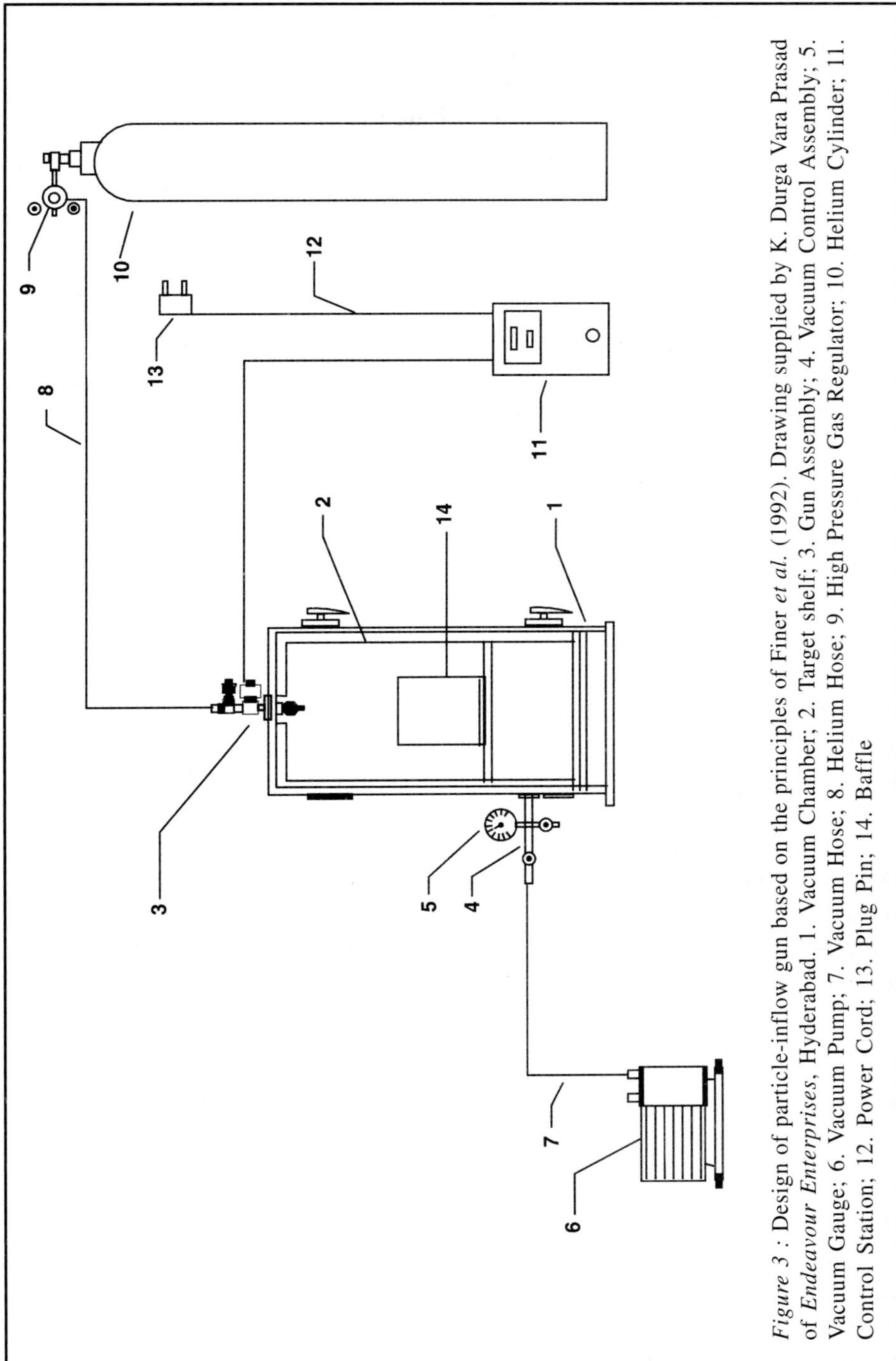

Figure 3 : Design of particle-inflow gun based on the principles of Finer *et al.* (1992). Drawing supplied by K. Durga Vara Prasad of *Endeavour Enterprises*, Hyderabad. 1. Vacuum Chamber; 2. Target shelf; 3. Gun Assembly; 4. Vacuum Control Assembly; 5. Vacuum Gauge; 6. Vacuum Pump; 7. Vacuum Hose; 8. Helium Hose; 9. High Pressure Gas Regulator; 10. Helium Cylinder; 11. Control Station; 12. Power Cord; 13. Plug Pin; 14. Baffle

the chamber has racked horizontal adjustable shelves (which are separated by 1.0 cm apart) to accommodate petridishes containing target tissue. An electronic control device, solenoid and metering valves were connected to the bombardment chamber to facilitate the conditions and acceleration of microprojectiles. The helium gas cylinder with pressure gauge is connected to the needle adapter (fitted at the top of the chamber) *via* a solenoid valve. On the other side of the chamber, a vacuum pump is connected *via* a vacuum gauge with an opening to release the vacuum to outside.

For each bombardment, concentrated particle suspension is placed in the center of the pre-autoclaved screen of the syringe filter unit, which is then screwed into the needle adapter in the PIG. Next, petri-plate containing the target tissue is placed on the adjustable shelf at the desired distance from the tip of the filter unit. Baffles made of nylon screens (1.0 mm) are placed above the petri-plates. The opening time for helium solenoid valve is set to 0.1 sec. Vacuum pump is turned on, and on raising the vacuum to the desired level (550-600 mm Hg) as indicated by the gauge, the solenoid-activating switch is pressed. This results in a burst of helium gas, which accelerates the tungsten particles towards the target material. Following firing, the bombardment chamber is vented to atmospheric pressure, and DNA coated particles are loaded for another round of bombardment.

3.4.2. Factors affecting microprojectile bombardment

Several parameters of particle bombardment such as preparation of DNA samples, pretreatment of explants, the flight distance of DNA coated tungsten particles, and helium pressure affect the efficiency of transformation. These factors are described by Rathus *et al.* (2001) and Rani *et al.* (2001). Standardization of protocols is not difficult and the parameters are independent of genotypes, and even explants (Rani *et al.* 2001).

3.4.3. DNA preparation, bombardment, and transient gene expression analysis

DNA mixture for bombardment is prepared as follows : 50 μL of tungsten suspension is transferred into sterile microcentrifuge tube and vortexed; to this, 10 μL of plasmid DNA, 50 μL of 2.5 M $CaCl_2$ and 20 μL of 100 mM spermidine are added in that order. After addition of each component, tube is vortexed briefly. This mixture is then placed on ice for 5 min. When the tungsten along with the coated DNA settled down, 100 µL of the supernatant is discarded and the rest of the aliquot is used for microprojectile bombardment. For each bombardment, 5 µL of the concentrated particle suspension is loaded into the syringe filter unit. Petridishes containing the explants is then placed on the adjustable shelf at 4, 8 or 15 cms distance from the tip of the filter unit. Explants are next bombarded under a helium gas pressure of 10, 15 or 18 kg cm^{-2} under partial vacuum (600 mm Hg).

Four hours before bombardment, the explants are placed on osmotic medium (containing 0.4 M mannitol and 0.4 M sorbitol) with 10 explants arranged over a sterile filter paper at the center of the petri-plate (3-cm diameter). After particle bombardment, explants are left on the osmoticum medium for the next 24 hours in the culture room. Later they are transferred to induction medium. Next, explants are analyzed for transient GUS expression according to the procedure of Jefferson *et al.* (1987) (Fig. 2B-D). In our experiments, average number of blue spots (GUS assay) on each explant mass was maximum with those derived from immature inflorescences. Therefore, we initially used those calli to standardize conditions for bombardment, and the results were verified with other explant types. Transient GUS expression was highest at a flight distance of 7 cm for all the explants tested. A pressure of 15 kg cm^{-2} produced maximum number of blue spots with all explant types than 10 kg cm^{-2} or 18 kg cm^{-2} helium pressure. The percentage of the explants showing transient expression was highest in isolated shoot meristems (Fig. 3).

3.4.4. Post-bombardment selection strategies

Kill curves were established and lethal doses determined for Basta and phosphinothricin using *in vitro* cultured explants (non-transformed). For this, explants were transferred onto LS medium (for immature embryos, shoot tips and immature inflorescence derived calli) or MS medium (for isolated shoot meristems) with different concentrations of filter-sterilized Basta (0, 0.1, 0.5, 1, 2, 3 and 5 mg L^{-1}) or phosphinothricin (0, 1, 2, 4, 8 and 12 mg L^{-1}). A concentration of 3.0 mg L^{-1} Basta was sufficient to kill all explants (LD_{100}) in callus cultures derived from the above explants. In the case of phosphinothricin, 50% of the explants died (LD_{50}) at 4.0 mg L^{-1}, and 100% at 12.0 mg L^{-1}. For shoot meristems, LD_{50} was 1.0 mg L^{-1} phosphinothricin, and LD_{100} was observed with 2.0 mg L^{-1}. However, post-bombardment recovery and regeneration rate was more with PPT based selection than that with Basta based selection. In addition, bombardment decreased the frequency of plant regeneration by 5-10 % when these bombarded calli were grown on medium with no selection agent. Bombarded calli resistant to the herbicide PPT were selected with relatively low frequency of abnormal structures, thus increasing the chances of recovery of transformants. In contrast, Basta based selection was not much effective as it increased abnormal structures under selection and hampered successful plant differentiation from such structures (Rani *et al.*, 2001). Further, when Basta was used as a selection agent, the frequency of abnormal roots was more with immature inflorescence and shoot tip derived calli. Once such abnormal roots appeared, it was very difficult to initiate shooting, and as a result, the calli failed to differentiate further. Post-selection (3 mg L^{-1} Basta) recovery was least in immature embryo derived callus, while it was maximum in immature inflorescence derive calli (Rani *et al.*, 2001). When compared to shoot tip or immature embryo derived calli, immature inflorescence-derived calli showed high frequencies of plant regeneration (Rani *et al.*, 2001).

3.5. *Agrobacterium*-mediated transformation

Wherever feasible, *Agrobacterim*-based methods are believed to be more practical as the success rates of transformation are greater than with biolistics. Further, unlike later, complex equipment is not involved. However, it is difficult to develop the protocol especially with cereals. A protocol has recently been developed at Kansas State University (Zhao *et al.*, 2000). Needles to say, several variations and modifications can be tried. The only disadvantage of the method is that both selectable marker and target genes are linked together (same plasmid), thus posing some problem in the acceptance of the transgenics. However, with improvement in technology, these can be overcome (Zuo *et al.*, 2001). Below, we will describe the salient features of the protocol that might be tried for sorghum.

3.5.1. Preparation of *Agrobacterium* culture

LBA4404 (pTOK 233) having 50 mg L^{-1} kannamycin and 50 mg L^{-1} hygromycin and pCAMBIA 1301 having 50 mg L^{-1} kannamycin are streaked on LB solid medium, and grow for 2 days at 28 °C to obtain single colonies. One such colony is picked and inoculated under aseptic conditions in 2 ml of culture of LB liquid medium having the above concentrations of antibiotics, and placed on a shaker at 200 rpm for 24 h at 28°C. Next, overnight liquid culture is diluted 1:5, 1:10 and 1:20 (v/v) with MS liquid medium having 200 ΩM acetosyringone and grown with continuous shaking. For infecting sorghum, a density of 0.5 – 1.0

OD units, (A_{660}) is essential.

3.5.2. Infection

Explants are immersed in *Agrobacterium* culture for 20 min following which, they are blot-dried with sterile filter paper and transferred to the MS co-cultivation medium containing 200 μM acetosyringone and incubated in dark for 2-3 days, at 26°C. After co-cultivation, explants are washed with MS liquid medium containing 400 mg L^{-1} cefatoxime to remove the bacterial growth, and retransfer onto culture medium (LS for callus; MS for shoot meristems), and cultured following standard protocol. For GUS staining, explants were transferred to X-glucuronide staining after thoroughly washing with cefatoxime.

3.6. Electroporation

Sorghum bicolor protoplasts were used successfully transformed *via.* electroporation to introduce kanamycin resistance gene and GUS reporter gene (Battraw and Hall, 1991). Protoplasts isolated from embryogenic suspension cultures were subjected to electroporation with the plasmid DNA carrying the above genes to obtain 77 resistant calli, but these calli failed to regenerate. Factors affecting transformation frequencies include DNA concentration and physical state of the plasmid (circular *vs.* linearized).

3.7. Analysis of the putative transformants

The common method of testing the transgenics is the histochemical localization of *GUS* activity in various organs such as spikelets, anthers, stigmas, microspores and immature embryos of the T_0 regenerants (Rani *et al.*, 2000, 2001) following the procedure of Jefferson *et al.* (1987) as shown in Figs. 2H-K. Next, PCR analysis is carried out using the GUS primers (Hamill *et al.*, 1991), and those suitable for amplification of target genes such as *Bt*. Following PCR, Southern analysis using genomic DNA of transgenics and control plants are carried. (Rani *et al.*, 2001) the results of which are shown in Fig. 2L.

3.8. Analysis of inheritance of selectable marker in T_1 transgenics

When *Basta* was painted on the fifth leaf of T_1 progeny, there was a clear demarcation between susceptible and resistant plants. From the control experiment, it is known that application of 0.5 % *Basta* results in > 90 % scorching of the treated leaf. Therefore in our studies it was assumed that, susceptible plants (that lack the transgene PAT) will show > 90 % leaf scorching, as do control plants, while the transgenic plants will withstand the herbicide (Figs. 2N-P).

3.9. Analysis of transgenic phenotypes for the desired trait and co-segregation of engineered traits

The effectiveness of transgenics can be evaluated even at T_0 generation, especially if the method involved is non-destructive (such as herbicide leaf paint assay and testing GUS gene expression in floral parts; Figs. 2N-P).

4. INHERITANCE STUDIES WITH PROGENY OF TRANSGENICS

It is important to study inheritance and levels of expression of transgene at least for two generations to verify expected ratios of segregation (depending on the number of copies inserted), and to detect any transgene silencing. Such a study with herbicide resistant transgenics is described below. Twenty progeny, from each of the four T_0 plants of the genotype 296B confirmed by southern analysis were grown in the P_2 greenhouse at ICRISAT. Genomic DNA was collected from the leaf samples of 15-day-old seedlings of such T_1 herbicide resistant transgenic plants. After leaf sampling, these plants were allowed to grow to maturity and at flowering, floral parts (anthers microspores and stigmas) were used for GUS histochemical staining. After confirming the segregation among the progeny for blue coloration, Southern hybridization was performed using GUS fragment as the probe ((Fig. 2M). The results of the southern analyses perfectly agreed with those from GUS staining (Rani *et al.*, 2001). The T_1 plants that showed blue coloration with X-Gluc substrate also showed the presence of GUS fragments when subjected to southern hybridization. To verify the above-obtained results from inheritance studies, additional 60 plants from each of the four T_0 plants were used.

4.1. Commercialization of transgenic sorghums and use of genetically modified foods

While transgenic crops have spread as the fastest of all agricultural technologies in the history, the debate on their wide-scale application is still raging especially in the developing countries where poor farmers grow crops like sorghum for their livelihood (Seetharama *et al.*, 2001). Besides public concerns, there are several technical considerations before transgenic are grown widely. Some of these issues are discussed here in the context of cultivation and utilization of sorghum.

4.2. Utilization of transgenics and biosafety considerations

While the debate on commercial cultivation of transgenics and safe use of genetically modified products is continuing, the general conclusion is that a case by case scientific study is warranted for each transgenic, and for specific use. Elsewhere, we have summarized special considerations related to deployment of transgenics with respect of developing countries (Seetharama *et al.*, 2001), and sorghum (Reddy *et al.*, 2001). In the near future, it is likely that major gains will come from insect resistant transgenic sorghums, as the conventional breeding is unable to build adequate level of resistance (Table 2). Obviously, this raises the possibility of insects becoming resistant of transgenics. The challenge can be addressed by use of multiple transgenes and better crop management (Sharma *et al.*, 2000; 2001). Gradually, we can expect the emphasis to shift towards production of specialized products (*e.g.*, better malting sorghums with introduction of β-amylase gene), and efficient whole plant utilization strategies following high-biomass production (*e.g.*, better quality stover by modifying cell wall constituents). In the end, it may be possible to further increase the resistance level of sorghum to physical stresses like water deficit, heat and nutrient stresses making it far more economical to produce even in the marginal areas. Use of transgenic sorghums for industrial purposes such as for biomass or alcohol production can also be envisaged in this decade because of unique qualities of sorghum. All such transgenics are likely to be less controversial than those used for direct consumption by humans, and hence readily acceptable.

4.3. Food safety and environmental considerations

By choosing appropriate gene promoters and processing technologies, one may be able to exclude the product of introduced exotic gene in transgenics from specific end-products such as sorghum grain used for food. However, still questions are raised on the presence of gene sequence itself in the transgenic plant. This is particularly true if antibiotic selectable markers are used during the production of transgenics. The co-bombardment of selectable marker and target gene described by us in this review enables us to eliminate markers simply by backcrossing with

Table 2 : Traits and candidate genes for transformation of Sorghum

	Gene	Gene product	Performance
(I) Against biotic stress			
A) Antifungal genes	- *vst1*	Phytoalexin synthesis	Resistance to Grain mold (*Fusarium* sp.), Grey leaf spot (*Cercospora sorghini*) /Ergot (*Claviceps sorghi*), Anthracnose (*Collitotrichum graminicola*).
	- *afp*	Antifungal protein	
	- *rip*	Ribosome inactivating protein	
	- *cht-1* & *cht-2*	Chitinase	
	- *gluc*	ß-glucanases	
B) Insect resistance	Bt genes - - *cry 1A(a)* - *cry 1A(c)*		Effective against Shoot fly (*Atherigona soccata*)/Stem borers (*Chilo partellus*)/Shoot bug (*Perigrinus maidis*), Midge (*Stenodiplosis*)/Armyworms (*Mythimna separata*), Head caterpillers (*Helicoverpa armigera; Heliothis zea*).
	- Enzyme inhibitors - (*CpTI*) (*SPIs*) - (*TPIs*) (*SKT I*)	Cowpea trypsin inhibitor, serine protease inhibitors, thiol protease inhibitors, soybean Kunitz trypsin inhibitor.	
	- (*WAAI*)/ (*BAAI*)	Amylase inhibitors from wheat and common bean.	
	- Plant lectins – *GNA*	snowdrop lectin gene	
	- *VIP 1* & *VIP 2*	Vegetative insecticidal protein	

Table 2. Continued

	Gene	Gene product	Performance
(II) For abiotic stress tolerance			
	Bet A	Choline dehydrogenase	Salinity tolerance.
	codA	Choline oxidase	Salinity and cold tolerance.
	HVA1	Group 3 LEA protein	Tolerance to water deficit and salinity tolerance
	MsFer	Ferritin	Tolerance to oxidative damage.
(III) Quality traits			
	- β-*amylase*	β-amylase enzyme in germinating grains	Better brewing quality
	- High-lysine protein genes	Grain protein	Nutritional quality
	- Lignin modifying enzymes	-	Digestibility of forage

the non-transgenic control plant and marker-assisted selection. To overcome this, better selection procedures employing safer selection methods such as use of cyanamide reductase (Weeks *et al.*, 2000) or mannose phosphate isomerase system (of *Novartis*) is available. Also, note the chemical-regulated, site-specific DNA excision in transgenic plants by Cre/loxP DNA recombination systems (Zuo *et al.*, 2001). Of particular interest is the elimination of marker genes like herbicide resistant gene in pollen from the point of environmental safety. In this regard, productions of transgenics in which exotic genes are introduced into the chloroplasts are very pertinent. With the successful development of protoplast regeneration system developed by us, this is an attractive proposition for sorghum.

4.4. Integration of transgenics with conventional breeding

The seed, especially the hybrid seed will be the final vehicle for delivery of transgenic technology to the farmers. This would mean effective integration of conventional breeding with transgenic research. A case study with special reference to sorghum has been described by Reddy and Seetharama (2001). The hybrid sorghum particularly offers an opportunity to separately pyramid two sets of traits in different parents; the efficiency of this approach is not yet adequately tested in any crop. For the time being, it is likely that most of the transgenes will be first introduced into the proper female parent (A line), and most of the natural resistance may be bred into the male (restorer) lines.

4.5. Public acceptance and economic constraints and opportunities

The questions related to public acceptance are not unique to sorghum transgenics. However, considering the fact that sorghum is a poor man's food crop, and only a small proportion of production enters global market, public institutions will have grater say and responsibility in this respect (Reddy *et al.*, 2001). On the other hand, if the cost of transgenic technology is too high, sorghum farmers many not be able to use it. Transgenics developed for industrial uses and those for export purposes will receive special consideration. All these issues have to be considered case by case on scientific lines, and ultimate decisions will be made after weighing the extent of economic and social benefits against acceptable magnitude of risk, if any.

5. CONCLUSIONS AND FUTURE PROSPECTS

The establishments of callus and cell cultures from various explants, which are competent to express totipotency, have become routine. To enhance the tissue culture response further, identification of genes controlling the *in vitro* response and mapping these is highly desirable. Additional research on some of the factors like the production of phenolics (pigments) and appearance of abnormal root and leaf like structures from the callus which drastically decreases the regeneration frequency can help in the progress of *in vitro* response of sorghum. Parallel developments are needed in improvement of other transformation methods like the *Agrobacterium* method and the identification and isolation of genes controlling agronomically important traits like disease resistance and stress tolerance. Access to better gene promoters and self-excising selector (Zuo *et al.*, 2001) regulatory systems, and suitable breeding programs combining efficient *in vitro* manipulation techniques could soon lead to the production of sorghum cultivars with novel traits.

Sorghum can be transformed with a number of candidate genes to increase resistance to biotic and abiotic stress factors as listed in the Table 2. The secondary gene pool of sorghum is a rich depository of resistance genes. Until the time these genes are isolated and fully characterized, it is still desirable to consider cellular techniques like somatic hybridization to incorporate novel resistance genes into the pool of cultivated sorghum germplasm. With the large-scale genome analysis of many organisms under way, in the long run we should be prepared to venture introducing novel sequences controlling plant growth and development to suit a variety of needs. Gradually it should be possible to pyramid multiple genes, *in vitro* within large vectors like binary bacterial artificial chromosomes (BI-BACs) that may be used for sorghum transformation. Such constructs will pave way for genetic engineering of metabolic pathways, and manipulation of such polygenic traits, there by enhancing the utility of transgenic approach.

REFERENCES

Battraw M and Hall TC (1991). Stable transformation of *Sorghum bicolor* protoplasts with chimeric neomycin phosphotransferase II and β-glucuronidase genes. *Theor. Appl. Genet.*, **82** : 161-168.

Bhaskaran S, Neumann AJ and Smith RH (1988). Origin of somatic embryos from cultured shoot tips of *Sorghum bicolor* (L.) Moench. *In Vitro Cell Dev. Biol.*, **24** : 947-950.

Boyes CJ and Vasil IK (1984). Plant regeneration by somatic embryogenesis from cultured young inflorescences of *Sorghum arundinaceum* (Desv) stapf. Var.

sudanese (Sudan grass). *Plant Sci. Lett.*, **35** : 153-157.

Brettel RIS, Wernicke W and Thomas E (1980). Embryogenesis from cultured immature inflorescences of *Sorghum bicolor*. *Protoplasma,* **104** : 141-148.

Cai T and Butler L (1990). Plant regeneration from embryogenic callus initiated from immature inflorescences of several high tannin sorghums. **20** : 101-110.

Cai T, Daly B and Butler L (1987). Callus induction and plant regeneration from shoot portions of mature embryos from high tannin sorghums. *Plant Cell Tiss. Org. Cult.,* **9** : 242-252.

Casas AM, Kononowicz AK, Zehr UB, Tomes DT, Axtell JD, Butler LG, Bressan RA, Hasegawa PM (1993). Transgenic sorghum plants via microprojectile bombardment, *Proc. Natl. Acad. Sci. USA,* **90** : 11212-11216.

Casas AM, Kononowicz AK, Zehr UB, Zhang L, Haan TG, Tomes DT, Bressan RA and Hasegawa PM. Approaches to the genetic transformation of sorghum In: *Maize Biotechnology: recent developments in maize transformation applications,* pp 89-93.

Chikwamba R and Godwin I (1992) Developing an *Agrobacterium* mediated gene transfer system for *Sorghum bicolor.* 32. *Abstracts of the Australian Society of Plant Physiologists 32nd Annual General Meeting.* 28 Sep 1992. Melbourne, Vic. (Australia). *Aust. Society of Plant Physiologists.* 1992. p8.

Chourey PS, Sharpe DJ (1985) Callus formation from protoplasts of Sorghum cell suspension cultures. *Plant Sci.,* **39** : 171-175

Chourey PS, RE Lloyd, DZ Sharpe and NR Isola (1986) Molecular analysis of hypervariability in the mitochondrial genome of tissue cultured cells of maize and sorghum. In: *The chondriome-chloroplast and mitochondrial genomes.* (Eds. Mantell SH, Chapman GP and Street PFS). Wiley and Sons, New York, pp 171-191.

Devi P, Zhong H and Sticklen M (2001). Production of drought tolerant transgenic sorghum plants using barley *HVA1* gene. In: *Sorghum Tissue Culture, Transformation & Genetic Engineering* (Eds. N. Seetharama & I.D. Godwin) ICRISAT & Oxford Publishers, India.

Eapen S, George L (1990) Somatic embryogenesis and plant regeneration in inflorescence segments of *Sorghum versicolor. Maydica,* **35** : 55-58

Finer JJ, Vain P, Jones MW and McMullen MD (1992). Development of the particle inflow gun for DNA delivery to plant cells. *Plant Cell Rep.,* **11** : 323-328.

Gamborg OL, Shyluk JP, Brar DS and Constable F (1977) Morphogenesis and plant regeneration from callus of immature embryos Sorghum. *Plant Sci. Lett.,* **10** : 64-67

George L and Eapen S (1988) Plant regeneration by somatic embryogenesis from immature inflorescence cultures of *Sorghum almum. Ann. Bot.,* **61** : 589-591.

Gray SJ, Rathus C and Godwin ID (2001). Enhancing transgene expression levels in sorghum: current status and future goals. In: *Sorghum Tissue Culture, Transformation & Genetic Engineering* (Eds. N. Seetharama & I.D. Godwin) ICRISAT & Oxford Publishers, India.

Hagio T, Blowers AD and Earle ED (1991). Stable transformation of sorghum cell cultures after bombardment with DNA coated microprojectiles. *Plant Cell Rep.,* **10** : 260-264.

Hamiel JD, Rounsley S, Spencer A, Todd G and Rhodes MJC (1991). The use of the polymerase chain reaction in plant transformation studies. *Plant Cell Rep.,*

10 : 221-224.

Harshavardhan D, Rani TS, Ulaganathan K and Seetharama N (2001). Rapid and high efficiency regeneration of *Sorghum bicolor* (L.) Moench from isolated shoot apices. *Plant Cell Tiss. Org. Cult.* (Submitted).

Jefferson RA, Kavnagh TA and Bevan M (1987). GUS fusions : β- *glucuronidase* as a sensitive and versatile gene fusion marker in higher plants. *EMBO J.,* **6** : 3907-3910.

Jeoung JM, Trick S, Krishnaveni S, Muthukrishnan S and Liang GH (2001). *Agrobacterium*-mediated transformation of grain sorghum. In: *Sorghum Tissue Culture, Transformation & Genetic Engineering* (Eds. Seetharama N and Godwin ID) ICRISAT & Oxford Publishers, India.

Klein TM, Fromm M, Weissinger A, Tomes D, Schaff S, Sletten M and Sanford JC (1988). Transfer of foreign genes into intact maize cells with high-velocity microprojectiles. *Proc. Nat. Acad. Sci. USA,* **85** : 4305-4309.

Kononowicz AK, Casas AM, Tomes DT, Bresan RA and Hasegawa PM (1995). New vistas are opened for sorghum improvement by genetic transformation. *Afr. Crop Sci. J.,* **3** : 171-180.

Ma H, Gu M and Liang GH (1987). Plant regeneration from cultured immature embryos of *Sorghum bicolor* (L.) Moench. *Theor. Appl. Genet.,* **73** : 389-394.

Murashige T and Skoog F (1962). A revised medium for rapid growth and bioassay with tobacco tissue cultures. *Plant Physiol.,* **15** : 473-497.

Mythili PK, Rao SV and Seetharama N (1997). Creating novel genetic variability through tissue culture for stress tolerance in sorghum and pearl millet. In: *Crop Improvement for Stress Tolerance* (Eds. Behl RK, Singh DP, Lodhi GP). CCHAU, Hisar and New Delhi: p 233-247.

Mythili PK, Seetharama N and Reddy VD (1998). Plant regeneration from embryogenic cell suspension cultures of wild sorghum *Sorghum dimidiatum* (Stapf.) *Plant Cell Rep.,* **18** : 424-428.

Rani TS, Harshavardhan D, Mythili PK and Seetharama N (2001). Factors affecting genetic transformation of *Sorghum bicolor* using microprojectile bombardment. *Plant Cell Rep.,* (submitted).

Rani TS, Mythili PK, Harshavardhan D and Seetharama N (2000). Sorghum genetic transformation research at ICRISAT. International Conference on Microbial Biotechnology Trade and Public Policy (15-17 July). Department of Microbiology, Osmania University Hyderabad, India. Pp: 93.

Rathus T, Nguyen T, Able JA, Gray SJ and Godwin ID (2001) Optimizing sorghum transformation technology via somatic embryogenesis. In: *Sorghum Tissue Culture, Transformation & Genetic Engineering* (Eds. Seetharama N and Godwin ID) ICRISAT & Oxford Publishers, India.

Rathus C and Godwin ID (2000). Transgenic sorghum (*Sorghum bicolor*) In: *Biotechnology in Agriculture and Forestry, Vol. 46, Transgenic crops I* (Eds. Bajaj YPS). Springer-Verlag. pp. 76-83.

Rathus C, Adkins AL, Henry RJ, Adkins SW and Godwin ID (1996). Progress towards transgenic sorghum. In: Foale MA, Henzell RG, Kneipp JF (Eds) Proceedings of the Third Australian Sorghum Conference, Tamworth, Feb 20-22, 1996. Occasional publication, no. 93. Australian Institute of Agricultural Science, Melbourne, pp 409-414.

Reddy BVS and Seetharama N (2001a). Sorghum improvement: a case study for integrating traditional breeding and transgenic research methods. In: *Sorghum Tissue Culture, Transformation & Genetic Engineering* (Eds. Seetharama N & Godwin ID) ICRISAT & Oxford Publishers, India.

Reddy P, Seetharama N and Rani TS (2001b). Considerations for adaptation of transgenic crops with special reference to agriculture in developing countries. In: *Sorghum Tissue Culture, Transformation & Genetic Engineering* (Eds. Seetharama N & Godwin ID) ICRISAT & Oxford Publishers, India.

Sairam RV and Seetharama N (1999). Regeneration from mesophyll protoplasts of *Sorghum bicolor* (L.) Moench. *Plant Cell Rep.,* **18** : 972-977.

Sairam RV, Seetharama N and Shyamala TR (1998). Plant regeneration from scutella of immature embryos of diverse sorghum genotypes. *Cer. Res. Comm.,* **28** : 279-285.

Sairam RV, Seetharama N, Murthy AK and Shyamala TR (1997). Comparison of tissue culture systems for genetic transformation in sorghum. Second International Crop Science Congress. Oxford & IBH Publication Co. Pvt. LTD. New Delhi. India.

Seetharama N, Sharma HC, Sharma KK, Reddy LJ, Rani TS and Ortiz R 2001. How irresistible are transgenic crops in developing countries? *Proceedings of the National Seminar on "Transgenic Crops and Foods"*, 7-8 November 2000. Organized by Acharya NG Ranga Agricultural University, Rajendranagar, Hyderabad 500 030.

Seetharama N, Sairam RV and Rani TS (2000). Regeneration of *Sorghum bicolor* (L.) Moench from shoot tip cultures. *Plant Cell Tiss. Org. cult.* **61** : 169-173.

Sharma HC, Sharma KK, Seetharama N, Ortiz R (2000) Prospects for using transgenic resistance to insects in crop improvement. EJB Electronic Journal of Biotechnology 3 (2). http://www.ejb.org/content/vol3/issue2/full/3/index

Sharma HC, Sharma KK, Seetharama N and Ortiz R (2001). Genetic transformation of crop plants: Risks and opportunities for the rural poor. *Curr. Sci.,* **80** : 1495-1508.

Smith CW and Frederiksen RA (Eds.) (2000). *Sorghum: Origin, History, Technology and Production,* John-Wiley & Sons, Inc. USA. P 824

Strogonov BP, Komizerko EI and Butenko RG (1968). Culturing of isolated glasswort, sorghum, sweet clover, and cabbage tissues for comparative study of their salt resistance. *Soviet Plant Physiol.,* **15** : 173-177.

Tadesse Y and Jacob M (1999). Nutritional quality improvement of sorghum through genetic transformation. In: *Sorghum Tissue Culture, Transformation & Genetic Engineering* (Eds. Seetharama N & Godwin ID) ICRISAT & Oxford Publishers, India (*In press*).

Thomas E, King PJ and Potrykus I (1977). *Naturwissenschaften.,* **64** : 587.

Vasil I and Philips R (2000) *Molecular improvement of cereal crops*, Kluwer Academic Publ., Dordrecht, The Netherlands.

Weeks JT, Koshiyama KY, Maier-Greiner U, Schaeffner T and Anderson OD (2000). Wheat transformation using cyanamide as a new selective agent. *Crop Sci.,* **40** : 1749-1754.

Wei ZM and Xu ZH (1990). Regeneration of fertile plants from embryogenic suspension culture protoplasts of *Sorghum vulgare. Plant Cell Rep.,* **9** : 51-53.

Wen FS, Sorensen EL, Barnett FL, Liang GH (1991) Callus induction and plant regeneration from anther and inflorescence culture of Sorghum. *Euphytica,* **52** : 177-81

Wernicke W, Brettell R (1982) Morphogenesis from cultured leaf tissue of *Sorghum bicolor* (L) culture initiation. *Protoplasma,* **111** : 19-27

Zhao Z, Cai T, Taglini L, Miller M, Wang N, Pang H, Rudert M, Schroeder S, Hondred D, Seltzer J and Pierce D (2000). *Agrobacterium*-mediated sorghum transformation. *Plant Mol. Biol.,* **44** : 789-798.

Zhong H, Srinivasan C and Sticklen M (1992). *In vitro* morphogenesis of corn (*Zea mays L.*) II. Differentiation of ear and tassel clusters from cultured shoot apices and immature inflorescences. *Planta,* **187** : 490-497.

Zhu H, Muthukrishnan S, Krishnaveni S, Wilde G, Jeoung JM and Liang GH (1998). Biolistic transformation of sorghum using a rice *chitinase* gene, *J. Genet. Breed.,* **52** : 243-252.

Zuo, Nin QW, Geir M and Chana NH (2001). Chemically regulated site-specific DNA excision system. *Nat. Biotechnol.,* **19** : 115-116.

Chapter 11

GENETIC TRANSFORMATION IN LEGUMES

Lingaraj Sahoo, N Dolendro Singh, Twinkle Sugla, Rana P Singh and Pawan K Jaiwal★

Department of Biosciences, M.D. University, Rohtak – 124 001, India

Summary

Legumes breeding programs, aim to produce improved varieties in order to broaden the range of phenotypes available for agriculture, are limited by the available narrow gene pool and restricted potential for hybridization between species. Improvement by the judicious application of biotechnological tools, for alleviating their coherent diseases has been the major focus. Genetic engineering offers the opportunity to modify the characteristics of plants by introduction of genes, from organisms across their genetic barriers. Significant progress has been made in transformation of forage legumes while grain legumes remained traditionally recalcitrant. The transformation of these crops vastly employed Agrobacterium tumefaciens, particle bombardment and electroporation. Most commonly used reporter gene, β-glucuronidase (gus), or genes for antibiotic resistance (nptII and hpt) have been used as selectable markers, and herbicide based selection (bar gene) also proved effective with recent success. The viable plants and seeds have been recovered showing stable integration of transgenes in many of the cases; however, progeny analysis not studied in details. A summary of the regeneration protocols, used for transformation with different procedures and marker genes, is presented. The major problems in developing reproducible transformation systems in grain legumes are the identification of transformation compatible regenerable cells and the threshold of recalcitrance these species exhibit. The vital

★Corresponding author : E-mail : pkjaiwal@yahoo.com

understandings associated to these problems are depicted and the future prospective crucial in this direction is highlighted.

Keywords : Genetic transformation, legumes, selectable markers, gene expression

1. INTRODUCTION

Legumes are of indispensable importance for the protein supply in human and animal nutrition, and form a vital niche of virtually all ecological and agricultural environments. Several disease outbreaks, insect pests and intense abiotic pressures severely affected their productivity. Plant breeding programs essentially seek the expansion of the range of phenotypes available to bring in improved varieties. Inevitably, these programmes are limited by the available narrow gene pool, further narrowing of the genetic base and restricted potential for hybridization between species. Plant biotechnology, a booming science, although bears the risk of being overestimated with respect to its immediate impact on plant breeding, judicious use of these techniques offer the potential to upgrade and/or impart the desired characteristics, in legumes, by the transfer of genes from exotic sources (Sahoo *et al.*, 2001*a*). The major differences between conventional breeding and biotechnology lie neither in goals nor processes, but rather in speed, precision, reliability and scope. Application of such tools largely depends on transformation compatible *in vitro* regeneration systems. Legumes, however, remained traditionally recalcitrant to regeneration as well as transformation. The striking successes in developing transformation protocols and/or transfer of desirable genes in legumes are restricted, so far, to very few crops of profound commercial importance (Kumar and Davey, 1991; Christou, 1994, 1997; Atkins and Smith, 1997; Sahoo *et al.*, 2002; Singh *et al.*, 2001; Sonia *et al.*, 2001). In majority of these crops, manipulation through genetic engineering is still a challenging task. This review focuses on the current status, limitations and future prospects of gene transfer in legumes highlighting the crucial understandings for successful transformation in these species.

2. *IN VITRO* REGENERATION

Efficient and reproducible regeneration systems, from tissues amenable to available transformation techniques, are pre-requisite for genetic transformation. Essentially, a transformation compatible regeneration system must have the following characteristics.

1. The regenerating tissue should ideally be at or close to the surface of an explant, to become easily accessible to the *Agrobacteria*. In case the regenerating cells are remote from the surface, a wound response is necessitated which is usually accomplished by various forms of wounding without hampering the regeneration. Incisions with scalpel blades and stabbing with needles are most frequent and now substituted by novel systems. The use of biolistic device to bombard the target tissue with naked particles and centrifugation of explants with glass beads to create micro-injuries facilitated *Agrobacterium* susceptibility (Bidney *et al.*, 1992). However, these have rarely been used for transformation in grain legumes. Use of ultrasound-sonication to assist *Agrobacterium*-mediated transformation (SAAT) is successful in development of transgenics in soybean (Trick and Finer, 1998).

2. The regenerating system should be rapid with the potential to produce many regenerants per explant.

3. Regenerants should arise from a single cell or a small group of cells. In case, the regenerants develops from many cells from which only a few cells are transformed, then a chimera is resulted. In such cases, application of a stringent selection pressure may be devised to avoid the escapants and chimeras. Alternatively, transformation can be targeted precisely to germlines for which development of suitable regeneration system is prerequisite.

4. Developing regeneration system and transformation protocol directly with elite varieties and extension of such protocols to other economically valuable varieties are extremely important and must be taken into consideration.

5. The culture process should not induce unintentional variability or infertility.

The development of regeneration through organogenesis, somatic embryogenesis, and protoplast culture, and evaluation of their practical compatibility to transformation techniques remain slow, particularly for the grain legumes, due to their extreme recalcitrant nature.

3. GENETIC TRANSFORMATION

3.1. Screenable and selectable markers for legume transformation

3.1.1. Screenable markers or reporter genes

Screenable markers or reporter genes code for products that detoxify selective agents or catalyze specific reactions whose products are detectable. Several reporter genes are used to monitor the efficiency of DNA delivery and expression pattern of transferred genes in cells as well as to analyze the activities of promoters in transformed tissues. Transient gene expression, which does not require the integration of the DNA into the host genome, allows evaluation of the activity of reporter genes within 24 to 48 h after gene delivery. An ideal reporter gene should have a low background activity; no detrimental effects on plant metabolism and its assay should be sensitive, versatile and simple to carry out. The β-glucuronidase *(uid/gusA)* gene (Jefferson *et al.*, 1987) is widely employed as reporter for transient and stable transformation studies in legumes due to its distinct advantage over the earlier used reporter genes viz. *cat* (chloramphenicol acetyl transferase) and *npt*II (neomycin phosphotransferase) whose assay are time consuming and involve the use of radioactive reagents. Localized *gus* activity in the intact tissue can be visualized as blue-spots or -areas within the tissue. For clearer visualization of GUS-positive regions, where chlorophyll interference is a problem, chlorophyll can be removed by transferring tissues to 70-100% ethanol once the blue coloration appears. However, the GUS assay is destructive in nature. The assays of the green fluorescent protein (GFP) gene of the jellyfish, a novel reporter gene for higher plants including the legumes (Ponappa *et al.*, 1999), and firefly luciferase gene (*luc*) are non-destructive in nature and do not require any cofactor or external substrate. In plants, GFP accumulation do not create any toxic effect and further, its variety of applications in higher plants to locate fusion protein, study intracellular protein traffic, characterization of cell-type specific gene regulation, in studies of virus invasion and spread, and monitoring of transgenic plants (Sonia *et al.*, 1998) has encouraged its use. Though the *luc* gene has also been used as reporter in legume transformation (Leon *et al.*, 1991), involvement of expensive CCD camera and substrate "luciferin" has prevented its wider use.

3.1.2. Selectable markers

The introduction of a selectable marker gene allows the transformed cells to proliferate in the presence of a selective agent while the non-transformed cells either do not grow or multiply at a slow rate.

The most commonly used selectable marker gene, in legumes, is neomycin phosphotransferase (*npt*II), which imparts resistance to the antibiotic, kanamycin or its analogues, geneticin or paromomycin by inactivating them by phosphorylation. However, effective selection of transformed cells with *npt*II gene was found difficult in some legumes due to their high level of inherent tolerance to kanamycin (Schroeder *et al.,* 1993). Grant *et al.* (1998) observed phenotypically three types of shoots in pea, that developed on kanamycin selection medium, *viz.* bleached, those that showed bleaching at the base and/or the leaf margins and green all over where kanamycin was very effective, as it neither impaired the normal shoot regeneration nor shoot vigor but bleached the emerging shoots from non-transformed cells. The choice of promoter to drive *npt*II, played a great role in developing efficient selection of transformants in pea as the double enhancer CaMV35S promoter was found ineffective (Davies *et al.*, 1993) whereas *nos* promoter proved effective (Grant *et al.*, 1998).

The hygromycin phosphotransferase (*hpt*) gene, which confers resistance to hygromycin, also provided efficient selection in pea as the untransformed cells were killed faster and the plant regeneration from transformed cells was not inhibited (Puonti-Kaerlas *et al.*, 1990). The successful selection of transformed plants using *hpt* gene has been accomplished in pea (Lulsdorf *et al.*, 1991), *Medicago sativa* (D'Malluin *et al.*, 1990) and *Glycine max* (Finer *et al.*, 1991; Dhir *et al.*, 1991).

The efficiency of transformation both in cereals and legumes has been improved by replacing antibiotic resistance genes with genes for herbicide tolerance, as selectable markers. The most widely used herbicide for selection is Basta (glufosinate or phosphinothricin). The *bar* gene, encodes for phosphinothricin acetyl transferase (PAT), provides resistance by acetylating phosphinothricin, the active ingredient in the herbicide Basta®. Bialaphos and glufosinate the active ingredient of the herbicide Liberty®, have allowed efficient selection of transformants in pea (Schroeder *et al.*, 1993; Bean *et al.*, 1997), soybean (Zhang *et al.*, 1999), lupinus (Pigeaire *et al.*, 1997), *Medicago sativa* (D'Malluin *et al.*, 1990) and *Trifolium subterraneum* (Khan *et al.*, 1994). In pea, both the CaMV35S promoter (Schroeder *et al.*, 1993) as well as *nos*-promoter (Bean *et al.* 1997) have been used to drive the *bar* gene.

The transfer of *CP4 EPSPS* (isolated from *Agrobacterium sp.* strain *CP4*) and *GOX* (Glyphosate oxidase reductase) genes into plants impart resistance to glyphosate, an active ingredient of herbicide Round up®, which inhibits the enzyme 5-enolpyruvylskimate-5-phosphate synthase, critical for the biosynthesis of aromatic amino acids (Phenylalanine, tyrosine and tryptophan) in plants and bacteria. *Petunia* mutated *EPSPS* (Hinchee *et al.*, 1988) as well as *CP4 EPSPS* (Padgette *et al.*, 1995) genes have also provided efficient selection in recovery of transgenic plants of soybean, that exhibited tolerance to foliar application of herbicide under field conditions (Padgette *et al.*, 1995).

3.2. Control of gene expression

The choice of effective promoters for high and tissue-specific expression of the introduced genes is critical for devising transformation systems in legume crops, considering their extreme recalcitrance nature (Sharma and Ortiz, 2002). For instance, GUS expression in transgenic plants can be highly specific in their cells or tissues depending on the promoter employed (Yang and Christou, 1990). The *35S* promoter of cauliflower mosaic virus (CaMV) has provided constitutive expression of not only selectable markers and GUS reporter genes, in majority of the legumes, but also for genes of agronomic interest. Constitutive promoters directs the expression of genes in almost all the tissues irrespective of the environmental or developmental conditions. However, the relative strength of this promoter varies greatly in vegetative tissues and seed. The synthetic methionine-rich 2S albumin gene of Brazil nut (*Bertholletia excelsa*) when transferred to *Vicia narbonensis,* under the control of *35S* promoter, expression was observed in all parts of plant including seeds albeit with low activity in seeds (Saalbach *et al.*, 1994). Whereas the seed-specific legumin B4-promoter (*leB4*) from *Vicia faba* (Baumlein *et al.*, 1991), provided the developmentally regulated tissue specific expression of 2S albumin gene in seeds (Saalbach *et al.*, 1995) with 5% accumulation of the foreign met-rich protein that lead to 2-fold increase in methionine content. High level of seed specific expression of GUS was observed when delivered under the control of *Con A* promoter (Kim and Minamikawa, 1996), of the gene for concanavalin A (*Con A*) from jackbean (*Canavalia gladiata*). Attempts were made to transfer the sunflower seed albumin (*sfa8*) gene (Kortt *et al.*, 1991) under the seed-specific USP (unknown seed protein) (Fiedler *et al.*, 1993) promoter to *V. narbonensis* (see review Pickardt and Saalbach, 2000) but the increase in methionine content was not determined. The *sfa8* gene had also been transferred into lupin (Molvig *et al.*, 1997), under *vicilin* promoter where the expression was associated with a 94% increase in methionine content. A comparative promoter analysis using the *leB4* promoter and the usp-5′-regulatory sequences to target the expression of the GUS gene to the seeds of *V. narbonensis* has been studied (Pickardt *et al.*, 1998). While the *usp*-GUS fusion showed GUS activity from the globular stage of seed development, the lines containing *leB4*-GUS fusion showed GUS expression, and transcript accumulation started at the mid cotyledon stage and an increase in GUS quantity continued until the end of the late cotyledon stage. Incidentally, in one of the two leB4-GUS lines, GUS activity was also detected in the seed coat.

3.3. Gene transfer methods

Among the available gene transfer techniques, the most widely used *Agrobacterium*-mediated transformation, biolistic, and protoplast-based methods have been employed for transformation of legumes.

3.3.1. *Agrobacterium*-mediated gene transfer

The ability of *Agrobacterium* to infect wounded plant cells leading to the induction of tumorous growth or hairy root formation at the site of infection has brought revolution in plant transformation. This tumor-inducing capability is due to the presence of Ti (tumor-inducing) plasmid in virulent strains of *A. tumefaciens* whereas hairy root formation is due to the Ri (root-inducing) megaplasmids found in virulent strains of *A. rhizogenes*. The detailed mechanism of transfer of T-DNA from *Agrobacterium* to plant cell and its integration into plant genome have been reviewed in great details (Zambryski, 1992; Hooykas and Bijersbergen, 1994; Zupan and Zambryski, 1995; Tinland, 1996; Gelvin, 2000). In both the cases, the foreign DNA gets integrated without any major rearrangement in the genome as a single copy. Transformation has been achieved for many leguminous plants either with tumorigenic strains or with disarmed vectors (Table 5 and 6).

3.3.2. *Agrobacterium tumefaciens* - mediated transformation

3.3.2.1. Wild type vector : The preliminary work on legume transformation involved the use of different wild oncogenic strains of *A. tumefaciens,* to determine the compatibility of different species of legumes for the efficient gene transfer. The susceptibility was incidentally variety-dependent and much of the specificity seems to be determined by the type of plasmid (octopine or nopaline encoding) carried by the bacteria or by hormonal status of the plant.

Several legume species have been reported to be susceptible to wild type *A. tumefaciens* and characteristic tumour phenotype noted in soybean (Owen and Cress, 1985; Pederson *et al.*, 1983; Byrne *et al.*, 1987), pea (Puonti-Kaerlas *et al.*, 1989; Hawers *et al.*, 1989; Hobbs *et al.*, 1989), bean (El-Khalifa and Lippincott, 1988; McClean *et al.*, 1991), chickpea (Islam *et al.*, 1994), pigeonpea (Rathore and Chand, 1997), lentil (Warkentin and McHughen, 1991), moth bean (Gill *et al.*, 1988), *Medicago* (Mariotti *et al.*, 1984; Hood *et al.*, 1986; Webb, 1986) and *Lotus corniculatus* (Armstead and Webb, 1987).

The transformation with wild type bacteria has assisted the identification of appropriate plant tissue for inoculation (e.g. hypocotyl, leaf, stem) as well as their optimum age for susceptibility. The cotyledons of young seedlings (7-d-old) and leaves excised from *in vitro* grown seedlings formed galls in culture of *Lotus corniculatus* (Armstead and Webb, 1987). The infection responses also observed to be tissue-, genotype- and accession-dependent in *Glycine clandestina* and *Glycine canescens* (Rech *et al.*, 1989) where hypocotyls of seedlings were more responsive to infection than cotyledons.

The oncogenic strains of *A. tumefaciens* incited tumours, in *Medicago sativa* (Mariotti *et al.*, 1984; Webb, 1986), *Trifolium* (Webb, 1986), *Glycine max* (Pedersen *et al.*, 1983) and in other legumes. In most of the cases, the tumors are disorganized *in situ* and in culture, in capable of regeneration except the tumours formed in *Lotus corniculatus* (Webb, 1986) that gave rise to teratomatous (shooty) phenotypes. Such shoots are phenotypically abnormal, non-rooting and infertile, when grafted. The molecular analysis of tumours by Southern hybridization showed the integration of T-DNA, and detection of opines in callusing tissue confirmed the expression of introduced genes (Mariotti *et al.*, 1984; Owens and Cress, 1985; Pedersen *et al.*, 1983; Hobbs *et al.*, 1989; McClean *et al.*, 1991).

3.3.2.2. ***Disarmed vector :*** The successful *Agrobacterium*-mediated plant transformation depends on efficient gene delivery method addressing cells for which a regeneration system exists. Thus such a protocol to be practically useful, the combination must show a sufficiently high overall efficiency. While wounded tissue of certain species, notably amongst the Solanaceae, are able to produce sufficient quantities of signal molecules to induce effective transformation (See Holford *et al.*, 1992 and references therein); with less amenable species like legumes, it is necessary to manipulate either the explant and/or the bacterium with cocultivation conditions to enhance the virulence so as to increase the transformation frequency. A number of variables involving host tissue notably explant type, mode of regeneration, wounding of explants, wounded cell extract, nurse cell or pH treatment, preculture of explants, ethylene antagonists and/or methylation inhibitor, and factors influencing *Agrobacterium* virulence viz. strain, vector type, cocultivation period, bacterial concentration, treatment with *vir* gene inducer (acetosyringone), use of appropriate antibacterial agent, vacuum infiltration and conditions influencing their interaction i.e. light and temperature status during cocultivation significantly affected the transformation frequencies, tissue viability and regeneration. Evaluation of these factors is critical while devising transformation protocols for recalcitrant plants like legumes. Transient expression of reporter gene (GUS) under the control of CaMV35S or plant specific promoters has allowed the identification of the conditions for efficient T-DNA transfer and integration. Interruption of the coding sequence by an intron (Vancenneyt *et al.,* 1990) essentially avoids false expression arise due to the presence of *Agrobacterium* in the plant tissue.

Explant type and mode of regeneration : The cultured cells that undergo *de novo* regeneration are usually more competent (i.e. transformable) to transformation, when accessed to *Agrobacterium* infection. *De novo* regeneration refers to the ability of somatic cells to dedifferentiate, proliferate and reorganise into shoot meristems (organogenesis) or embryos with shoot and root meristems (somatic embryogenesis) which ultimately develop into viable plants. The *de novo* regeneration pathway has so far been most successful in generation of homogenously transformed plants due to the nature of the pathway. But recent advances have precluded that *de novo* regeneration is not essential for transformation as protoplasts, suspension cultured cells, callus cells or cells forming part of an organ or tissue can also serve as target for transformation. The small seeded forage legumes, due to the ease of their regeneration, are easy to transform compared to the large seeded grain legumes where regeneration from protoplasts, suspension-cultured cells or via an unorganized callus phase is limited. The direct adventitious shoot regeneration from various explants, cotyledon (Hinchee *et al.*, 1988; Muthukumar *et al.*, 1996; Sharma and Anjaiah, 2000), cotyledonary nodes (Davies *et al.*, 1993; Jordan and Hobbs, 1993; Di *et al.*, 1996;

Bean *et al.*, 1997; Meurer *et al.*, 1998; Jaiwal *et al.*, 2001), epicotyl (Sato *et al.*, 1993; Saini and Jaiwal, 2002), apical meristem (Zhang *et al.*, 1999; Geetha *et al.*, 1999) excised from developing, mature or germinated embryonic axes, decapitated embryonic axes (Fontana *et al.*, 1993; Kar *et al.*, 1996; Krishnamurthy *et al.*, 2000) and nodal thin cell layer (Nauerby *et al.*, 1991) have been used for transformation of large seeded grain legumes (Table-1).

Agrobacterium-mediated transformation coupled with regeneration of transformed plants via direct or indirect somatic embryogenesis has been reported in a few grain legumes, i.e. soybean (Parrot *et al.*, 1989), *Vicia narbonensis* (Pickardt *et al.*, 1995; Saalbach *et al.*, 1995), *Cicer arietinum* (Ramana *et al.*, 1996), *Medicago truncukata* (Wang *et al.*, 1996; Chabaud *et al.*, 1996), *M. varia* (Deak *et al.*, 1996) and *M. sativa* (D'Malluin *et al.*, 1990; Trinh *et al.*, 1998).

The limitation of direct or indirect somatic embryogenesis to plant transformation is that the regeneration is highly cultivar specific, lengthy and labour intensive, and conversion of embryos into plantlets is difficult. In *Medicago*, only the genotypes with high embryogenic potential are transformable. Moreover, the somatic embryos arising from callus or directly from an explant are multicellular in origin that result in development of chimeric somatic embryos. The process of repetitive somatic embryogenesis circumvents the problem of chimeric embryos by allowing recovery of completely transformed secondary embryos from transformed sector within a primary somatic embryo. Though, chimeric embryos are recovered from the first cycle of repetitive embryogenesis, continued cycling in the presence of a selective agent eventually result in embryo consisting entirely of transformed cells (Parrott *et al.*, 1994; Liu *et al.*, 1992, 1996).

The transformation of soybean using embryogenic suspension culture with particle bombardment has been discussed in other section.

Majority of the explants, regenerating through direct organogenesis pathway, contain only a few regenerating cells whose capacity for regeneraiton is short lived. Further, the efficiency of these regenerable cells to transformation is not optimum. Thus, wounding is made in the tissue for bacterial infection, in such a way that the regeneration is not hampered by the wounding response. Wounding is usually accomplished by scalpel, needle, bombardment with naked particles and ultrasonication prior to *Agrobacterium* inoculation. As a consequence, the susceptibility

Table 1. Transformation of legumes with disarmed strains of *A. tumefaciens*

Genotype/ cultivar	Explant/ Regeneration	*Agrostrain* and plasmid/vector	Reporter/ Marker gene(s)	Selection (mg/L)	Transformation efficiency (%)	Viable plant or seed	Molecular Analysis	Reference
Soybean	Cotyledon	-	*nptII*	-	-	-	-	Owens & Smigocki (1988)
Soybean cv, Peking	Mature cotyledon/ Organogenesis	pTiT37-SE pTiT37-SE pMCN894	*npt*II, *gus* *EPSPS*	- -	- -	Yes Transgenric R_0, R_1 Peking only	GUS Southern analysis, kanamycin and glyphosate	Hinchee *et al.*, 1988
Soybean cv. Peking	Immature cotyledon/ Embryogenesis	*Agrobacterium* 15kD Zein	-	-	-	Transgenic R_0	-	Parrott *et al.*, 1989
Soybean cv. Peking	Mature coty ledonary node/ Organogenesis	A.t. Ac transposon	-	-	-	Transgenic plants	-	Zhon and Atherly, 1990
Soybean cv. ADOC	Seedling cotyledon	-	marker genes	-	-	Transgenic R_0		Chang and Chan, 1991
Soybean cv. ADOC	Embryogenic suspension/ Embryogenesis, immature cotyledons	A.t. + sonication EHA105 (pVec035) EHA105 (pIG121Hm)	*hpt* *hpt*	hygromycin (20 mg/L)	- -	Transgenic clones	GUS, Southern blot	Trick and Finer, 1998; Santarem *et al.*, 1998

Table 1. Continued

Genotype/ cultivar	Explant/ Regeneration	*Agrostrain* and plasmid/vector	Reporter/ Marker gene(s)	Selection (mg/L)	Transformation efficiency (%)	Viable plant or seed	Molecular Analysis	Reference
Soybean cv. Fagette	Cotyledonary nodes of germinating seeds/ Embryogenesis	7707(pMON530)	coat protein precursor (CP-P) of BPMV	kanamycin (200mg/L)	-	Transgenic R_0, R_2	ELISA, Southern analysis	Di *et al.*, 1996
Soybean A3237	Cotyledonary node	EHA105 or EHA101 (pPTN125)	*bar*	glufosinate (3.3 or 5.0 mg/L)	- assay.	T0 plant	Southern blot and leaf paint	Zhang *et al.*, 1999
Pea	Embryogenic axis, hypocotyl segments	GV3101(pGV150 3)	*hpt* or *npt*II	hyg (15mg/L) or kan (75 mg/L)	-	Transformed plant	Southern analysis of calli & plant	Puonti-Kaerlas *et al.*, 1989, 1990
Pea cv. Rando, cv. Green feast	Embryo axis of developing seeds	AGLI (pSLJ15C1)	*npt*II *and bar*	PPT (15 mg/L)	1.5-2.5%	Herbicide resistant pea plant	NPTII and *PAT assay* 3:1 mendelian inheritance of both genes	Schroeder *et al.*, 1993
Pea cv. p1244253	Stem explant	EHA101 (pBI1042)	*nptII* or *hpt*	kan (50 mg/L) hyg (50 mg/L)	77%	Transformed	- calli only	Lulsdorf *et al.*, 1991
Pea cv. madria	Nodal and epicotyl explants	GV2260 (p35SGUSInt), GV3850 HPT	*npt*II, *gus, hpt*	-	-	-	-	de Kathen and Jacobsen, 1990
Pea	Root explant protoplasts seed	C58C1(pGV2260, pGA482)	*npt*II	-	-	-	-	Schaerer and Pilet, 1991

Table 1. Continued

Genotype/ cultivar	Explant/ Regeneration	*Agrostrain* and plasmid/vector	Reporter/ Marker gene(s)	Selection (mg/L)	Transformation efficiency (%)	Viable plant or seed	Molecular Analysis	Reference
Pea cv. puget	Cotyledonary moristem of germinating seed	C5813(pSLJ1911)	*npt*II and *gus*	-	1.44%	Transgenic fertile plant	PCR, GUS, Southern analysis	Davies *et al.*, 1993
Pea	Immature cotyledon	AGL1(pLN27)	*bar/npt*II	-	-	Transgenic herbicide resistant plant	PAT, Southern stable inheritance	Grant *et al.*, 1995
Pea Bodel, Trille Puget Rondo	Nodal thin cell layer	C58C1(pGV2260, pGSGLUC1) C58C1(pGV3850, pGSGLUC1)	*npt*II and *gus*	-	-	GUS assay of leaves of transformed shoots	GUS assay	Nauerby *et al*, 1991
Pea cv. Laser and Heiga	Immature cotyledon	EHA105 p35SGUSINT, C58C1 pGPTV-kan LBA4404 pGPTV-*hpt* pGPTV-*bar* pGPTV-*dhfr*	*hptII* *hpt* *bar* *dhfr*	kan (50 mg/L) hyg (100 mg/L) PPT (2 mg/L) Methotrexate (1 mg/L)	2.6% - 9%	T_0 & T_1 plants	PCR, Southern	Nadolska-Orczyk and Orczyk, 2000
Vigna radiata	Cotyledonary node	LBA 4404 (pTOK233)	*npt*II and *gus*	kan (75 mg/L)	0.9%	T_0 plant	GUS, PCR, Southern	Jaiwal *et al.*, 2001
V. unguiculata	Mature embryo	C58C1(pGV2260 and p35SGUSInt)	*gus*	-	-	Chimeric GUS (+) shoots	GUS, DNA dot blot	Penza *et al.*, 1991

Table 1. Continued

Genotype/ cultivar	Explant/ Regeneration	*Agrostrain* and plasmid/vector	Reporter/ Marker gene(s)	Selection (mg/L)	Transformation efficiency (%)	Viable plant or seed	Molecular Analysis	Reference
V. mungo	Primary leaf segments	pGA472	*npt*II, *gus*	kan (50mg/L)	Transgenic calli	No plant regnerated	NPTII and Southern analysis	Karthikeyan et al., 1996
V. narbonensis	Epicotyl explant via callus/ Embryogenesis	C58C1 (pGV3850 HPT)	Nopaline synthase gene as reporter and *hpt*	-	-	Studied transgenes stability upto 9 generations	Southern blot analysis	Pickardt *et al.*, 1991 Brinkmann *et al.*, 1997
V. *narbonensis*	Epicotyl explant via callus/ Embryogenesis	EHA101 pGSGluc1/BNA pGA472/23 EHA101 pGA472/LeB4 GUS	*npt*II *npt*II	2S brazil nut Legumin B4-GUS, seed specific	-	Low activity of 35S promoter in seeds, X2 x net content; Legumin B4 promoter activity	GUS and Southern analysis	Saalbach *et al.*, 1994 1995 Pickardt *et al.*, 1998
Arachis hypogaea	Leaf section explant	-	*gus/npt*II	-	-	Yes	-	Chang *et al.*, 1996
Lens culinaris	Shoot apex of germinating embryonic axis	-	*gus/bar*	-	-	-	-	Barton *et al.*, 1996
Lupinus angustifolius	Shoot apex of germinating seed	-	*gus/bar*	PPT	0.4 to 2.8%	Transgenic plants, seed progeny	Southern analysis herbicide resistant palnt	Atkin *et al.*, 1994; Pigeaire *et al.*, 1996

Table 1. Continued

Genotype/ cultivar	Explant/ Regeneration	*Agrostrain* and plasmid/vector	Reporter/ Marker gene(s)	Selection (mg/L)	Transformation efficiency (%)	Viable plant or seed	Molecular Analysis	Reference
Lupinus mutabilis cv. Potosi	Shoot apical meristem	LBA4404 with a binary vector together with the supervirulent pTOK47	*gus* and *npt*II	-	-	Kanamycin resistant plants	GUS, NPTII Southern	Babaoglu *et al.*, 2000
Cicer arietinum	Sterile seedlings	-	-	-	-	Tumor/callus formation	Southern analysis	Srinivasan *et al.*, 1992
C. arietinum	Embryonic axis	LBA4404 (pBI121)	*npt*II and gus	-	4%	Transformed plants	GUS, *npt*II Southern blot	Fontana et al., 1993
C. arietinum	Sterile Seeding	A281, C58, A6	-	-	-	Tumor/callus formation	Southern blot and opine analysis	Riazzudin and Husnain, 1993
C. arietinum	Sterile Seedling	-	-	-	-	Tumor/callus formation	Opines assay	Islam *et al.*, 1994
C. arietinum	Immature cotyledon	-	-	-	-	Transfromed somatic embryo	-	Ramana *et al.*, 1996
C. arietinum various cvs.	Embryonic axes	C58C1/GV2260 (p35GUSINT), EHA101 (pBGUS)	*npt*II/*bar*	kan (100 mg/L)	-	Transgenic plant	GUS, Southern KAN and PPT resistant plant	Krishnamurthy *et al.*, 2000
Phaseolus acutifolius cv. Gray	Buds	C58C1 (carrying *npt*II *gus* and arcelin gene)	*npt*II	-	-	Transformed plants	GUS Southern analysis	Dillen *et al.*, 1997

Table 1. Continued

Genotype/ cultivar	Explant/ Regeneration	*Agrostrain* and plasmid/vector	Reporter/ Marker gene(s)	Selection (mg/L)	Transformation efficiency (%)	Viable plant or seed	Molecular Analysis	Reference
Phaseolus vulgaris cv. dark red kidney	Leaf and hypocotyl segment	EHA101 (pKYL x 71GUS)	*gus* and *npt*II	-	-	No	NPTII, ELISA and Southern analysis	Franklin *et al.*, 1993
P. vulgaris cv. various lines	Nodal region of intact seedlings	-	*gus*	-	-	No	GUS	Levis and Bliss, 1994
P. vulgaris various cultivars	Cotyledonary node and epicotyl explants	-	*gus* and *npt*II	-	-	Chimeric roots	GUS activity	Banos *et al.*, 1997
Lens	Shoot apex	GV2260	*gus*	-	-	No	GUS assay	Warkentin and McHughen, 1992
Pigeonpea	Cotyledonary node and shoot apex	-	*gus* and *npt*II	kan (50 mg/L)	-	Transgenic plant	GUS, PCR, Southern analysis	Geetha and Lakshmi Sita, 1999
Lathyrus sativa	Immature leaflet and nodal segments	-	*gus*	-	-	-	GUS expression	Barna and Mehta, 1995
Stylosanthes humilis	Leaf section and stolons	GV3101(pGV385 0:1103 neo)	*npt*II and *nos*	kan (100 mg/L)	-	R_2 generation	NPTII Southern	Manners, 1988

Table 1. Continued

Genotype/ cultivar	Explant/ Regeneration	*Agrostrain* and plasmid/vector	Reporter/ Marker gene(s)	Selection (mg/L)	Transformation efficiency (%)	Viable plant or seed	Molecular Analysis	Reference
S. guianesis	Leaf section via callus	EHA101 (pGV1040)	*npt*II *bar*	PPT (1 mg/L)	-	Yes	Southern blot, mendelian inheritance	Sarria *et al.*, 1994
Medicago sativa	Stem cutting somatic embryogenesis	A.t. bo42 (pGA471)	*npt*II	-	-	Yes	NPTII, Southern	Deak *et al.*, 1996
Medicago sativa	Leaf segmens	-	*Chicken ovalbumin*	-	-	Yes	-	Schoreder *et al.*, 1991
Medicago sativa	Mature stem segments	-	*nos/npt*II	-	-	Yes	-	Shahin *et al.*, 1986
Medicago sativa	Stem & Petiole discs	CS8C1(pGSFR780A and pGSFR761A	*bar, npt*II and *hpt*	kan (50 or 100 mg/L) hyg (75 mg/L) ppt (5 or 10 mg/L)	-	Yes	NPTII, PAT,	D'Halluin *et al.*, 1990 Southern
M. truncatula Jamalong 2 HA	Leaf/Shoot regeneration from calli	LBA4404 (pBI121)	*npt*II, *gus*	kan (50 mg/L)	-	Transgenic R_0, R_1	GUS Southern blot	Wang *et al.*, 1996

Table 1. Continued

Genotype/ cultivar	Explant/ Regeneration	*Agrostrain* and plasmid/vector	Reporter/ Marker gene(s)	Selection (mg/L)	Transformation efficiency (%)	Viable plant or seed	Molecular Analysis	Reference
M. truncatula	Leaf, somatic embryogenesis	LBA4404 (p35SGUSINT or PLP100.pr.Mt12)	*Mt ENOD 12/gus*	-	-	Yes	GUS	Chabaud *et al.*, 1996
M. truncatula	Embryogenic calli from leaf explants	EHA101(PMT2)	*gus*	-	-	Yes	-	Thomas *et al.*, 1992; Rose and Nolan, 1995
M. varia	Stem cuttings	-	*npt*II	-	-	Yes	-	Deak *et al.*, 1986
Lotus japonicum	cotyledonary node *in planta*	-	*npt*II	-	-	Yes	-	Oger *et al.*, 1996
Trifolium repen	Stolen internode segments/ callus formation and plant regeneration	LBA4404 pBin19	*npt*II, *nos*	kan	-	Transgenic plant	-	White and Greenwood, 1987
Trifolium repen	Cotyledonary axil/direct organogenesis	LBA4404 pPE64	*npt*II, *gus*	Kanamycin	-	Transgenic plant obtained from 1% treated cotyledon	-	Voisey *et al.*, 1994

Table 1. Continued

Genotype/ cultivar	Explant/ Regeneration	*Agrostrain* and plasmid/vector	Reporter/ Marker gene(s)	Selection (mg/L)	Transformation efficiency (%)	Viable plant or seed	Molecular Analysis	Reference
Trifolium repen	Internode stolon segments	LBA4404 p03692	Pea albumin (*PAI*)	Kanarmycin	-	Yes	-	Ealing *et al.*, 1994
Trifolium repen	Cotyledon of mature imbibed seed /direct shoot organogenesis	AGLI(pTAB10 derivatives pJJ430 AGL1(pTAB10 derivative PCP1001 AGL1(pKYLX35 S2) derivative with AMV subgenomic RNA4	*GH3-gus bar*	PPT Kanamycin	-	Transgenic plant recovered from 50% of treated cotyledons	-	PiHock *et al.*, 1997; Larkin *et al.*, 1996; Garrett and Chen, 1997
T. subterran eum	Hypocotyl segments explants from seed/ adventitious shoot regeneration	-	*gus/bar*	PPT	-	Transgenic plant	-	Khan *et al.*, 1994

Table 1. Continued

Genotype/ cultivar	Explant/ Regeneration	*Agrostrain* and plasmid/vector	Reporter/ Marker gene(s)	Selection (mg/L)	Transformation efficiency (%)	Viable plant or seed	Molecular Analysis	Reference
T. pratense	Petiole segments/ somatic embryo via callus	-	-	Kanamycin	70% of petiols produced transgenic plant	Transgenic plant	-	Quesenbergy *et al.*, 1996

Ab : A.t. Agrobacterium tumefaciens; Kan – Kanamycin; hyg – hygromycin; PPT – phosphinothricin of wounded plant cells to *Agrobacterium* is enhanced due to the secreted phenolic compounds that induce *vir* genes (Potrykus, 1991; Bidney *et al.*, 1992). Legumes, traditionally, showed poor wound response. A mild treatment of explants with a pectinase, an enzyme that disorganises the plant cell wall, is also effective as it facilitates bacterial assess to the host cells (Alibert *et al.*, 1999). The dividing cells, such as those in cotyledon, can be transformed in the presence of *vir* inducing compounds (acetosyringone) where a wound response may not be necessary (Hiei *et al.*, 1997). Moreover, wounding of plant tissue prior to inoculation is not always vital for *Agrobacterium*-mediated transformation (Escudero and Hohn, 1996) as bacteria can use stomata openings to access target cells.

Addition of acetosyringone during cocultivation and/or *Agrobacterium* culture has proven to be beneficial in some of the legume crops (Godwin *et al.*, 1991; Voisey *et al.*, 1994; Khan *et al.*, 1994) while transformation rates could not be enhanced even after adding acetosyringone in pea (de Kathen and Jacobsen, 1990; Grant *et al.*, 1995). Subsequent studies observed that acetosyringone had no apparent influence on efficiency of transformation of pea with hypervirulent strain EHA105, but affected the transformation efficiency when moderate virulent strains C58C1 were used (Nandolska–Orczyk and Orczyk, 2000). Even pretreating the explants with acetosyringone also enhanced transformation. Efficient *vir* gene induction requires an acidic medium (Joersbo *et al.*, 1999), a monosaccharide concentration (sugar) (Khan *et al.*, 1994), presence of wounded cell extracts and feeder (nurse cells) (Lulsdrof *et al.*, 1991) in cocultivation medium.

The presence of plant growth regulators (Khan *et al.*, 1994; Schroeder *et al.*, 1993), cell cycle regulators (de Kathen and Jacobsen, 1995) and silver thiosulfate (Joersbo *et al.*, 1999) in the co-cultivaiton and selection medium enhanced the recovery of transformed shoots. The inclusion of NAA or IAA, and abscisic acid (de Kathen and Jacobsen, 1992, 1995), 2,4-D and kinetin (Schroeder *et al.*, 1993), in the cocultivation medium increased the transformation competency of plant cells in pea whereas TDZ and NAA in *Trifolium subterraneum* (Khan *et al.*, 1994) and BAP in azukibean (Yamada *et al.*, 2001) influenced transformation frequency. The presence of TDZ and silver thiosulfate in selection medium significantly increased the transformation frequency in guar by enhancing the selection efficiency of antibiotics (Joersbo *et al.*, 1999).

Type of cocultivation medium : The effect of state of co-cultivation media, liquid or semi-solid was examined on infection rates by transient GUS expression (Zang *et al.*, 1997). Though, there was no difference in survival rates, the percentage of explants cocultivated on semi-solid medium expressed significantly higher GUS activity (Zang *et al.*, 1997; Jaiwal *et al.*, 2001), but the survival of explants of mungbean greatly affected when cocultivated in liquid medium (Jaiwal *et al.*, 2001). On contrary cocultivation on solid medium decreased the frequency of GUS expression in explants of common bean than those co-cultivated on filter paper. Though the reason for enhanced GUS expression on filter paper is not clear, the phosphate starvation of *Agrobacterium cells* due to the lack of nutrient supply on the filter paper may be the reason for enhancing DNA transfer (Zang *et al.*, 1997). Application of cell cycle and DNA replication inhibitors, aphidicolin, colchicines and nalidixic acid drastically reduced *Agrobacterium*-mediated DNA transfer in pea (de Kathen and Jacobsen, 1995).

Preconditioning and preculture of explants : Germination of seeds, and preculture of explants in dark prior to co-cultivation with *A. tumefaciens*, increased infection rates as compared to those grown under light conditions (Zang *et al.*, 1997; Jaiwal *et al.*, 2001). The dark treatment might have influenced the permeability of seedlings as etiolated ones have thinner cell walls and cuticles.

Age/maturity of explants : The younger explants were more susceptible to *Agrobacterium* than the older ones in mungbean (Jaiwal *et al.*, 2001) whereas in *Phaseolus vulgaris*, older and mature ones were more responsive (Zhang *et al.*, 1999).

In some legumes, plant *Agrobacterium* interaction cause necrosis beyond the inoculation site. The necrosis seemed to be oxygen dependent and correlated with elevated levels of peroxidases. Very short exposure with diluted cultures of *Agrobacterium*, and use of antioxidants such as polyvinylpyrrolidone and dithiothreitol (Perl *et al.*, 1996), ascorbic acid, and cysteine and silver nitrate (Enrizuez-obregon *et al.*, 1999), inhibitors of the plant ethylene sensing system during cocultivation improved plant survival and recovery of stable transgenic plants.

Addition of inhibitor of cytosine methylation, 5-azacytidine, to the regeneration medium enhanced the recovery of transgenics and the expression of introduced genes in pea (Nadolska-Orczyk and Orczyk, 2000). Demethylated constructs have been advanced for their preferential integration into transcriptionally active chromatin (Palmgren *et al.*, 1993), which increased stable expression and frequency of transformation (Mandal *et al.*, 1993).

***Agrobacterium* strain and vectors :** Virulence of the *Agrobacterium* against a particular plant genotype (species) is specific. Hence, the choice of suitable *Agrobacterium* strain and Ti plasmid with appropriate natural or modified virulence is important for the higher efficiency of transformation for a particular cultivar. Identification of transformable germplasm by preliminary screening, reported in pea (Nadolska-Orczyk and Orczyk, 2000) and soybean (Donaldson and Simmonds, 2000). In most of such studies, the comparison of susceptibility was based mainly on tumor formation (see section 5.1.1.). A comparison of susceptibility of disarmed *Agrobacterium* strains with same Ti plasmid in different leguminous species showed EHA101 (pBI1042) (Lulsdorf *et al.*, 1991) and EHA105 (pGPT4-Bar) (Nadolska-Orczyk and Orczyk, 2000) in pea, KYRTI (pTichry5) in soybean (Torisky *et al.*, 1997; Meurer *et al.*, 1998), AGLO in *Lupinus angustifolia* (Pigeaire *et al.*, 1997), AGL1 in *Galega orientalis* (Collen and Jarl, 1999), GV3101 (pMP90) in *Robinia pseudoacacia* (Igasaki *et al.*, 2000) and LBA4404 (pTOK233) in *Vigna radiata* (Jaiwal *et al.*, 2001) resulted greater transformation response. Further, the use of a hyper virulent strain of *A. tumefaciens* was suggested to overcome the genotypic resistance observed (Torisky *et al.*, 1997).

The supervirulence of the strain A281 containing pTiBO542 is attributed to hyperactive *virG* and some *virB* genes (Jin *et al.*, 1987). Additional copies of *virG* and *virE* genes not only facilitated the transfer of high molecular weight T-DNA (150 kb) into the plant genome but also enhanced transformation efficiency (Park *et al.*, 2000). *Agrobacterium* with additional copies of *vir* genes (Hiei *et al.*, 1994; Ishida *et al.*, 1996) or with constitutively expressing virulence genes placed on the helper plasmid, the binary vector or a ternary plasmid (Hansen *et al.*, 1994) increased the efficiency of plant transformation. The use of a "super binary" cointegrative vector containing an extra set of virulence genes (*virG, virC* and *virB*), from the supervirulent strain A281, appeared to perform better in many recalcitrant plants including legumes (Komari *et al.*, 1996; Jaiwal *et al.*, 2001). Co-integrative vector performed better than the binary vector in pea (Lulsdorf *et al.*, 1991). The low transformation efficiency associated with binary vectors may be attributed to the problems encountered with stability and small size of this vector.

Bacterial concentration : Increase in concentration of *Agrobacterium* cells in the co-cultivation process increased the number of plants cells infected (Lin *et al.*, 1994). A concentration of *Agrobacterium* ($1\text{-}5\times10^8$ cells/ml) was appropriate for pea and most of the legumes (Bean *et al.*, 1993). Higher concentrations cause hypersensitive action on explant, aggregation of *Agrobacterium* cells or post co-cultivation difficulties in elimination of *Agrobacteria*. Growing *Agrobacterium* cells in YM broth prevented aggregation (Lin *et al.*, 1994).

Duration of cocultivation time : The length of cocultivation required for achieving maximum gene transfer was genotype dependent, ranging from 2 to 5-d with most plant species (Chabaud *et al.*, 1988; Lulsdorf *et al.*, 1991; White and Greenwood, 1997; de Kathen and Jacobsen, 1990; Jaiwal *et al.*, 2001) with an exception of longer (6-d) cocultivation optimal in *Phaseolus vulgaris* (Zang *et al.*, 1997).

Temperature : T-DNA transfer to plants drastically decreased at temperature of 25°C and above (Dillen *et al.*, 1997). The thermo-sensitivity of T-DNA transfer possibly related to the formation of conjugal pilus and not to the *vir* gene induction.

Anti bacterial agents : The choice of an efficient antibiotic to eliminate the *Agrobacterium* after cocultivation of plant tissues is an important consideration. An ideal antibiotic, for eliminating *Agrobacterium* species, should be highly effective, inexpensive with no negative effect on plant growth and regeneration, and stable in culture (Cheng *et al.*, 1998). Though many antibiotics are described, carbenicillin and cefotaxime are most preferred due to their least toxicity on most plant tissues and their capacity to efficiently eliminate the *Agrobacterial* cells (Lin *et al.*, 1995; Nauerby *et al.*, 1997). The carbenicillin and cefotaxime also shown to exert positive effect on shoot regeneration in some species (Bilings *et al.*, 1997; Nauerby *et al.*, 1997; Ling *et al.*, 1998). However, the inactivation of carbenicillin and ticarcillin by the bacterial β-lactamases and the inhibition of plant regeneration by cefotaxime, the potential inhibitor of β-lactamases, discourage their use (Hammerschleg *et al.*, 1995, 97). Negative effects of antibiotics, especially cefotaxime, on rooting have also been observed (Nauerby *et al.*, 1997). Further, plant hormone like effects of carbenicillin and cefotaxime on tissue (Mathias and Boyd, 1986) may complicate the hormone ratio in the culture medium and thereby influence the *Agrobacterium*-mediated transformation efficiency by reducing the efficiency of regeneration (Lin *et al.*, 1995). The phytotoxicity of antibiotics can vary considerably between plant species and also depends on their concentration. Timentin, a mixture of ticarcillin (a penicillin derivative) and clavulanic acid (Nauerby *et al.*, 1997) has been found resistant to β-lactamases as well as stimulates *in vitro* shoot or root organogenesis (Nauerby *et al.*, 1997; Costa *et al.*, 2000). Thus the probable cytotoxic effect of antibiotics on plant regeneration should be evaluated prior to their use in any transformation experiments.

3.3.3. *Agrobacterium rhizogenes*-mediated transformation

The development of roots on the susceptible host plant results from the transfer of T-DNA carrying oncogenes, the so-called "rol genes" of Ri plasmid, to the plant cell and their stable integration into plant genome and expression. The hairy roots synthesize unusual amino acid derivatives called opines. The expression of the *rol B* gene is critical for hairy root production while other *rol* genes, *iaaM* and *iaaH* genes seem to have an additional role in establishing complete hairy root symptoms and in broadening the host range of infection.

A. rhizogenes-mediated transformation depends on plant genotype, type of tissue and its age, genetic make-up and concentration of the strain, co-cultivation period and conditions (Rech *et al.*, 1989; Ramsay and Kumar, 1990). The interactions were invariably species as well as tissue-specific. The hypocotyls and petioles are generally more responsive, with wider exceptions (Table 2).

Hairy roots, formed due to *A. rhizogenes* infection, are well differentiated probably due to their origin from single transformed cell, and therefore essentially uniformly transformed which is in contrast to the crown galls that consists of undifferentiated rapidly growing mass of transformed and untransformed cells. Hairy roots are hormone-autotrophic, and regenerated easily into whole plants, in a few grain legumes, either spontaneously or from the root cultures in *Lotus corniculatus* (Stougard *et al.*, 1986; Petit, 1987) or upon transfer of roots to media with plant growth hormones in *Medicago sativa* (Spano *et al.*, 1987), *Glycine argyrea* (Kumar *et al.*, 1991) and *G. canescens* (Rech *et al.*, 1989). The plants regenerated from hairy roots display characteristic phenotype with wrinkled leaves, shortened internodes, decreased apical dominance, altered flower morphology etc. (Jauanin *et al.*, 1987). The resulting differed phenotypes between species with some of their features, obtained in *A. rhizogenes* transformed legumes are highlighted (see Table-2). However, transgenic plants without or with minimal Ri phenotype effects (viz. *Stylosanthes* plants with normal

Table 2. Transformation of legumes via *Agrobacterium rhizogenes*

Species	Explant	A.r. strain/vector	Gene(s) introduced	Shoot regeneration	Phenotype/Genotype observed	Reference
Cicer arietinum	-	8 strains of *A. rhizogenes* Ri plasmid	-	-	Hairy roots 10-15 mm in length and < than 1.0 mm wide	Siefkes-Boer *et al*, 1995
Pisum sativum	intact root segment	A4, 8196	Wild	H, callus	Fast grown rate Lack of gravitropism Agropine, Mannopine	Schaerer and Pilet, 1991
Phaseolus vulgaris	-	A.r. 8196	*nptII, GUS*	H	-	Brasileiro *et al*, 1996 Barros *et al*, 1997.
P. acutifolius	-	A.R. 8196	-	-	-	Brasileiro *et al*, 1996
Vigna radiata	Hypocotyl and cotyledon	LBA9402 (pRi1855) pBin9GusInt	*nptII, GUS*	*H*, callus	roots grow faster hairless, extensive lateral branches hormone autotrophic GUS expression Southern analysis	Jaiwal *et al.*, 1998
Glycine argyrea	Hypocotyl	Wild LBA9402, A4T, R1601, A4T (pRiTL DNA)	*nptII*	H, S (through callus)	Opines, NPTII, RiTL DNA	Kumar *et al*, 1991
G. canescens	Hypocotyl	R1601 (pRiA4b) *npt*II cointegrate into pRiTL-DNA	*nptII*	H, S (through callus)	Fast growth, lateral branching, plagiotropic, no change in internodes of plant, integration of TL-DNA	Rech *et al*, 1989

Table 2. Contd.

Species	Explant	A.r. strain/vector	Gene(s) introduced	Shoot regeneration	Phenotype/Genotype observed	Reference
Lotus angustrissimum	Plantlet	LBA9404 (Ri1855 and pBI121)	*npt*II, *GUS*	S (through callus)	Transgenic plant flowering	Nenz *et al*, 1996
Lotus corniculatus	Stolon segment	Wild type A.r. C58C1 (pRil5834)	-	H, S	Root morphogenesis, leaf parameter, flower and seed production	Petit *et al*, 1987; Stougaard *et al*, 1986; Webb *et al*, 1990.
	Seedling	LBA9402 (pRi1855 pJit73)	*GUS npt*II/*hpt*	H, S	-	Webb *et al*, 1994
L. corniculatus *L. tenuis*	Plantlets	LBA9402 (pRi1855) pAG125	*hpt*	H, S	Reduced internodes, leaf size leaf wrinkling prostrate growth habit fibrous root	Dmiani *et al*, 1993.
Medicago arborea	-	-	*hpt*	H	-	Damiani and Arcioni, 1991 Spano *et al*, 1987; Golds *et al*, 1991
M. sativa	-	-	*Wild type*	H	-	Sukhapinda *et al*, 1987
M. truncatula	-	-	*npt*II	H	-	Thomas *et al*, 1992

Table 2. Contd.

Species	Explant	A.r. strain/vector	Gene(s) introduced	Shoot regeneration	Phenotype/Genotype observed	Reference
Onobrychis viciifolia	Sainfoin	-	*Wild life*	S	-	Golds *et al*, 1992
Stylosanthes humilis	leaf, stem explants	A.r. NCPPB1855 pGA492	*npt*II	H, S	One with normal or with minor morphological abnormal type opine, NPTII assays, transfer binary vectors and TL-DNA	Manners and Way (1989)
Robinia pseudoacacia	-	-	*npt*II	H,S	Abnormal shoot	Hans *et al*, 1993.
Vicia faba	Cotyledon, stem	LBA9402 with Ri Plasmid and pBin19	*npt*II	H	Hormone autotrophy, NPTII dot bot, no root hair	Ramsay and Kumar, 1990

A.r. – *Agrobacterium rhizogenes;* H – Hairy root; S-shoot phenotype) have also been obtained (Manners and Way, 1989). Molecular analysis revealed, all plants with normal phenotype contained the binary vector T-DNA but one lacked the TL-DNA. Minor changes in plant morphology were noted in transgenic *Lotus corniculatus*, even though two or more copies of TL-DNA were present (Webb *et al.*, 1990). In plants where Ri phenotypic effects are undesirable, phenotypically normal plants can also be obtained. When binary vectors are used, the independent insertion of the Ri T-DNA and binary vector T-DNA may result in segregation of T-DNA's at meiosis in subsequent generation.

The several advantages of *A. rhizogenes*-mediated transformation over other methods including *A. tumefaciens*, are the spontaneous shoot regeneration directly from hairy roots, avoiding a callus phase eliminates the risk of somaclonal variation in the resulting plants, the rapid procedure, and efficiency for the production of non-chimeric plants. The transfer of root cultures from dark to light induced spontaneous shoot regeneration in *Lotus japonicus* (Petit *et al.*, 1987). Transgenic plants were produced at a high frequency by *A. rhizogenes*-mediated transformation as compared to *A. tumefaciens* methods in *S. humilis* (Manners and Way, 1989). In *Robinia,* hairy roots appeared within a week of co-cultivation and transgenic shoots regenerated within four weeks. Hairy root morphology is also used for the primary selection of transgenic plants. Thus, the use of selective agent, which may inhibit shoot regeneration, is essentially avoided. The maintenance of transgenic hairy root cultures as organ cultures allow long-term culture and subsequent shoot regeneration, without the encountered problems associated with cytological abnormalities in long term callus. However, cytological analysis of transformed root clones showed variation in ploidy level up to the octaploid and structural rearrangement of chromosomes in *Vicia faba* (Ramsay and Kumar, 1990). The main disadvantages of *A. rhizogenes*-mediated transformation are the limited host range of *A. rhizogenes* and the production of transgenic plants with altered phenotype.

Oncogenic wild type strains of *A. rhizogenes* have been used to transform *Sesbania rostrata* (Vlachova *et al.*, 1987) and *Stylosanthes sp.* (Manner, 1987) while plants were successfully regenerate from transgenic hairy roots using disarmed strains in several legume crops of economic importance (Table-2). While production of transgenic *Lotus corniculatus* (Jensen *et al.*, 1986), *Glycine canescens* (Rech *et al.*, 1989) and *G. argyrea* (Kumar *et al.*, 1991) employed co-integrative Ri-plasmid; a high frequency of co-transfer of genes through binary vector has been reported in the absence of selection in *Medicago arborea* where 70% of hairy roots were co-transformed (Damiani and Arcioni, 1991). Integration of Ri T-DNA and binary T-DNA at independent loci resulted in segregation of T-DNA's at meiosis in subsequent generation leading to normal transgenic plants without Ri-T-DNA phenotype. Such plants carried only T-DNA of binary vector.

Homogenously transformed plants were regenerated in *Medicago arborea*, *Lotus corniculatus* and *Stylosanthes*, employing *A. rhizogenes*-mediated transformation and thus provide opportunity to transfer agronomically important traits to these crop plants.

3.3.4. Protoplasts-based methods

Transformation of isolated protoplasts has been successful in a few legumes, either by coculturing protoplasts with *Agrobacterium* or by microinjection of *Agrobacterium* Ti-plasmid (Table-3) or direct DNA uptake by using PEG (Table-4) or electric current (Table-5). Direct DNA uptake into protoplasts has also provided quick system for monitoring gene expression. Such transient gene expression studies are useful in evaluating constructs and gene promoters with readily assayable reporter genes, such as CAT and β-glucuronidase.

3.3.4.1. Agrobacterium-protoplasts cocultivation or introduction of Ti-plasmid into protoplasts : The transformed clones resulted from cocultivation of protoplasts with the *A. tumefecians* strain T37, with a frequency of 10^{-3} (Mi and Zhon, 1986), and C58 (Baldes *et al.*, 1987) in soybean showed integration of full length DNA by Southern analysis (Baldes *et al.*, 1987). The transformed clones of *Vigna aconitifolia* generated from protoplasts, using Ti plasmid carrying *npt*II gene, showed expression of *npt*II and nopaline synthase gene in twenty-three percent of the clones as confirmed by Southern analysis. However, the transformation rate depended on the cultivar used (Kohler *et al.,* 1987b). *Agrobacterium* protoplast cocultivation technique has also been used in *Lotus corniculatus* (Nisbet, 1987).

On contrary, a variety of Ti-plasmid were microinjected into the nucleus of protoplasts of *Medicago sativa* (Reich *et al.*, 1986) with a transformation frequency of 15-26%, as determined by nopaline synthase activity assay. Several transformed lines with pTiC58 showed integration that avoided the T-DNA transfer mechanism typical for crown gall tissue. The transformation frequency was not improved even when the functional T-DNA carried by small plasmid or the coinjection of functional *vir* regions on a separate plasmid. The use of *Agrobacterium*–protoplast

Table 3. Agrobacterium-protoplasts co-cultivation or microinjection of Ti-plasmid into protoplasts of legumes

Species	Explant/Tissue	Agrostrain	Results	Reference
Glycine max	Protoplast	*A. tumefaciens* strain T37	Two clones out of 1226 found resistant to tetracycline and produce nopaline	Mi and Zhon, 1986
G. max	Protoplast	*A. tumefaciens* nopaline strain C58 PTiC8: pMON200-pfB2	Transformed callus resistant to kanamycin Southern analysis confirmed integration	Baldes *et al*, 1987
Vigna aconitifolia	Protoplast	*A. tumefaciens* PgV36/850::1130*neo* with selectable marker *npt*II and nopaline synthase genes	Kanamycin resistant transformed clones confirmed by Southern analysis	Eapen *et al.*, 1987
Medicago sativa	Protoplst	Microinjection of plasmid into protoplast	Nopaline synthase activity in transformed calli	Reich *et al.*, 1987

Table 4. Transformation of legumes using PEG-mediated DNA uptake

Species	Explant	Plasmid/ Construct	Reporter or Selectable marker	Plant	Mol. Analysis	Reference
Vigna aconitifolia 2 genotypes	Protoplasts	-	Kan.	Plants resistant to kanamycin	Southern bloting	Kohler *et al.*, 1987
Psophocarpus tetragonoloba	Epicotyl derived protoplasts	pABDI or pHP23	Kan.	transformed calluses, no shoot regeneration	NPTII activity and Southern	Gill *et al.*, 1990
Phaseolus vulgaris	Cell-suspension culture protoplasts	-	Transient Luciferase or GUS expression	-	-	Leon *et al.*, 1991
Pisum sativum	Protoplasts	-	-	Plant with reduced fertility and phenotypic abnormality	-	de Kathen *et al.*, 1998

Table 5. Transformation of legumes using electroporation technique

Species/Cultivar	Explant	Plasmid/ Construct	Selection or Reporter	Regeneration (Yes/No)	Molecular analysis	Reference
Medicago borealis	Shoot and leaf segment protoplasts	pGA472	*npt*II	Transgenric plant resistant to kanamycin	-	Kuchuk *et al.*, 1990
Glycine max	Immature cotyledon protoplasts	-	*gus* and *CAT*	No transgenic plant	Transient expression	Dhir *et al.*, 1991a
	-do-		*hpt*	Calli, shoots	Southern for *gus* and *hpt* in calli and shoots	Dhir *et al.*, 1991b
Phaselous vulgaris var, Bush Bean Tendergreen	Immature cotyledon protoplast	p795	*gus* and *CAT*	No	Transient	Bustos *et al.*, 1991
Glycine max cv. raiden	Immature cotyledon protoplast	pFF19G	*gus*	No	Transient	Ishikawa *et al.*, 1993
Vigna unguiculata cv. Blackeye	Embryonic axis	pBPG1 p35SGUS INT	*gus*	Yes	Seedling with *gus* activity	Akella and Lurquin, 1993
Phaseolus vulgaris	Embryonic axis (EA)	pDE435SG USInt.	*gus*	No	Transient expression in various tissue of EA	Dillen *et al.*, 1994, 1995
Phaseolus valgaris cv. Giza3	Callus	pDPG165	*bar*	Shoots regenerated	PCR and Southern	Sarkar and Kuhne, 1998
Pisum Sativum, Lens culinaris Glycine max Cicer arietimum	Intact nodal meristem	-	*gus*	Plants regenerated	-	Chowira *et al.*, 1995

cocultivation technique or microinjection of Ti plasmid into protoplast is not exploited in other legumes.

3.3.4.2. PEG-mediated DNA uptake : PEG-mediated DNA uptake assisted in the development of an extremely simple transformation technique for some species (Birch, 1997) as PEG-6000 at a concentration of 15-20% (w/v), in the presence of divalent cations, minimizes charge repulsion between DNA and membrane, protecting DNA from nuclease activity and stimulating endocytosis. Kohler *et al.* (1987a) reported production of transgenic (kanamycin resistant) calli and regenerants of mothbean (*V. aconitifolia*) when heat shocked protoplasts were treated with polyethylene glycol and plasmid DNA containing *npt*II. Transformation was confirmed by Southern analysis. Leon *et al.* (1991) studied transient expression of luciferase and *gus* genes in protoplasts, isolated from cell suspension culture of *Phaseolus vulgaris* that used PEG, and a plasmid containing chimaric genes comprising CaMV35 promoter fused to luciferase and *gus* genes. The transient expression of these genes increased by 10-fold when RNA leader sequence omega (Ω), derived from tobacco mosaic virus, and intron from the maize alcohol dehydroglycerase gene were introduced in the untranslated leader sequence of the chimeric genes.

Gill (1990) generated kanamycin resistant-transformed calli from epicotyl derived protoplasts of *Psophocarpus tetragonoloba*, following uptake of plasmid DNA assisted by PEG treatment. The transformed calli expressed NPTII activity and also exhibited the integration of gene. However, regeneration of shoot buds could not be accomplished from the transformed clones. de Kathen *et al.* (1998) recovered transgenic plants of pea after PEG-mediated DNA transfer into protoplasts. But the fertility of these plants was drastically reduced and phenotypic abnormalities were observed (Table-4).

3.3.4.3. Electroporation : Electroporation involves the application of high-voltage electric pulses to cells so as to induce transient membrane pores, allowing DNA to move into the cells. Electroporation has been preferred over PEG as the apparatus delivers DNA effectively at appropriate electric pulses. The process is cost-effective and, treatment is both reproducible and simple to apply. Electroporation has proved to be broadly applicable gene transfer technique, facilitating foreign gene expression in protoplasts as well as intact plant cells, both in suspension cultures and as part of organized tissues (Table-5). However, efficient permeability to cells of different types and sizes required different electrical conditions. Both transient expression and stable gene integration have been reported by this method.

Bustos *et al.* (1991) developed a protocol for the introduction of foreign DNA by electroporation into protoplasts, isolated from immature bean cotyledons. High expression of the GUS and CAT was reported when cognate genes were driven by the promoter and upstream sequences of a bean β-phas gene. Dhir *et al.* (1991) and Ishikawa *et al.* (1993) optimized the electroporation conditions for the transient expression of reporter genes, CAT and/or β-glucuronidase (GUS) respectively in protoplasts of soybean immature cotyledons. Protoplast viability and reporter enzyme activity depended on the field strength employed, which varied with genotype. Treatment of PEG (5.6%) and heat shock (5 min. at 45°C) to protoplasts prior to addition of DNA further enhanced the transformation efficiency (Dhir *et al.*, 1991). Use of calf thymus DNA as carrier DNA at concentration between 20-40 μg/ml and presence of divalent cation, Ca^{2+} (0.1 mM $CaCl_2$) in electroporation buffer increased the expression of GUS (Ishikawa *et al.*, 1993). The protoplasts, isolated from immature cotyledon of soybean, were stably transformed by electroporation of plasmid DNA, carrying chimaric genes encoding GUS and *hpt*. Transformed colonies were selected after growing 15-d-old protoplasts in the presence of hygromycin (40 μg/ml) and shoots were regenerated from the transformed calli at a frequency of 0.8% that showed the presence of GUS and *hpt* genes as confirmed by Southern hybridization (Dhir *et al.*, 1991).

Electroporation carries the same drawbacks like PEG method as both relies mainly on the use of protoplasts for introduction of foreign genes, and the technique for regeneration of plant from protoplast in most of the legumes is not perfected. An innovative procedure to deliver DNA into intact plant cells, both in suspension cultures and as a part of organized tissues has been demonstrated by Akella and Lurquin (1993) that showed imbibitions of cowpea (*Vigna unguiculata*) embryos in the presence of p35GUSINT/pUC8 DNA and the generated seedlings expressed the *gus* transgene quite efficiently in a variety of tissues including apical meristem after passive DNA uptake. Embryo electroporation in the presence of DNA and protectants such as spermine and cationic liposome (Lipofectin) increased both the proportion of embryo-derived seedlings expressing the chimeric gene and the level of gene expression. However, the transgene expression could not be accomplished when intact embryonic axes of *P. vulgaris* were simply incubated with DNA (Dillen *et al.*, 1994). Whereas DNA introduced into cells of intact embryonic axes of *Phaseolus vulgaris* and related species through electroporation (Dillen *et al.*, 1995), showed transient expression of *gus* reporter gene, in hypocotyl and epicotyl tissues, in section of variable size and penetration. Though the transgene expression was observed within a wide range of electrical conditions, the expression was maximum when a single pulse of 260 ms applied, at field strength of 225 v cm^{-1} (Dillen *et al.*, 1995). The protocol can be useful to test gene constructs in transient assays on whole tissue of grain legumes.

Chowira *et al.* (1995) produced transgenic plants of pea, lentil, soybean and chickpea by electroporation of intact nodal meristem of mature plants that expressed *gus* gene. Subsequent attempts in electroporation of axillary buds of pea with a construct containing the coat protein gene of Pea Enation Mosaic Virus (PEMV) lead to the generation of transformed plants resistant to PEMV (Chowira *et al.*, 1996).

Collen and Jarl (1999) compared the efficiency of plant materials of different kind, and mature embryos of 1-week-old *in vitro* grown seedlings and 8-week-old greenhouse grown plants of the legume, *Galega orientalis* to electroporation. Transformed chimeric shoots were recovered at a frequency of 33%, by electroporation of apical meristem of 8-week-old green house-grown plants. Similar transformation rate was obtained in electroporation of intact tissue of other legumes (Chowira *et al.*, 1996). The disadvantage with electroporation of whole plants is the requirement of high concentration of DNA (600 μg plasmid used for 40 plants) and generation of chimeric transformants in the first generation. In order to obtain complete transformants, the seeds of the chimeric plants need to be selected on selection medium. Electroporation of whole plant tissue might result in unstable pattern of inheritance of the introduced trait. However, the mechanism underlying the transfer of DNA molecules through the cell wall, during this procedure remains to be investigated (Table-5).

3.3.5. Biolistic method

Direct gene transfer by biolistics involves acceleration of high velocity DNA coated particles (gold or tungsten, 1-4 μm) to regenerable intact cells or tissues, and regeneration of transformants in a genotype-independent manner bypassing *Agrobacterium*-host specificity and tissue-culture associated regeneration difficulties. Introduction of multiple genes located on a single plasmid or separate plasmids (co-transformation) has been possible. At least three different types of particle bombardment devices have been constructed in order to accelerate these microprojectiles. All these systems are based on the generation of a shock wave with enough energy to move the microprojectiles. All the three biolistic guns have been used to deliver DNA into cells of legumes to develop transgenic plants using different explants (Table-6). Electric discharge particle accelerating device ACCELL® (McCabe and Christou, 1993) can accelerate DNA coated gold particles to any velocity by varying the voltage while the biolistic PDS-1000/He (Kikkert, 1993) device is powered by a burst of gas that accelerates a macrocarrier, on which DNA coated particles are uniformly placed, towards a perforated stopping screen and the microprojectiles continue at high velocity, usually under vacuum, into the target tissue. A modified version of the biolistic PDS1000 device was developed i.e. particle Inflow gun (PIG) (Vain *et al.*, 1993), where bombardment caused less injury to tissues than the PDS-1000/He, allowing regeneration and transformation of the more recalcitrance varieties. The DNA is dissociated from the particles within the target cells and gets integrated into plant genomic DNA, although the exact mechanism involved in this process remain complex.

The *gus*, widely used reporter gene, has been driven by CaMV35S promoter in majority of the legumes transformed with exception to concanavalin (ConA) promoter in *Phaseolus vulgaris* (Kim and Minamikawa, 1996). The untranslated leader sequence from RNA 4 of alfalfa mosaic virus significantly increased gus activity when included between the promoter and the *gus* coding region. The *gfp* (Ponappa *et al.*, 1999) assisted monitoring gene expression as well as screenings shoots at the preliminary stage, which is beneficial particularly in case of high degree of chimeras. Light stimulated GFP fluorescence does not require any co-factors, substrates, or additional gene products and moreover, GFP accumulation in transgenic crops produced, so far, did not appear to have a toxic effect (Maximova *et al.*, 1998). The *gfp* has been effectively used as reporter in *Vigna angularis* (Yamada *et al.*, 2001). The *npt*II has invariably used as selectable marker in majority of the cases with exception of *hpt* (Hadi *et al.*, 1996; Ozias-Akin *et al.*, 1993; Yang *et al.*, 1998) or herbicide genes (Aragao *et al,* 2000).

The generation of transgenic legumes by particle bombardment, so far, depended on two regenerating systems, one that used shoot apical meristem of mature seeds or whole embryos as target tissues in *Glycine max* (McCabe *et al.*, 1988; Sato *et al.*, 1993), *Phaseolus vulgaris* (Russell *et al.*, 1993; Kim and Minamikawa, 1996; Aragao *et al.*, 1997), *Arachis hypogea* (Schnall and Weissinger, 1993; Ozias-Akin *et al.*,

Table 6. Transformation of legumes using biolistic technique

Species	Cultivar/ Variety	Method	Explant/culture	Plasmid	Gene Construct	Reporter or Market gene/ Selective agent	T_0 plant regenerated or not	Mol. Analysis of T_0/ Transformation efficiency	T_1/T_2	Reference
Glycine max	Williams 82 Mandarin Ottawa Hardin	-	Callus	pCMC1022	35S NPTII CDSNOS	Kan	Yes	-	-	Christou *et. al.*, 1988
G. max	Williams 82 Mandarin Ottawa	-	Mertistems	pCMC1022	35S NPTII NOS35GUS	Kan	Yes	2% Yes Southern *nptII* analysis	R_1	McCabe *et al.*, 1988
G. max	Fayette	Microproj-ectile-He	Embryogenic suspensions Embroygenesis		-	Market Genes	Transgenic R_0 and R_1	-	R_1	Finer and McMullen 1991
G. max	William 82	Microproj-ectile-He	Shoot tips, Embroyonic suspension	pMON1002 6 pMON1371	35S *nptII* *nos* GUS	*npt*II, GUS	Yes	0.4%	-	Sato *et al.*, 1993
G. max	Pecking	Microproj-ectile-He	Embryogenic suspension and Embryogenesis	Bt. Insect-resistant	-	-	Transgenic R_0 and R_1	-	R_1	Parrott *et al.*, 1994
G. max	-	Microproje ctile	Immature embryo axes, organogenesis	-	*lysC* and *dap A* genes (to enhance seed lysine)	-	Transgenic R_1 and R_2	-	-	Falco *et al.*, 1995

Table 6. Continued

Species	Cultivar/ Variety	Method	Explant/culture	Plasmid	Gene Construct	Reporter or Market gene/ Selective agent	T_0 plant regenerated or not	Mol. Analysis of T_0/ Transformation efficiency	T_1/T_2	Reference
Glycine max	-	Microprojectile-He	Somatic embryo-hypocotyl Somatic embryo-cycling	-	Mannmalian desaturase	-	Transgenic R_0 abnormal flower	-	-	Liu *et al.*, 1996
G. max	Jack	Microprojectile-He	Embryogenic suspension Embryogenesis	-	Bt. *cryIAc* gene	-	Transgenic R_0 and R_1	Yes	-	Stewart *et al.*, *al.*, 1996
G. max	Fayette	BPG Biolistic	Embryogenic suspension/ Embryogenesis	umc29, 34, 38, 39, 82, 84, 107, 115, 119, pC 1B709, pUCGUS pbnl114.28	35S HPT35S GUS	Hygromycin, GUS	-	-	-	Hadi *et al.*, 1996
Arachis hypogea	Florunner Georgia Runner MARC-1	-	Embryogenic tissue	pPBIN19	-	*hph.* TSWV-N	Yes	RT-PCR, ELISA, Southern of T_0	T_1/T_2	Yang *et al.*, 1998
A. hypogea	-	PIG	Embryogenic tissue	-		*hpt*	Yes	-	-	Ozias-Akins *et al.*, 1993
A. hypogea	Floruner, Florigent	PIG	Shoot meristem of mature EA	-	-	GUS, BAR and tswv-np	-	6-8 percent transient *gus* assay	Southern R_0, R_1	Brar *et al.*, 1994

Table 6. Continued

Species	Cultivar/ Variety	Method	Explant/culture	Plasmid	Gene Construct	Reporter or Market gene/ Selective agent	T_0 plant regenerated or not	Mol. Analysis of T_0/ Transformation efficiency	T_1/T_2	Reference
Arachis hypogea	Virginia Tamnut 74 UPL Pn4 Florunner	PDS-1000/He	Zygotic Embryos	pRT99	35SGUS	Hyg.	-	-	-	Schnall and Weissinger, 1993
Phaseolus vulgaris	Carioca	Electric particle gun	Embryonic axis	p35SBN	35S2S and 35SGUS	-	-	-	GUS, Western and ELISA assays, Expression of 2S gene	Aragao *et al.*, 1992
P. vulgaris	-	-	Apical meristem from seeds	pWRG2204	*gus, bar*	PPT	-	GUS, Southern northern analysis, herbicide resistant plants	-	Russel *et al.*, 1993
P. vulgaris	Gold Star	GIE-III	Embryonic axis	pSOG1016	ConAGUS	Kan	Yes	GUS assay, PCR, Southern	-	Kim and Minakawa, 1996
P. vulgaris	Olathe	-	Embryonic axis	pBI426 pEA23 pAC123B C1-BZ	35S:35SGUS *neo tnos*	Kan	Yes	GUS, PCR, Southern	$R_1R_2R_3$	Aragao *et al.*, 1996

Table 6. Continued

Species	Cultivar/ Variety	Method	Explant/culture	Plasmid	Gene Construct	Reporter or Market gene/ Selective agent	T_0 plant regenerated or not	Mol. Analysis of T_0/ Transformation efficiency	T_1/T_2	Reference
Phaseolus vulgaris	Olathe	-	Embryonic axis	pMDI pBI426	35S-*gus*-*noster*	-	-	-	-	Ribeiro *et al.*, 1998
Medi-cago sativa	-	-	Pollen	pBI121	*nos*pro *npt*II *nos-ter-CaMV 35S-GUS nos*-ter	Kan	Yes	PCR, Southern GUS of T0	-	Romaiah; and Skinner, *1997*
M. sativa	C2-4	PDS 1000-He	Calli (derived from petiole and stem section	pKANGUS pFF19K	35S NPTII 35S GUS *nos* ter	Kan	Yes	Southern	Progeny analysis	Pereira and Erickson, 1995
Lathyrus sativus	-	-	Immature leaflets and nodal segments	-	-	GUS	-	-	-	Barna and Mehta, 1995
Vigna aconitifoli-a V.mungo V.radiata	-	PIG	Mature embryo without cotyledon	pBI121	-	-	-	Transient GUS assay	-	Bhargava and Smigocki, 1994

1993; Brar *et al.*, 1994) and with limited success in *Vigna* species (Bhargava and Smigocki, 1994). Whereas the embroyogenic cell suspensions were targeted in soybean (Finer and McMullen, 1991; Sato *et al.*, 1993; Parrott *et al.*, 1994; Stewart *et al.*, 1996; Ponappa *et al.*, 1999) and embryogenic cultures in *Medicago sativa* (Pereira and Erickson, 1995) and peanut (Yang *et al.*, 1998). The pollen grain has also been targeted for transferring genes through particle bombardment in *Medicago sativa* (Ramaiah and Skinner, 1997).

3.3.5.1. Shoot apical meristem : The shoot apical meristem of mature seed or whole embryos, with their high regeneration potential, regenerates to complete fertile plant with minimal tissue culture manipulations (Smith and Murashige, 1970) as well as in a genotype independent fashion. The plants originating from the apical or axillary meristem cultures are considered to show low somaclonal variability. Introduction of DNA into shoot meristem cells frequently results in the production of chimeric plants, consisting of both transformed and untransformed sectors. However, transgenic offspring can be obtained from such chimeric plants provided that the original transformation events involved the progenitor cells of the gametes. These progenitor cells are located in the sub-epidermal layer of the apical meristem, the L2 layer (Satina *et al.*, 1940). With most particle gun types, the particles penetrate to a depth of only one or two cell layers. However, the electric gun (ACCELL) gauge and control the penetration by DNA-carrying particles by accurately controlling voltage (McCabe *et al.*, 1988). The bombardment of shoot meristematic tissue of immature soybean seeds followed by tissue culture multiplication, resulted in production of transgenics that transmitted the introduced DNA to progenies (McCabe *et al.*, 1988). Christou *et al.* (1990) developed a procedure for introduction of genes into virtually any genotype of soybean by combining a simple genotype-independent regeneration protocol (based on proliferation of multiple shoots from the general area of meristem of embryonic axis) with electric discharge particle acceleration. Independently transformed soybean lines were obtained representing a wide phenotypic spectrum of transformation events. Parameters such as bombardment conditions, intensity of DNA and gold beads were optimized to recover phenotypes associated with germline transformation. The expression of *gus* reporter in specific tissue types (stem, leaf and petiole sections) of R_0 chimeric soybean transformants, at various stages of development was correlated with transmission of the introduced genes to progeny to identify the germline transformation (Christou, 1994). A large number of plants recovered, virtually expressed the introduced genes in all their cells which inherited to subsequent generations (Christou, 1997). A similar approach was followed in bean (Russell *et al.*, 1993) and peanut (Brar *et al.*, 1994) for recovery of fertile transgenic plants. The various parameters affecting the gene delivery such as the pressure of helium shock wave, distance between the particle plate and the sample, amount of particles and DNA per bombardment were optimised using transient GUS expression (Russell *et al.*, 1993; Aragao *et al.*, 1993; Brasileiro *et al.*, 1996).

Aragao *et al.* (1996) exploited the biolistic process to regenerate stable transgenic bean plants with linked or unlinked foreign genes with a co-transformation frequency of 40-50% and 100% respectively. The introduced genes inherited in Mendelian fashion in most of the transgenic bean lines.

3.3.5.2. Embryogenic suspension tissue : In soybean, embryogenic suspension cultures were established by the proliferation of new embryoids from the surface of older somatic embryos (Finer, 1988; Finer and Nagasawa, 1988). Finer and McMullen (1991) used the particle inflow gun to transform the tissue from embryogenic suspension culture. Surface origin of the embryoids proliferation makes this system attractive for transformation as the regeneration competent surface embryogenic cells are numerous and thus, can easily be targeted. A few *in vitro* embryogenic proliferate cycles of the bombarded tissue under selection pressure lead to the recovery of non-chimeric transformants. This system has been most successful in soybean (Sato *et al.*, 1993; Parrott *et al.*, 1994; Hadi *et al.*, 1996; Stewart *et al.*, 1996; Liu *et al.*, 1996). The major drawback of this system is the difficult and lengthy efforts required to establish soybean liquid culture system. Further, establishment of soybean embryogenic suspension culture seem highly genotype dependent and the plants regenerated from older culture may exhibit partial or full sterility and abnormalities. Therefore, it is imperative to establish and transform young (<1 yr. old) cultures (Liu *et al.*, 1996).

Hadi *et al.* (1996) co-transformed 12 different plasmids into suspension cultures of soybean via particle bombardment. All the co-transforming plasmids were found present in most of the transgenic soybean clones and there was no preferential uptake and integration of any of the plasmids. Plasmid amplification, concatamer formation (indicative of homologous recombination) and ligation of plasmid fragments also occurred in transformed clones. This indicates that co-transformation can be used to study the process of recombination and to introduce large number of genes present in different plasmid without intensive and inefficient process of repeated transformation.

3.3.5.3. Other systems : Christou *et al.* (1988) generated transformed calli on kanamycin selection medium from the protoplasts of immature embryos, bombarded with pCMC1022 that contained *npt*II under the control of 35S promoter and *nos* polyadenylation region. Enzyme assays and Southern hybridization confirmed the expression of transgene and its stable integration in soybean genome.

Ramaiah and Skinner (1997), for the first time, targeted pollens by microprojectile bombardment in *Medicago sativa* by introducing pBI121 bearing GUS reporter gene. The bombarded pollens expressed GUS activity and produced viable transgenic seeds by direct pollination of the flowers of male-sterile plants. Thirty percent of the plants derived from the fertile seeds showed integration of GUS as confirmed by PCR and Southern analysis. However, some transgenics (T_0) apparently lost the integrated *gus* after 10 vegetative generations. The possible cause for the loss could not be ascertained.

Many agronomically important genes have been introduced in legumes employing the particle bombardment technique, which include resistance to herbicides (Delanney *et al.*, 1995; Padgette *et al.*, 1995), insect (Stewart *et al.*, 1996), viruses (Aragao *et al.*, 1998), and improvement in protein quality (Aragao *et al.*, 1992, 1996).

Though gene delivery by biolistic has been most successful in three major legumes i.e. soybean, peanut and *Phaseolus vulgaris*, certain drawbacks associated with this technique viz. the copy number and high rearrangement of the introduced DNA, that lead to gene silencing, complicated its merits. The other major drawback is the restricted accessibility of the device due to its high cost.

3.3.6. *In Planta* transformation

Although legumes are susceptible to *Agrobacterium*, difficulties in regenerating whole plants from tissues competent for transformation, necessitated the development of alternative transformation methods that essentially avoid tissue culture and regeneration procedures. Further, complications associated with the time consuming and labor intensive regeneration steps, and somaclonal variation or morphological abnormalities exhibited by the regenerated plants (van de Bulk *et al.*, 1990; Evans and Sharp, 1986; Larkin and Scowcroft, 1991) and reduced fertility (Scholl *et al.*, 1981) necessarily demands development of plant transformation procedures that obviate the *in vitro* regeneration steps. There are only a few species for which transformation systems avoiding tissue culture based regeneration system are available which includes the model plant *Arabidopsis thaliana* for which a variety of *in planta* methods have been developed. Transgenic plants regenerated from germinated seeds (Feldman, 1992; Feldman and Marks, 1987) or wounded plants (Chang *et al.*, 1994) inoculated with *Agrobacterium*, were grown to maturity in the absence of any selection, progeny seeds were collected and germinated on antibiotic containing medium to identify the transformed plants. The procedure was difficult to reproduce consistently and transformants were selected at a low frequency. Bechtold *et al.* (1993) inoculated flowering plants of *A. thaliana* by vacuum infiltration of *Agrobacterium* suspension. Plants were transplanted back to soil, seeds were collected and stable transformants were selected in the next generation using the antibiotic or herbicide. This method proved reproducible and now routinely used in many laboratories. Even, uprooting the plants for *Agrobacterium* inoculation and replantation is not required. Transformants could be obtained by dipping protruding inflorescence in *Agrobacterium* suspension with a strong non-toxic surfactant (Bent, 2000). It is demonstrated that ovules are the site of productive *Arabidopsis* transformation.

The *in planta* transformation has been successful in model legumes, *Medicago truncatula* and *Lotus japonica* as well as in soybean and peanut (Table-7). Chee *et al.* (1989) developed a technically simple and tissue culture-independent method for soybean transformation by inoculating plumule, cotyledonary node and adjacent cotyledonary tissue of germinating seeds with *Agrobacterium tumefaciens* strains C58C1, with pGA482G which contained *npt*II. Sixteen R_0 plants were identified positive for *npt*II out of total 2200-tested. About 10% of these R_0 plants yielded transformed R_1 progeny. The presence of *nptII* gene was demonstrated in R_0 and R_1 plants by genomic blot hybridization and PCR, and one R_1 soybean line demonstrated the integration of this gene. The overall transformation frequency was low, i.e. 0.07%, which involved extensive labour, greenhouse space and time to obtain a reasonably large number of stable transgenic plants. The low frequency of this method may be attributed to the failure in identifying the exact location of those cells in the plumule and cotyledonary nodal region contributing

Table 7 ***In planta* transformation**

Species	Cultivar	Explant	Agrostrain	Construct	Selectable Marker	Plant regnerat-ion (Yes/No)	Progeny (ies)	References
Glycin max	A0949	germinated seeds	C58Z707	pGA482G	*nptII*	Yes Southern & PCR analysis	nos-nptII gene in 0.7% R_0 0.07% R_1	Chee *et al.*, 1989
Lotus japonicum	Gifu	Seedlings	C58C1	pGU3850:: 1103, pR;18196; pGV3850::1103	*nos* and *npt*II	Yes, PCR Southern analysis	F_1 65-66%	Oger *et al.*, 1996
		Infiltration of flower plants	ASE1 EHA 105	pSLJ525 pSK1006 pBIN *mgfp. ER*-bar or *bar*	*bar* *bar*	Southern analysis	2.9 to 27.6% 4.7 to 76.6% T2 transgene inherited in Mendelian fashion	Trieu *et al.*, 2000
Medicago truncatula	Jemalong (Line A17)	Vacuum infiltration of seedings	GV3101 EHA105	pSK1015 pB1121-bar pGA482-bar pKYLX7GUS pBINmgfp-ER-*bar*	- *bar* *bar* - *bar*			
Arachis hypogaea	TMV-2	Embryonic axis of mature seed with one cotyledon	LBA4404	pKiW1105	GUS and *nptII*	Yes gus assay and PCR analysis	T_1 & T_2 Sothern analysis	Rohini and Rao (2000)

to the ovule/pollen cells. The identification of these cells could lead to improve targeting for transformation.

Oger *et al.* (1996) developed a simple method that based on *in planta* inoculation, to spontaneously regenerate shoots of *Lotus japonicus* at cotyledon attachment site of the seedlings. The plantlets (7-15 d-old) were detached at the cotyledon attachment site and inoculated with disarmed *Agrobacterium* strain C58C1 (pGV3850:1103) whose T-DNA harbored screenable marker gene i.e. nopaline synthase (nos) and neomycin phosphotransferase (*npt*II), that allowed easy detection. The shoots that appeared from this site on hormone free medium were nopaline producing (transformed) and probably chimeric. Shoots (R_0) were propagated to produce mature plants, which were selfed to produce seeds. Seeds were germinated and yielded non-chimeric transgenic *Lotus* plants (R_1) at a frequency of 75%. Integration of *nos* gene in plant genome was confirmed by PCR and Southern blot analysis.

Trieu *et al.* (2000) reported *in planta* transformation in *Medicago truncatula* by vacuum infiltration of flowering plants and young seedlings. The transformation frequencies ranged from 4.7 to 76% for flower infiltration and 2.5 to 27% for young seedlings. The analysis of T_2 transformants showed transgene inheritance in Mendelian fashion. Further studies are required to unravel the mechanism underlying transformation in *M. truncatula*. This transformation system will make feasible large scale insertional mutagenesis or gene tagging, and will also provide a wide range of opportunities for the analysis of the unique aspects of legumes.

A non-tissue culture based approach (*in planta* transformation) was followed for generating transgenic peanut plants (McKently *et al.*, 1995; Rohini and Rao, 2000) employing embryos and *Agrobacterium tumefaciens*. However, such methods are limited by high input requirement in terms of the number of treatments to obtain very small number of independent transformants. Additionally, such approach is labour intensive requiring *vir* gene induction treatment such as the use of wounded tobacco leaf extract. Only three independent transformation events involving 150 embryos resulted in 2% transformation frequency (Rohini and Rao, 2000) and a high degree of chimeric shoots resulted with only a sector of T_0 transformants containing the T-DNA (McKently *et al.*, 1995).

Alternatively, it may be possible to deliver *Agrobacterium* by microinjection to ovaries or shooting *Agrobacterium* in flowers using microprojectile (Mesa *et al.*, 2000) or high-pressure air guns (Bent, 2000). These transformation approaches can be adopted for recalcitrant legumes.

4. CONCLUSIONS AND FUTURE PROSPECTS

The revolution in genetic engineering of crop plants with pioneering technologies and with the array of genes introduced, is limited to the model plants. The transfer of these technologies and their practical application to legume crops has not been facilitated vastly due to the difficulties in identifying regenerable cells accessible to transformation, high recalcitrance nature, and lack of significant fund involvement in the research programs undergoing in developing countries, the major producers. Preliminary progress in developing transformation system in legume crops of commercial importance, by employing the most promising gene transfer techniques, *Agrobacterium*-mediated and particle bombardment, are encouraging but their further improvement is not accelerated due to the restriction of intellectual property right issues that prevents availability of novel genes and effective promoters, for high and tissue-specific expression of the introduced genes. Most of the currently established or promising plant genetic engineering strategies are covered by patents owned by private biotechnologies companies (see, e.g., Sanford *et al.*, 1990, 1992; Schilperoort *et al.*, 1990; Hiei *et al.*, 1994; Coffee and Dunwell, 1995; Maliga and Maliga, 1995; Pazkowski *et al.*, 1995) which include many isolated genes, promoters and techniques for plant gene manipulation, thereby leading to serious implications (Birch, 1997). Further, link amongst corporate research organizations, academic institutions and international organizations need to be strengthened to avoid duplication of effort and to maximize efficient utilization of limited resources (Christou, 1997). The need of pragmatic approaches for effective handling either by global initiatives on intellectual property right (IPR) issues or through the development of expertise in gene identification and cloning in the developing world, is realized. Future field trials of genetically modified legume crops and their subsequent release must address the present lack of proper biosafety regulations in developing countries and associated ethical, sociocultural and political equations. Endeavors and meaningful coordinations should be on priority to enhance public and government perceptions on transgenic crops and their applications to sustainable agriculture.

REFERENCES

Akella V and Lurquin PF (1993) Expression of cowpea seedlings of chimeric transgenes after electroporation into seed-derived embryos. *Plant Cell Rep.,* **12** : 110-117.

Alibert B, Lucas O, Gall VL, Kallerhoff J and Alibert G (1999) Pectolytic enzyme treatment of sunflower explants prior to wounding and co-cultivaiton with *Agrobacterium tumefaciens*, enhances efficiency of transient β-glucuronidase expression. *Physiol. Plant.,* **106** : 232-237.

Armstead IP and Webb KJ (1987) Effect of age and genotype of tissue on genetic transformation of *Lotus corniculatus* by *Agrobacterium tumefaciens. Plant Cell Tiss. Org. Cult.,* **9** : 95-101.

Aragao FJL, Grossi de Sa M, Almeida ER, Gander ES and Rech EL (1992) Particle bombardment-mediated transient expression of a Brazil nut methionine-rich albumin in bean (*Phaseolus vulgaris*). *Plant Mol. Biol.,* **20** : 357-359.

Aragao FJL, Grossi-de-Sa MF, Davey MR, Brasileiro ACM, Faria JC and Rech EL (1993) Factors influencing transient gene expression in bean (*Phaseolus vulgaris*) using an electrical particle acceleration device. *Plant Cell Rep.,* **12** : 483-490.

Aragao FJL, Barros LMG, Brasileiro AC, Ribeiro SG, Smith FD, Sanford JC, Faria JC and Rech EL (1996) Inheritance of foreign genes in transgenic bean (*Phaseolus vulgaris* L.) co-transformed via particle bombardment. *Theo. Appl. Genet.,* **93** : 142-150.

Aragao FJL and Rech EL (1997) Morphological factors influencing recovery of transgenic bean plants (*Phaseolus vulgaris* L.) of a carioca cultivar. *Intl. J. Plant Sci.,* **158** : 157-163.

Aragao FJL, Ribeiro SG, Barros LMG, Brasileiro ACM, Maxwell DP, Rech EL and Faria JC (1998) Transgenic bean (*Phaseolus vulgaris* L.) engineered to express viral antisense RNAs show delayed and attenuated symptoms to bean golden mosaic geminivirus. *Mol. Breed.,* **4** : 491-499.

Aragao FJL, Sarokin L, Vianna GR and Rech EL (2000) Selection of transgenic meristematic cells utilizing a herbicidal molecule results in the recovery of fertile transgenic soybean (*Glycine max* (L.) Merril) plants at high frequency. *Theor. Appl. Genet.,* **101** : 1-6.

Atkins CA and Smith PMC (1997) Genetic transformation and regeneration of legumes. In: *Biological fixation of nitrogen for ecology and sustainable agriculture.* (Eds Legocki A, Bothe H, Puler A). Springer, Berlin Heidelberg, New York, pp, 283-304.

Baldes R, Moos M and Geider K (1987) Transformation of soybean protoplasts from permanent suspension cultures by cultivation with *Agrobacterium*-cell of *Agrobacterium tumefaciens. Plant Mol. Biol.,* **9** : 135-146.

Baumlein H, Boerjan W, Nagy I, Panitz R, Inze D and Wobus U (1991) Upstream sequences regulating legumin gene expression in heterologous transgenic plants. *Mol. Gen. Genet.,* **225** : 121-128.

Bean SJ, Gooding PS, Mullineaux P and Davies DR (1997) A simple system for pea transformation. *Plant Cell Rep.,* **16** : 513–516.

Bechtold N, Ellis J and Pelletier G (1993) *In planta Agrobacterium*-mediated gene transfer by infiltration of adult *Arabidopsis* plants. *C.R. Acad Sci. Paris Life Sci.,* **316** : 1194-1199.

Bent AF (2000) *Arabidopsis in planta* transformation, uses, mechanisms, and prospects for transformation of other species. *Plant Physiol.,* **124** : 1540-1547.

Bhargava SC and Smigocki AC (1994) Transformation of tropical grain legumes using particle bombartment. *Curr. Sci.,* **66** : 439-442.

Bidney D, Scelonge C, Martichs J, Burrus M, Sims L and Huffman G (1992) Microprojectile bombardment of plant tissue increases transformation frequency by *Agrobacterium tumefaciens. Plant Mol. Biol.,* **18** : 301-313.

Billings S, Jelenkovic G, Chin CK and Eberhardt J (1997) The effect of growth regulators and antibiotics on eggplant transformation. *J. Am. Soc. Hort. Sci.,* **122** : 158-162.

Birch RG (1997) Plant transformation: Problems and strategies for practical application. *Annu. Rev. Plant Physiol. Mol. Biol.,* **48** : 297-326.

Brar GS, Cohen BA, Vick CL and Johnson GW (1994) Recovery of transgenic peanut (*Arachis hypogaea* L.) plants from elite cultivars utilizing ACCELL technology. *Plant J.,* **5** : 745-753.

Brasilleiro ACM, Aragao FJL, Rossi S, Dusi DMA, Barros LMG and Rech EL (1996) Susceptibility of common and tepani bean to *Agrobacterium* spp. strains and improvement of *Agrobacterium*-mediated transformation using microprojectile bombardment. *J. Amer. Soc. Hort. Sci.,* **121** : 810-815.

Bustos MM, Battraw MJ, Kalkan FA and Hall TC (1991) Transient gene expression in electroporated bean cotyledon protoplasts. *Plant Mol. Biol. Reporter,* **9** : 322-332.

Byrne MC, Mcdonnell RE, Wright MS and Carnes MG (1987) Strain and cultivar specificity in the *Agrobacterium* soybean interaction. *Plant Cell Tiss. Org. Cult.,* **8** : 3-15.

Chabaud M, Passiatore JE, Cannon F and Buchanan-Wallaston V (1988) Parameters affecting the frequency of kanamycin resistant alfalfa obtained by *Agrobacterium tumefaciens* mediated transformation. *Plant Cell Rep.,* **7** : 7, 512-516.

Chabaud M, Larsonneau C, Marmoget C and Huguest T (1996) Transformation of bassel medic (*Medicago truncatula* Gaertn) by *Agrobacterium tumefaciens* and regeneration via somatic embryogenesis of transgenic plants with the MtENOD12 nodulin promoter fused to the gus reporter gene. *Plant Cell Rep.,* **15** : 5, 305-310.

Chang SS, Park SK, Kim BC, Kang BJ, Kim DU and Nam HG (1994) Stable genetic transformation of *Arabidopsis thaliana* by *Agrobacterium* inoculation *in planta. Plant J.,* **5** : 551-558.

Chee PP, Forber KA and Slinghtom JL (1989) Transformation of soybean (*Glycine max*) by infecting germinating seeds with *Agrobacterium tumefaciens. Plant Physiol.,* **91** : 1212-1218.

Cheng ZM, Schnurr JA and Kapaun JA (1998) Timentin as an alternative antibiotic for suppression of *Agrobacterium tumefaciens* in genetic transformation. *Plant Cell Rep.,* **17** : 646-649.

Chowira GM, Akella V and Lurquin PF (1995) Electroporation-mediated gene transfer into intact nodal meristems in planta. *Mol. Biotechnol.,* **3** : 17-23.

Chowira GM, Akella V, Fuerst PE and Lurquin PF (1996) Transgenic grain legumes obtained by *in planta* electroporation-mediated gene transfer. *Mol. Biotechnol.,* **5** : 85-96.

Christou P, McCabe DE, Martinell BJ and Swain WF (1990) Soybean genetic engineering–Commercial production of transgenic plants. *Trends Biotech.,* **8** : 145.

Christou P (1994) Biotechnology of crop legumes. *Euphytica*, **74** : 165-185.

Christou P (1997) Biotechnology applied to grain legumes. *Field Crops Res.*, **53** : 187-204.

Coffee RA and Dunwell JM (1995) Transformation of plant cells. US Pat. No. 5, 464, 765.

Collen AMC and Jarl CI (1999) Comparison of different methods for plant regeneration and transformation of the legume *Galega orientalis* Lam (goat's rue) *Plant Cell Rep.*, **19** : 1, 13-19.

Costa MGC, Nogueira FTS, Figueira ML, Otoni WC, Brommonschenkel SH and Cecon PR (2000) Influence of timentin on plant regeneration of tomato (*Lycopersicon esculentum* Mill.) cultivars. *Plant Cel Rep.*, **19** : 327-332.

Damiani F and Arcioni S (1991) Transformation of *Medicago arborea* L with an *A. rhizogenes* binary vector carrying hygromycin resistance gene. *Plant Cell Rep.*, **10** : 300-303.

Davies DR, Hamilton J and Mullineaux P (1993) Transformation of peas. *Plant Cell Rep.*, **12** : 180-183.

de Kathen A and Jacobson HJ (1990) *Agrobacterium tumefaciens*-mediated transformation of *Pisum sativum* L. using binary and co-integrative vectors. *Plant Cell Rep.*, **9** : 276–279.

de Kathen A and Jacobsen HJ (1992) Introduction of competence for transformation in *Pisum sativum* L. Proceedings of European conference on grain legumes, Angers, 117-118.

de Kathen A and Jacobsen HJ (1995) Cell competence for *Agrobacterium*-mediated DNA transfer in *Pisum sativum* L. *Transgenic Res.*, **4** : 184-191.

de Kathen A, Waglin T, Kiesecker H, Meyer B and Jacobsen HJ (1998) Transgenic grain legume from protoplasts. *Proc. 3rd European Conf. on grain legumes*. Valladolid, Spain, pp 370-371.

Deak M, Kiss GB, Konca C and Dudits D (1986) Transformation of *Medicago* by *Agrobacterium*-mediated gene transfer. *Plant Cell Rep.*, **5** : 97.

D'Halluin K, Botterman J and DeGreef W (1990) Cell biology and molecular genetics: engineering of herbicide-resistant alfalfa and evaluation under field conditions. *Crop Sci.*, **30** : 866-871.

Dhir SK, Dhir S, Hepburn A and Widholm JM (1991) Factors affecting transient gene expression in electroporated *Glycine max* protoplasts. *Plant Cell Rep.*, **10** : 106-110.

Di R, Purcell V, Collins GB and Ghabrial SA (1996) Production of transgenic soybean lines expressing the bean pod mottle virus coat protein precursor gene. *Plant Cell Rep.*, **15** : 746-750.

Dillen W, Montagu MV and Angenon G (1994) Electroporation-mediated DNA delivery to embryos of leguminous species. In: Cell Biology. A Laboratory Handbook. Academic Press. New York pp 72-76.

Dillen W, Engler G, Montagu MV and Angenon G (1995) Electroporation-mediated DNA delivery to seedling tissues of *Phaseolus vulgaris* L. (common bean). *Plant Cell Rep.*, **15** : 119-124.

Dillen W, Clercq J De, Kapila J, Zambre M, Montagu MV and Angenon G (1997) The effect of temperature on *Agrobacterium tumefaciens*-mediated gene transfer to plants. *Plant J.*, **12** : 1459-1463.

D' Malluin, Botterman J and DeGreef (1990) Cell biology and molecular genetics: engineering of herbicide resistant alfalfa and evaluation under field conditions. *Curr. Sci.*, **30** : 866-871.

Donaldson PA and Simmonds DH (2000) Susceptibility to *Agrobacterium tumefaciens* and cotyledonary node transformation in short-season soybean. *Plant Cell Rep.,* **19** : 478-484.

El Khalifa MD and Lippincott JA (1988) The influence of plant growth factors on the initiation and growth of crown-gall tumours on primary pinto bean leaves. *J. Exp. Bot.,* **19** : 61-74.

Enriquez-Obregon GA, Prieto-Samsonov DL, de la Riva G, Perez M, Selman-Housein G and Vazquez-Padron RI (1999) *Agrobacterium*-mediated japonica rice transformation: a procedure assisted by an antinecrotic treatment. *Plant Cell Tiss. Org. Cult.,* **59** : 159-168.

Escudero J, Neuhaus G, Schlappi M and Hohn B (1996) T-DNA transfer in meristematic cells of maize provided with intracellular *Agrobacterium. Plant J.,* **10** : 355-360.

Evans DA and Sharp WR (1986) Application of somaclonal variations. *Bio/Technol.,* **4** : 528-532.

Feldmann KA and Mark MD (1987) *Agrobacterium*-mediated transformation of germinating seeds of *Arabidopsis thaliana*: a non-tissue culture approach. *Mol. Gen. Genet.,* **208** : 1-9.

Feldmann K (1992) T-DNA insertion mutagenesis in *Arabidopsis*: sedd infection transformation In: *Methods in Arabidopsis Research* (Eds Koncz C, Chua N-H, Schell J) World Scientific, Singapore, pp 274-289.

Finer JJ (1988) Apical proliferation of embryogenic tissue of soybean (*Glycine max* (L.) Merrill). *Plant Cell Rep.,* **7** : 238-241.

Finer JJ and Nagasawa A (1988) Development of an embryogenic suspension culture of soybean (*Glycine max*). *Plant Cell Tiss. Org. Cult.,* **15** : 125-136.

Finer JJ and McMullen MD (1991) Transformation of soybean via particle bombardment of embryogenic suspension culture tissue. *In Vitro Cell Dev. Biol.,* **27** : 175-182.

Fontana G, Santini L, Caretto S, Frugis G and Moriotti D (1993) Genetic transformation in the grain legume *Cicer arietinum* L. (chickpea). *Plant Cell Rep.,* **12** : 194-198.

Gelvin SB (2000) *Agrobacterium* and plant genes involved in T-DNA transfer and integration. *Annu. Rev. Plant Physiol. Plant Mol. Biol.,* **51** : 223-256.

Geetha N, Venkatachalam P and Laxmi Sita G (1999) *Agrobacterium*-mediated genetic transformation of pigeonpea (*Cajanus cajan* L.) and development of transgenic plants via direct organogenesis. *Plant Biotech.,* **16** : 213-218.

Gill R (1990) Direct gene transfer in *Psophocarpus tetragonoloba* resistance to Kanamycin. *Ann. Bot.,* **66** : 31-39.

Gill R, Eapen S and Rao PS (1988) Transformation of the grain legume *Vigna aconitifolia* Jacq. Marechal by *Agrobacterium tumefaciens. Plant Sci.,* **98** : 495.

Godwin I, Gordon T, Ford–Lloyd B and Newbury HJ (1991) The effects of acetosyringone and pH on *Agrobacterium*-mediated transformation vary according to plant species. *Plant Cell Rep.,* **9** : 671–675.

Grant JE, Copper PA, McAra AE and Frew TJ (1995) Transformation of peas (*Pisum sativum* L) using immature cotyledons *Plant Cell Rep.,* **15** : 254-258.

Grant JE, Cooper PA, Gilpin BJ, Hoglund SJ, Reader JK, Pither Joyce MD and Timmerman-Vaughan GM (1998) Kanamycin is effective for selecting transformed peas. *Plant Sci.,* **139** : 159-164.

Hadi MZ, McMullen MD and Finer JJ (1996) Transformation of 12 different plasmids into soybean via particle bombardment. *Plant Cell Rep.,* **15** : 500-505.

Hammerschlag FA, Zimmeman RH, Yadava UL, Hunsucker S and Gercheva P (1997) Effect of antibiotics and exposure to an acidified medium on the elimination of *Agrobacterium tumefaciens* from apple leaf explants and on shoot regeneration, *J. Am. Soc. Hortic. Sci.,* **122** : 758-763.

Hansen G, Das A and Chilton MD (1994) Constitutive expression of the virulence genes improves the efficiency of plant transformation by *Agrobacterium. Proc. Natl. Acad. Sci. USA,* **91** : 7603-7607.

Hawers MC, Robbs SL and Pueppke SG (1989) Use of a root tumorigenesis assay to detect genotypic variation in susceptibility of thirty-four cultivars of *Pisum sativum* to crown gall. *Plant Physiol.,* **90** : 180.

Hiei Y, Ohta S, Toshishiro K and Komari T (1994) Efficient transformation of rice (*Oryza sativa* L.) mediated by *Agrobacterium* and sequence analysis of the boundaries of the T-DNA. *Plant J.,* **6** : 271-282.

Hiei Y, Komari T and Kubo T (1997) Transformation of rice mediated by *Agrobacterium tumefaciens. Plant Mol. Biol.,* **35** : 205-218.

Hinchee MAW, Conner-Ward DV, Newell CA, McDonnell RE, Sato SJ, Gasser CS, Fischhoff DA, Re DB, Fraley RT and Horsch RB (1988) Production of transgenic soybean plants using *Agrobacterium*-mediated DNA transfer. *Bio/ Technol.,* **6** : 915-922.

Hobbs SLA, Jackson JA and Mahon JD (1989) Specificity of strain and genotype in the susceptibility of pea to *Agrobacterium tumefaciens. Plant Cell Rep.,* **8** : 274-277.

Holford P, Hernandez N and Newbury HJ (1992) Factors influencing the efficiency of T-DNA transfer during cocultivation of *Antirrhinum majus* with *Agrobacterium tumefaciens. Plant Cell Rep.,* **11** : 196–199.

Hood EE, Helmer GL, Fraley RI and Chilton MD (1986) The hypervirulence region of *Agrobacterium tumefaciens* A281 is encoded in a region outside of T-DNA. *J. Bacteriol.,* **168** : 1291-1301.

Hooykaas PJJ and Beijersbergen AGM (1994) The virulence system of *Agrobacterium tumefaciens. Ann. Rev. Phytopathol.,* **32** : 157-179.

Igasaki T, Mohri T, Ichikawa H and Schinohara (2000) *Agrobacterium tumefaciens*-mediated transformation of *Robinia pseudoacacia. Plant Cell Rep.,* **19** : 448-453.

Ishida Y, Saito H, Ohta S, Hiei Y, Komari T and Kumashiro T (1996) High efficiency transformation of maize (*Zea mays*) mediated by *Agrobacterium tumefaciens. Nature Biotechnol.,* **14** : 745-750.

Ishikawa M, Harda K, Sakai F, Ohashi Y and Naito T (1993) Transient gene expression system using protoplasts from developing soybean cotyledons. *Japan J. Breed.,* **43** : 53-60.

Islam R, Malik T, Husmain T and Riazuddin S (1994) Strain and cultivar specificity in the *Agrobacterium* and chickpea interaction. *Plant Cell Rep.*, **13** : 561-563.

Jaiwal PK, Sautter C and Potrykus I (1998) *Agrobacterium rhizogenes*-mediated gene transfer in mungbean (*Vigna radiata* L. Wilczek). *Curr. Sci.,* **75** : 41-45.

Jaiwal PK, Kumari R, Ignacimuthu S, Potrykus I and Sautter C (2001) *Agrobacterium tumefaciens*-mediated genetic transformation of mungbean (*Vigna radiata* L. Wilczek) – a recalcitrant grain legume. *Plant Sci.,* **161** : 239-247.

Jauanin L, Vilaine F, Tourneur J, Tourneur C, Pautot V, Muller JE and Caboche M (1987) Transfer of a 4.3 Kb fragment of the TL-DNA of *Agrobacterium rhizogenes* strain A4 confers the hairy root phenotype to regenerated tobacco plants. *Plant Sci.,* **53** : 53-56.

Jefferson RA, Kavanagh TA and Bevan MW (1987) GUS fusions: β-glucuronidase as a sensitive and versatile gene fusion marker in higher plants. *EMBO J.,* **6** : 3901-3907.

Jensen JS, Marcker KA, Othen L and Shell J (1986) Nodule-specific expression of chimeric soybean leghaemoglobin gene in transgenic *Lotus corniculatus*. *Nature,* **321** : 669-674.

Jin S, Komari T, Gordon MP and Nester EW (1987) Genes responsible for the supervirulence phenotype of *Agrobacterium tumefaciens* AA281. *J. Bacteriol.,* **169** : 4417-4425.

Joersbo M, Paterson SG and Okkels FT (1999) Parameters interacting with mannose selection employed for the production of transgenic sugerbeet. *Physiol. Plant.,* **105** : 109-115.

Jordan MC and Hobbs SLA (1993) Evaluation of a cotyledonary node regeneration system for *Agrobacterium*-mediated transformation of pea (*Pisum sativum*). *In Vitro Cell. Dev. Biol.,* **29** : 77-82.

Kar S, Johnson TM, Nayak P and Sen SK (1996) Efficient transgenic plant regeneration through *Agrobacterium*–mediated transformation of chickpea (*Cicer arietinum* L.). *Plant Cell Rep.,* **16** : 32-37.

Khan MRI, Tabe LM, Heath LC Spencer D and Higgins TJV (1994) *Agrobacterium tumefaciens*–mediated transformation of subterranean clover (*Trifolium subterraneum* L). *Plant Physiol.,* **105** : 81-88.

Kikkert JR (1993) The Biolistic PDS-1000/He device. *Plant Cell Tiss. Org. Cult.,* **33** : 221-226.

Kim JW and Minamikawa T (1996) Transformation and regeneration of French bean plants by the particle bombardment process. *Plant Sci.,* **117** : 131-138.

Kohler F, Golz C, Eapen S, Kohn H and Schieder O (1987*a*) Stable transformation of moth bean (*Vigna aconitifolia*) via direct gene transfer. *Plant Cell Rep.,* **6** : 313-317.

Kohler F, Golz C, Eapen S and Schieder O (1987*b*) Influence of plant cultivar and plasmid DNA on transformation rates in tobacco and moth bean. *Plant Sci.,* **53** : 87-91.

Komari T, Hiei Y, Saito Y, Murai N and Kumashiro T (1996) Vectors carrying two separate T-DNAs for co-transformation of higher plants mediated by *Agrobacterium tumefaciens* and segregation of transformants free from selection markers. *Plant J.,* **10** : 165-174.

Kortt AA, Caldwell JB, Lilley GG and Higgins TJV (1991) Amino acid and cDNA sequence of a methionine rich 2S protein from sunflower seed (*Helianthus annuus* L). *Eur. J. Biochem.,* **195** : 329-334.

Krishanamurthy KV, Suhasini K, Sagare AP, Meixner M, de Kathen, Pickardt T and Schieder O (2000) *Agrobacterium*–mediated transformation of chickpea (*Cicer arietinum* L.) embryo-axes. *Plant Cell Rep.,* **19** : 235-240.

Kumar V, Jones B and Davey MR (1991) Transformation by *Agrobacterium rhizogenes* and regeneration of transgenic shoots of the wild soybean *Glycine argyrea*. *Plant Cell Rep.,* **10** : 135-138.

Larkin PJ and Scowcroft WR (1991) Somaclonal variation- a novel source of variability from cell cultures for plant improvement. *Theor. Appl. Genet.,* **60** : 197-214.

Leon P, Planckaert F and Walbot V (1991) Transient gene expression in protoplasts of *Phaseolus vulgaris* isolated from a cell-suspension culture. *Plant Physiol.,* **95** : 968-972.

Lin JJ, Garcia–Assad N and Kuo J (1994) Effect of *Agrobacterium* cell concentration on the transformation efficiency of tobacco and *Arabidopsis thaliana. Focus,* **16** : 72–77.

Lin JJ, Garcia-Assad N and Kuo J (1995) Effect of antibiotics on the plant transformation efficiency of plant tissues by *Agrobacterium tumefaciens. Plant Sci.,* **109** : 171-177.

Ling HQ, Kriseleit D and Granal MW (1998) Effect of ticarcillin/potassium clavulanate on callus growth and shoot regeneration in *Agrobacterium*-mediated transformation of tomato (*Lycopersicum esculentum*). *Plant Cell Rep.,* **17** : 843-847.

Liu CN, Li XO and Gelvin SB (1992) Multiple copies of *virG* enhance the transient transformation of celery carrot and rice tissues by *Agrobacterium tumefaciens. Plant Mol. Biol.,* **20** : 1071-1082.

Liu W, Torisky RS, McAllister KP, Avdiushko S, Hikdebrand DF and Collins GB (1996) Somatic embryo cycling: evaluation of a novel transformation and assay system for seed-specific gene expression in soybean. *Plant Cell Tiss. Org. Cult.,* **47** : 33-42.

Lulsdorf MM, Rempel H, Jackson JA, Baliski DS and Hobbs SLA (1991) Optimizing the production of transformed pea (*Pisum sativum* L.) callus using disarmed *Agrobacterium tumefaciens* strains. *Plant Cell Rep.,* **9** : 479–483.

Mandal A, Lang V, Orczyk W and Palva ET (1993) Improved efficiency for T-DNA-mediated transformation and plasmid rescue in *Arabidopsis thaliana. Theor. Appl. Genet.,* **86** : 621-628.

Manners JM (1987) Transformation of *Stylosanthes* spp. using *Agrobacterium tumefaciens. Plant Cell Rep.,* **6** : 204-207.

Manners JM (1988) Transgenic plants of the tropical pasture legume *Stylosanthes humilis. Plant Sci.,* **55** : 61-68.

Manners JM and Way H (1989) Efficient transformation with regeneration of the tropical pasture legume *Stylosanthus humilis* using *A. rhizogenes* and a Ti-plasmid-binary vector system. *Plant Cell Rep.,* **8** : 341-345.

Mariotti D, Davey MR, Droper J, Freean JP and Cocking EC (1984) Crown gal tumorigenesis in the forage legume *Medicago sativa* L. *Plant Cell Physiol.,* **25** : 473.

Mathias RJ and Boyd LA (1986) Cefotaxime stimulates callus growth, embryogenesis and regeneration in hexaploid bread wheat (*Triticum aestivum* L. Thell). *Plant Sci.,* **46** : 217-223.

Maximova SN, Dandekar AM and Guiltinan MJ (1998) Investigation of *Agrobacterium*-mediated transformation of apple using green fluorescent protein: high expression and low stable transformation suggest that factors other than T-DNA transfer are rate-limiting. *Plant Mol. Biol.,* **37** : 549-559.

McCabe DE and Christou P (1993) Direct DNA transfer using electric discharge particle acceleration (ACCELL™ technology). *Plant Cell Tiss. Org. Cult.,* **33** : 227-236.

McCabe DE, Swain WF, Martinell BJ and Christou P (1988) Stable transformation of soybean (*Glycine max*) by particle acceleration. *Bio/Technol.,* **6** : 923-926.

McClean P, Chee P, Held B, Simental J, Drong RF and Salghtom J (1991) Susceptibility of dry bean (*Phaseolus vulgaris* L.) to *Agrobacterium* infection and transformation of cotyledonary and hypocotyl tissues. *Plant Cell Tiss. Org. Cult.,* **24** : 131.

McKently AH, Moore GA, Doostdar H and Niedz RP (1995) *Agrobacterium*-mediated transformation of peanut (*Arachis hypogaea* L.) embryo axes. *Plant Cell Rep.,* **14** : 699-703.

Meurer CA, Dinkins RD and Collins GB (1998) Factors affecting soybean cotyledonary node transformation. *Plant Cell Rep.,* **18** : 180-186.

Mesa MC, Jimenez-Bermudez S, Pliego-Alfaro F, Quesada MA and Mercado JA (2000) *Agrobacterium* cells as microprojectile coating: a novel approach to enhance stable transformation rates in strawberry. *Aust. J. Plant Physiol.,* **27** : 1093-1100.

Mi JJ and Zhou H (1986) *In vitro* transformation of protoplasts from soybean by *Agrobacterium tumefaciens* T37. *J. Agri. Sci. China,* **2** : 109-114.

Muthukumar B, Mariamma M, Veluthambi K and Gnanam A (1996) Genetic transformation of cotyledon explants of cowpea (*Vigna unguiculata* L. Walp) using *Agrobacterium tumefaciens. Plant Cell Rep.,* **15** : 980–985.

Molvig L, Tabe LM, Eggum BO, Moore AE, Craig S, Spencer D and Higgins TJV (1997) Enhanced methionine levels and increased nutritive value of seeds of transgenic lupins (*Lupinus angustifolius* L) expressing a sunflower seed albumin gene. *Proc. Natl. Acad. Sci. USA,* **94** : 8393-8398.

Nadolska-Orczyk A and Orczuk W (2000) Study of the factors influencing *Agrobacterium*-mediated transformation pea (*Pisum sativum* L.) *Mol. Breed.,* **6** : 185-194.

Nauerby B, Madsen M, Christiansen J and Wyndaele R (1991) A rapid and efficient regeneration system for pea (*Pisum sativum*) suitable for transformation. *Plant Cell Rep.,* **9** : 676-679.

Nauerby B, Biling K and Wyndaele R (1997) Influence of antibiotic timentin on plant regeneration compared to carbenicilin and cefotaxime in concentrations suitable for elimination of *Agrobacterium tumefaciens. Plant Sci.,* **123** : 169-177.

Nisbet GS (1987) Transformation of *Lotus corniculatus* and regeneration of transgenic plants. M.Sc. Thesis, Univ. Wales.

Oger P, Petit A and Dessaux Y (1996) A simple technique for direct transformation and regeneration of the diploid legume species *Lotus japonicus. Plant Sci.,* **116** : 159-168.

Owens LD and Cress DE (1985) Genotypic variability of soybean response to *Agrobacterium* strains harbouring the Ti or Ri plasmids. *Plant Physiol.,* 77-87.

Ozias-Akins P, Schnall JA, Anderson WF, Singsit C, Clemente TE, Adang MJ and Weissinger AK (1993) Regeneration of transgenic peanut plants from stably transformed embryogenic callus. *Plant Sci.,* **93** : 185-194.

Padgette SR, Kolacz KH, Delannay X, Re DB, LaVallee BJ, Tinius CN, Rhodes WK, Otero VI, Barry GF, Eichholz DA, Peschke VM, Nida DL, Taylor NB and Kishore GM (1995) Development, identification and characterization of a glyphosate-tolerant soybean line. *Crop. Sci.,* **35** : 1451-1461.

Palmgren G, Mattson O and Okkels FT (1993) Transient of *Agrobacterium* or leaf discs with 5-azacytidine increases transgene expression in tobacco. *Plant Mol. Biol.,* **21** : 429-435.

Park SH, Lee BM, Salas MG, Srivatanakul M and Smith RH (2000) Shorter T-DNA or additional virulence genes improves *Agrobacterium*-mediated transformation. *Theor. Appl. Genet.,* **101** : 1015-1020.

Parrott WA, Hoffman LM, Hildebrand DF, William EG and Collins GB (1989) Recovery of primary transformants of soybean. *Plant Cell Rep.,* **7** : 615-617

Parrott WA, All JN, Adang MJ, Bailey MA, Boerma HR and Stewart Jr CN (1994) Recovery and evaluation of soybean (*Glycine max* (L.) Merr.) plants transgenic for a *Bacillus thuringiensis* var. kurstaki insecticidal gene. *In Vitro Cell Dev. Biol.,* **30** : 144-149.

Paszkowski J, Potrykus I and Hohn B (1995) Transformation of hereditary material in plants. US Pat. No. 5, 453, 367.

Pederson HC, Christiansen J and Wyndaele R (1983) Induction and *in vitro* culture of soybean crown gall tumours. *Plant Cell Rep.,* **2** : 201-204.

Pereira LF and Erickson L (1995) Stable transformation of alfalfa (*Medicago sativa* L.) by particle bombardment. *Plant Cell Rep.,* **14** : 290-293.

Petit A, Stougaard J, Kuble A, Marcker KA and Tempe J (1987) Transformation and regeneration of the legume *Lotus corniculatus*: a system for molecular studies of symbiotic nitrogen fixation. *Mol. Gen. Genet.,* **207** : 245-250.

Pickardt T, Perales HE and Schieder O (1995) Seed specific expression of the 2S albumin gene from Brazil nut (*Bertholletia excelsa)* in transgenic *Vicia narbonensis. Mol. Breed.,* **1** : 295-301.

Pickardt T and Saalbach I (2000) Genetic transformation of Narbon bean (*Vicia narbonensis*) In: *Biotechnology in Agriculture and Forestry Vol. 46. Transgenic Crops* (Ed. Bajaj YPS) Springer-Verlag, Berlin, Heidelberg, pp 373-389.

Pigeaire A, Abernethy D, Smith PM, Simpson K, Fletcher N, Lu Chin Yi, Atkins CA and Cornish E (1997) Transformation of a grain legume (*Lupinus angustifolius* L.) via *Agrobacterium tumefaciens* mediated gene transfer to shoot apices. *Mol. Breed.,* **3** : 341-349.

Ponappa T, Brzozowski AE and Finer JJ (1999) Transient expression and stable transformation of soybean using jellyfish green fluorescent protein. *Plant Cell Rep.,* **19** : 6-12.

Potrykus I (1991) Gene transfer to plants: Assessment of published approaches and results. *Annu Rev. Plant Physiol. Plant Mol. Biol.,* **42** : 205-225.

Puonti-Kaerlas J, Stabel P and Eriksson T (1989) Transformation of pea (*Pisum sativum* L.) by *Agrobacterium tumefaciens. Plant Cell Rep.,* **8** : 321.

Puonti-Kaerlas J, Erikson T and Engstrom P (1990) Production of transgenic pea (*Pisum sativum* L.) plants by *Agrobacterium tumefaciens*-mediated gene transfer. *Theor. Appl. Genet.,* **80** : 246-252.

Ramaiah SM and Skinner DZ (1997) Particle bombardment: a simple and efficient method of alfalfa (*Medicago sativa* L.) pollen transformation. *Curr. Sci.,* **73** : 674-682.

Ramana RV, Venu C, Jayasree T and Sadanadam A (1996) Direct somatic embryogenesis and transformation in *Cicer arietinum* L. *Indian J. Exp. Biol.,* **34** : 716-718.

Ramsay G and Kumar A (1990) Transformation of *Vicia faba* cotyledon and stem tissues by *Agrobacterium rhizogenes*: Infectivity and cytological studies. *J. Exp. Bot.,* **41** : 841-847.

Rathore RS and Chand L (1997) *In vitro* transformation of pigeonpea genotypes by wild strains of *Agrobacterium tumefaciens. Intl. Chickpea and pigeonpea Newsletter,* **5** : 10-11.

Rech EL, Golds TJ, Husnain T, Vainstein MH, Jones B, Hammatt N, Mulligan BJ and Davey MR (1989) Expression of a chimaric kanamycin resistance gene introduced into the wild soybean *Glycine canescens* using a cointegrate Ri Plasmid vector. *Plant Cell Rep.,* **8** : 33-36.

Reich TJ, Iyer VN and Miki BL (1986) Efficient transformation of alfalfa protoplasts by the intranuclear microinjection of Ti-plasmids. *Biotechnology,* **4** : 1001-1004.

Rohini VK and Rao KS (2000) Transformation of peanut (*Arachis hypogaea*): a non-tissue culture based approach for generating transgenic plants. *Plant Sci.,* **150** : 41-49.

Russell JA (1993) The Biolistic© PDS-1000/ He device. *Plant Cell Tiss. Org. Cult.,* **33** : 221-226.

Saalbach I, Pickardt T, Waddell DR, Hillmer S, Scheider O, Muntz K, Cassells AC (ed) and Jones PW (1995) The sulphur-rich Brazil nut 2S albumin is specifically formed in transgenic seeds of the grain legume *Vicia narbonensis. Eucarpia Genetic Manipulation & Plant Breeding Section meeting held in Cork,* Irish Republic, 11-14, September 1994.

Saalbach I, Pickardt T, Machemehl F, Saalbach G, Schieder O and Muntz K (1994) A chimeric gene encoding the methionine rich 2S albumin of the Brazilnut *Bertholletica excelsa* is stably expressed and inherited in transgenic grain legumes. *Mol. Gen. Genet.,* **242** : 226-236.

Sahoo L, Singh ND, Sonia, Sugla T, Singh RP and Jaiwal PK (2001) Genetically Modified Crops: A Bane or Boon to Green Revolution. *Physiol. Mol. Biol. Plants,* **7** : 1-2.

Sahoo L, Sugla T and Jaiwal PK (2002) *In vitro* regeneration and genetic transformation of mungbean, urdbean, cowpea and azuki bean. In : *Biotechnology for the improvement of legumes* (Eds. Jaiwal PK, Singh RP), Kluwer Acad. Publ., The Netherlands (in press).

Satina S, Blakeslee AF and Avery AG (1940) Demonstration of the three germ layers in the shoot apex of *Datura* by means of induced polyploidy in periclinal chimeras. *Am. J. Bot.,* **27** : 895-905.

Sato S, Newell C, Kolacz K, Tredo L, Finer J and Hinchee M (1993) Stable transformation via particle bormbardment in two different soybean regeneration systems. *Plant Cell Rep.,* **12** : 408-413.

Saini R and Jaiwal PK (2002) Age, position in mother seedlings, orientation and polarity of the epicotyl segments of blackgram (*Vigna mungo* L. Hepper) determines its morphogenic response. *Plant Sci.,* **163** : 101-109.

Sato T (1995) Basic study of biotechnology in adzuki bean (*Vigna angularis* Ohwi & Ohashi). *Report of Hokkaido Pref. Agri. Exp. Station,* **87** : 1-68.

Sato T, Asaka D, Harada T and Matsukawa I (1993) Plant regeneration from protoplasts of Adzuki bean (*Vigna angularis* Ohwi & Ohashi). *Jap. J. Breed.,* **43** : 183-191.

Schnall JA and Weissinger AK (1993) Culturing peanut (*Arachis hypogaea* L.) zygotic embryos for transformation via microprojectile bombardment. *Plant Cell Rep.,* **12** : 316-319.

Scholl RL, Keathley DE and Baribault TJ (1981) Enhancement of root formation and fertility in shoots regenerated from anther and seedling-derived callus cultures of *Arabidopsis thaliana. Z. Pflanzephysiol. Bd.,* **104** : 225-231.

Schroeder HE, Schotz AH, Wardley-Richardson T, Spencer D and Higgins TJ (1993) Transformation and regeneration of two cultivars of pea (*Pisum sativum* L.) *Plant Physiol.,* **101** : 751-757.

Sharma KK and Anjaiah V (2000) An efficient method for the production of transgenic plants of peanut (*Arachis hypogaea* L.) through *Agrobacterium tumefaciens*-mediated genetic transformation. *Plant Sci.,* **159** : 7-19.

Sharma KK, Sharma HC, Seetharama N and Ortiz R (2002) Development and deployment of transgenic plants: Biosafety considerations. *In vitro Cell. Dev. Biol.-Plant,* **38** : 106-115.

Singh ND, Sonia L, Sahoo, SM Singh and PK Jaiwal (2000) Biotechnological approaches for the improvement of Pigeonpea (*Cajanus cajan* (L) Millsp): A review In: *Recent advances in Biotechnology* (Ed. Trivedi PC) Panima Pub. Co., New Delhi. pp 154-173.

Smith RH and Murashige T (1970) *In vitro* development of isolated shoot apical meristems of angiosperms. *Am. J. Bot.,* **57** : 562-568.

Spano L, Mariotti D, Pezzotti M, Damiani F and Arcioni S (1987) Hairy root transformation in alfalfa (*Medicago sativa* L.) *Theor. Appl. Genet.,* **73** : 523-530.

Sonia, Jaiwal PK, Ahad A and Sahoo L (1998) Green fluorescent protein: A novel reporter gene. *Curr. Sci.,* **74** : 402-405.

Sonia, Sharma P, Pabbi P, Ragini and Jaiwal PK (2001) Applications of biotechnology for the improvement of chickpea (*Cicer arietinum* L) In : *Recent advances in Biotechnology* (Ed. Trivedi PC) Panima Pub. Co., New Delhi.

Stewart CN Jr., Adang MJ, All JN, Boerma HR, Cardineau G, Tucker D and Parrott WA (1996) Genetic transformation, recovery, and characterization of fertile soybean transgenic for a synthetic *Bacillus thuringiensis cry1Ac* gene. *Plant Physiol.,* **112** : 1, 121-129.

Tinland B (1996) The integration of T-DNA into plant genomes. *Trends Plant Sci.,* **1** : 178-184.

Torisky RS, Kovaes L, Adviushko S, Newman JD, Hund AG and Collins GB (1997) Development of a binary vector system for plant transformation based on the supervirulent *Agrobacterium tumefaciens* strain chry 5. *Plant Cell Rep.,* **17** : 102-108.

Trick HN and Finer JJ (1998) Sonication-assisted *Agrobacterium*-mediated transformation of soybean (*Glycine max* (L.) Merrill) embryogenic suspension culture tissue. *Plant Cell Rep.,* **17** : 482-488.

Trieu AT, Burleigh SH and Kardailsky IV (2000) Transformation of *Medicago truncatula* via infiltration of seedlings or flowering plants with *Agrobacterium. Plant J.,* **22** : 531-541.

Trinh TH, Ratet P, Kondorosi E, Durand P, Kamate K, Bauer P and Kondorosi A (1998) Rapid and efficient transformation of diploid *Medicago truncatula* and *M. sativa* spp falcata lines improved in somatic embryogenesis. *Plant Cell Rep.,* **17** : 345-355.

Vain P, Keen N, Murillo J, Rathus C, Nemes C and Finer JJ (1993) Development of the Particle Inflow Gun. *Plant Cell Tiss. Org. Cult.,* **33** : 237-246.

Vancanneyt G, Schmidt R, O'Connor Sanchez A, Willmitzer L and Rocha-Sosa M (1990) Construction of an intron-containing marker gene: splicing of the intron in transgenic plants and its use in monitoring early evens in *Agrobacterium*-medicated plant transformation. *Mol. Gen. Genet.,* **220** : 245-250.

van der Bulk RW, Loffler HJM, Lindhant WH and Koornneef M (1990) Somaclonal variation in tomato: effect of explant source and a comparision with chemical mutagenesis. *Theor. Appl. Genet.,* **80** : 817-825.

Vlachova M (Vlakhova-M), Metz BA, Schell J, Bruizn FJ de and Vlakhova M (1987) The tropical legume *Sesbania rostrata*: tissue culture, plant regeneration and infection with *Agrobacterium tumefaciens* and *rhizogenes* strain. *Plant Sci. Irish-Republic,* **50** : 3, 213-223.

Voisey CR, White DWR, Dudas B, Appleby RD, Ealing PM and Scott AG (1994) *Agrobacterium*-mediated transformation of white clover using direct shoot organogenesis. *Plant Cell Rep.,* **13** : 309-314.

Wang J, Rose RJ, Donaldson BI and Wang JH (1996) *Agrobacterium*-mediated transformation and expression of foreign genes in *Medicago truncatula. Australian J. Plant Physiol.,* **23** : 3, 265-270.

Warkentin TD and McHughen A (1991) Crown gall transformation of lentil (*Lens culinaris* Medik.) with virulent strains of *Agrobacterium tumefaciens. Plant Cell Rep.,* **10** : 489-493.

Webb KJ (1986) Transformation of forage legumes using *Agrobacterium tumefaciens. Theor. Appl. Genet.,* **72** : 53-58.

Webb KJ, Jones S, Robbins MP and Minchin FR (1990) Characterization of transgenic root cultures of *Trifolium repens*, *Trifolium pratense* and *L. corniculatus* and transgenic plants of *L. corniculatus. Plant Sci.,* **70** : 243-254.

White DWR and Greenwood D (1987) Transformation of of the forage legume *Trifolium repens* L. using binary *Agrobacterium* vectors. *Plant Mol. Biol.,* **8** : 461-469.

Yamada T, Teraishi M, Hattori K and Ishimoto M (2001) Transformation of azuki bean by *Agrobacterium tumefaciens. Plant Cell Tiss. Org. Cult.,* **64** : 47-54.

Yang H, Singsit C, Wang A, Gonsalves D and Ozias-Akins P (1998) Transgenic peanut plants containing a nucleocapsid protein gene of tomato spotted wilt virus show divergent level of gene expression. *Plant Cell Rep.,* **17** : 693-699.

Zambryski PC (1992) Chronicles from the *Agrobacterium* plant cell DNA transfer story. *Annu. Rev. Plant Physiol. Plant Mol. Biol.,* **43** : 465-490.

Zang ZY, Coyne DP and Mitra P (1997) Factors affecting *Agrobacterium*-mediated transformation of common bean. *J. Am. Soc. Hort. Sci.,* **122** : 300-305.

Zhang Z, Xing A, Staswick P and Clemente TE (1999) The use of glufosinate as a selective agent in *Agrobacterium*-mediated transformation of soybean. *Plant Cell Tiss. Org. Cult.,* **56** : 37-46.

Zupan JR and Zambryski P (1995) Transfer of T-DNA from *Agrobacterium* to the plant cell. *Plant Physiol.,* **107** : 1041-1047.

Zupan J and Zambryski P (1997) The *Agrobacterium* DNA transfer complex. *Crit. Rev. Plant Sci.,* **16** : 279-295.

Chapter 12

GENETIC TRANSFORMATION OF *BRASSICA RAPA* (syn. *BRASSICA CAMPESTRIS*)

Viktor Kuvshinov* **and Eija Pehu**

Research Project in Agricultural Biotechnology, Biotechnology Institute, Helsinki Science Park, Viikinkaari 6, 00014 University of Helsinki, Finland

Summary

The species Brassica rapa includes an important oil crop B. rapa ssp. oleifera and several vegetable crops, the major ones being Chinese cabbage (B. rapa ssp. pekinensis), pakchoi (B. rapa ssp. chinensis) and turnip (B. rapa ssp. rapa) widely used especially in Asia. While a related species B. napus has been one of the pioneering crops in plant biotechnology, B. rapa has turned out to be a very recalcitrant species complex in tissue culture. Most efforts have focused on B. rapa ssp. oleifera because of its economic importance as an oil crop of the cool climates in Northern Europe and Canada. Successful transformation protocols are available for this species and the work has shifted to the application of these tools for traits of economic importance. For the vegetable crops of B. rapa biotechnology efforts have been rather scarce, but advances in plant tissue culture and transformation methods have been achieved and are reviewed.

Keywords : *Brassica rapa, Brassica campestris* ssp. *oleifera,* ssp. *pekinensis*, transformation, *in vitro*.

Abbreviations : BAP : 6-benzylaminopurine; IAA : indole-3-acetic acid; IBA : indole-3-butyric acid; NAA : α-naphthaleneacetic acid; 2,4D : 2,4-dichlorophenoxy-acetic acid.

---

*Corresponding author : E-mail : Viktor.Kuvshinov@Helsinki.fi

1. INTRODUCTION

The genus *Brassica,* including about 40 species, of the family *Brassicaceae* is one of the most versatile *geni* in plant taxonomy in terms of different crops species derived from it by humankind. The two major classes of these crops are the oil-seed Brassicas and the annual and biannual foliage and root crops. The majority of the cultivated forms belong to six species, the monogenomic *B. nigra, B. oleracea, B. rapa,* and the amphidiploid *B. carinata, B. juncea and B. napus.* The present chapter reviews the advances in plant biotechnology in the species *B. rapa* (2n=10). The major sub-species used as crops within the taxon *Brassica rapa* (*B. campestris*) are summarized in the following table (Table 1).

Brassica rapa ssp. *oleifera*, turnip rape, is an important oil seed crop in northern regions, because of its adaptability to cool climates and because of the high nutritional quality of the oil. The seed meal is high in protein and used as animal feed. Another related cultivated form of the same species is Brown sarson produced in the Indian sub-continent. The main lines of biotechnology research of *Brassica rapa* comprise improvement in oil composition and productivity, adjustments in seed storage quality for improved feeds, tolerance to biotic and abiotic stresses, male sterility for hybrid seed production and environmental risk assessment (Kuvshinov *et al.*, 1997).

Chinese cabbage (*B. rapa* ssp. *pekinensis*) is one of the most important vegetable crops is China, Japan and Korea. It suffers from severe virus diseases and insect pests and transgenic improvement has

Table I. Major crops of the species *Brassica rapa*

Crop species	Major form of use
Brassica rapa ssp. *oleifera* (turnip rape)	Oil seed crop
Brassica rapa ssp. *pekinensis* (Chinese cabbage)	Leafy vegetable, used in salads
Brassica rapa ssp. *rapifera* (*rapa*) (turnip)	Roots used as cooked vegetable, leaves as herbs, canned or frozen
Brassica rapa ssp. *chinensis* (pakchoi)	Leaves and petioles used fresh
Brassica rapa ssp. *parachinensis* (false pakchoi)	Leaves and petioles used fresh

Source: USDA, NRCS, 1999.

been viewed as a potential tool to improve biotic stress resistance in this crop. Pakchoi (*B. rapa* ssp. *chinensis*) is an important leafy vegetable in China occupying an area of 600 000 ha (Cao *et al.*, 1994). As in Chinese cabbage virus diseases and insect pests are the main attractions for transgenic research also in pakchoi.

In spite of the economic importance of the sub-species of *B. rapa* genetic transformation techniques have only been developed for a few species. There are only two reports on transformation of turnip rape, one on seed oil modification (Knutzon *et al.*, 1992) and the other on self-incompatibility modification (Shiba *et al.*, 1995). Pathogen derived resistance has been used to improve virus resistance in Chinese cabbage (Jun *et al.*, 1995). As discussed in this chapter *B. rapa* has proven to be a very recalcitrant species in tissue culture.

2. OVERVIEW OF TISSUE CULTURE, REGENERATION AND TRANSFORMATION METHODS OF *B. RAPA*.

Successful transformation procedure of any plant comprises effective shoot regeneration and a successful transformation event. As a rule, young or fast growing meristems are the tissues most capable of cell differentiation into somatic embryos or shoot primordia. The most amenable plant tissues for regeneration of *Brassica rapa* are cotyledons, hypocotyls, inflorescence carrying stems or young leaves. Microspore and protoplast culture have also been applied to in *in vitro* manipulation of *B. rapa*. It is important to note, however, that compared to other *Brassica* species, *B. rapa* is more recalcitrant in shoot regeneration. While shoot regeneration procedures were developed for most other *Brassica* species rather early in the history of plant biotechnology, *B. rapa* shoot regeneration remained unsuccessful until mid-nineties.

There were a few critical factors that finally solved the problem of *B. rapa* regeneration. It has been found that ethylene inhibitors like silver nitrate significantly enhance regeneration capacity of *B. rapa*. The growing conditions of the source plants for *in vitro* explants is another important factor for regeneration. Experiments have shown that hypocotyl or cotyledon explants should be excised from 4 - 5 days old seedlings. High shoot regeneration can also be achieved from inflorescence carrying stem segments of turnip rape. The highest shoot regeneration was achieved from the segments excised from the two upper internodes at the stage of fast elongation of the stem. Experience supports the fact that meristems of *B. rapa* become capable of morphogenesis only in particular organs and tissues and usually for just a short developmental period. As in many other species shoot regeneration from *B. rapa* explants is genotype dependent and requires precise hormone and media composition. Besides the limitations to the efficiency of primordia formation, there are problems with vitrification, rooting or spontaneous early flowering of recovered shoots. Through extensive experimentation, regeneration techniques have been successfully developed for vegetative tissues, microspores, anthers and protoplasts of *B. rapa*.

While regeneration capacity of *B. rapa* is the lowest of the species in the genus *Brassica*, its responsiveness to *Agrobacterium tumefaciens* is much better than that of *B. oleracea*, but the transformation efficiency is rather genotype specific. However, average *A. tumefaciens* transformation levels range from a percent to tens of percents, that makes *A. tumefaciens* transformation protocols reliable for *B. rapa*. *Agrobacterium rhizogenes* is even more effective in transformation of *B. rapa* than *A. tumefaciens*. The negative side of Ri-plasmid transformation is the overproduction of auxins, which leads to the 'rooty' phenotype and sterility of the regenerants. Although, direct transformation requires high regeneration capacities of transformed explants, particle bombardment and protoplast transformation protocols have also been developed for some subspecies of *B. rapa*. Various regeneration and transformation procedures have been developed especially for the two economically most important subspecies of *B. rapa*: the oilseed turnip rape *B. rapa* ssp. *oleifera* and the vegetable, Chinese cabbage *B. rapa* ssp. *pekinensis*. Below is a comprehensive review of regeneration and transformation protocols separately for the subspecies *oleifera* and *pekinensis* followed by an overview on the techniques developed for the other subspecies of *B. rapa*.

2.1. *Brassica rapa* ssp. *oleifera*

B. rapa ssp. *oleifera* (syn. *B. campestris*) is an important food crop, which includes two closely related varieties. The first one named 'turnip rape' is grown in Northern Europe and North America (primarily in Canada), mainly as an annual, spring sown oil crop. Turnip rape is adapted to a relatively short growing season and low temperatures and thus has occupied a niche where oilseed rape, *B. napus* (originating as a diploid hybrid of *B. rapa* and *B. oleracea*) cannot grow because of too short a growing season. The second is brown sarson, also called mustard. It is cultured in warm climatic regions, mainly in the Indian subcontinent. Regeneration and transformation protocols were developed for both of these cultivated forms.

2.1.1. *In vitro* culture

Any transformation technique is based on successful shoot regeneration. Despite the great interest in developing transformation procedures for *B. rapa*, in the eighties, only evidences of recalcitrance of this subspecies and failure in shoot regeneration from callus (Dietert, 1982; Murata and Orton, 1987), leaf discs (Dunwell, 1981), cotyledons (Jain *et al.*, 1988; Narasimhulu *et al.*, 1988), isolated protoplasts (Glimelius, 1984) or anther culture (Keller and Armstrong, 1979) were reported. It was pointed out that, in contrast to other *Brassica* species, cotyledon explants of *B. rapa* required auxins in the culturing media for shoot regeneration (Jain *et al.*, 1988). It was also shown that, in comparison to rapeseed, *B. napus*, cotyledon explants of turnip rape exhibited very low callusing and shoot morphogenesis frequency (Javed and Hassan, 1992). First reports on successful shoot regeneration of *B. rapa* appeared early in the nineties after the negative effect of ethylene on regeneration was identified (Lentini *et al.*, 1988). While the ethylene effect in general is critical for *Brassica rapa* regeneration, it is interesting, that the first remarkable (70%) shoot regeneration rate was achieved on cotyledon explants without the application of ethylene inhibitors (Hachey *et al.*, 1991). The importance of ethylene inhibition was blurred by the differences in impact by different types of inhibitor compounds. Winnie *et al.* (1997) reported that silver nitrate can reduce shoot regeneration on cotyledon explants. Authors noted that the age of the seedlings (4 days) and light intensity also had a profound effect on regeneration potential. The hormones used in the media in this experiment were 2 mg/l BAP and 1 mg/l NAA. In contrast to this, Palmer (1992) reported a strong enhancement of shoot regeneration on cotyledons of *B. rapa* ssp. *oleifera* by addition of silver nitrate in the medium. Cotyledon (including petioles) explants were excised from 4 to 6 days old seedlings of which the 4 days old seedlings showed maximal shoot regeneration on MS medium supplemented with the same hormone composition as in the previous report. Addition of silver nitrate at a concentration of 60 μM enhanced shoot regeneration from 6% to 83%. Hypocotyl segments were unresponsive to any treatments reported by Palmer (1992). In another study it was pointed out that the addition of BAP or NAA to the growth medium during the *in vitro* culture of seedlings can significantly enhance subsequent shoot regeneration from the cotyledon explants (Burnett *et al.*, 1994). Summarizing the data on regeneration experiments with explants of *B. rapa* var. *oleifera* seedlings, it can be concluded that cotyledons of four days old seedlings were the most responsive to shoot regeneration. Agar solidified MS (Murashige and Skoog, 1962) medium with 2 mg/l BAP, 1 mg/l NAA and 60 μM $AgNO_3$ has proven to be optimal to regeneration.

Investigations on different ethylene inhibitors showed that other than silver nitrate chemicals can remarkably enhance shoot regeneration. For example, the use of aminoethoxyvinylglycine (1-2 μM) in the regeneration medium of cotyledon explants was more effective than silver nitrate (Burnett *et al.*, 1994; Winnie *et al.*, 1997). Shoot regeneration was successfully developed from peduncle segments excised from the top 5 cm of an inflorescence-carrying stem of turnip rape (Eapen and George, 1997). The regeneration rate ranged from 16 to 33%. The combination of the ethylene inhibitors, silver nitrate and silver thiosulfate, enhanced shoot regeneration 4 - 5 times. Other putative ethylene inhibitors, such as 1-aminocyclopropane-1-carboxylic acid, norbornadiene (Songstad *et al.*, 1988), cobalt chloride and salicylic acid (Palmer, 1992) or methylglyoxal-bis-guanylhydrazone (Burnett *et al.*, 1994) had no positive effect on shoot regeneration.

The important method for crop improvement to produce homozygous plants from isolated microspores was developed for *B. rapa* ssp. *oleifera*. Microspores excised from buds of plants grown at low temperatures (10/5°C) formed embryos in three weeks of culturing on high sucrose containing (17 - 10%) media (Baillie *et al.*, 1992). Microspore culture with elevated temperatures for source plant growth (22°C) and for the first day of microspore culturing (35°C) resulted in effective plant regeneration and seed production (Burnett *et al.*, 1992). The most efficient microspore culture was achieved using fully matured donor plants grown in low temperature in a medium supplemented with activated charcoal (Guo and Pulli, 1995; Guo and Pulli, 1996). There is also a report on methods for *B. rapa* pollen germination and preservation (Sato *et al.*, 1998). Unfortunately, the development of a microspore culture of *B. rapa* ssp. *oleifera* has not yet resulted in any direct or *Agrobacterium*-mediated transformation protocols for embryos.

An important tool for direct gene transfer and cellular manipulations is protoplast culture. The first report on *B. rapa* protoplast culture has been published by Schenk and Hoffmann as early as in 1979. The early investigations were performed on different *Brassica* species. Comparing to other *Brassicae B. rapa* protoplasts had low regeneration capacity (Ulrich *et al.*, 1980; Glimelius, 1984; Jourdan and Earle, 1989; Loudon *et al.*, 1989). Only the use of silver nitrate made shoot regeneration from 6-day old hypocotyl protoplasts possible at a level of 20% (Pauk *et al.*, 1991a). Basing on the reports of sucessful shoot regeneration from callus derived from isolated protoplasts (Pauk *et al.*, 1991a; Zhao *et al.*, 1995), the following general features of *B. rapa* ssp. *oleifera* protoplast culture can be indicated. The most amenable source material for protoplasts were hypocotyls and cotyledons of 4-6 days old in dim light or darkness germinated seeds. The hypocotyls or cotyledons were cut in small pieces of 1-2 mm in length treated overnight with 1% Cellulase R-10 and Macerozyme R-10 enzymes dissolved in K3 (Kao and Michayluk, 1975; Nagy and Maliga, 1976) medium with osmotic pressure of 0.714 M mannitol. Isolated protoplasts were washed the next day several times with W5 salt solution (Menczel *et al.*, 1981) and cultured in 0.4 M K-75 or K8P (Kao and Michayluk, 1975) or NN (Nitsch and Nitsch, 1969) at 25-28°C in the dark. Alternatively, microcalli can be embedded the second day into low melting temperature agarose medium. Callus growth and shoot regeneration from calli was carried out on K3 or MS medium supplemented with 2% sucrose, 0.6% agarose, 0.5 -2 mg/l BAP, 0.5 mg/l Zeatin, 0.1 mg/l IAA and, alternatively, 100 mg/l glutamine and 250 mg/l casein amino acid. Silver nitrate (20 - 60 μM) increased shoot regeneration rate from 0 to 22 %. Whole cycle from protoplast isolation to shoot regeneration took 2 -3 months. Rooting was difficult and had to be continued for two weeks on MS medium supplemented with 0.1 mg/l IBA or 0.1-0.2 mg/l NAA without silver nitrate. The successful protoplast culture development sparked the interest in applying the protoplast fusion technique and cybrid development (Cardi and Earle, 1997).

2.1.2. Genetic transformation

The most reliable way to transfer genes into the plant genome, especially of dicots, is transformation by *Agrobacterium tumefaciens*. Development of a transformation protocol consists of three major parts - *Agrobacterium* infection, selection of transformed cells and tissues and shoot regeneration with subsequent rooting. *B. rapa* ssp. *oleifera* is a rather recalcitrant plant species at all of these three stages of a genetic transformation process. The plant-microbe interaction of turnip rape and *A. tumefaciens* was investigated on different strains of the bacteria. The first investigations on the variation in inoculation capacities of different *Agrobacterium* strains showed that Ach5 strain did not transform turnip rape, N2/73 caused moderate tumorigenesis and A281 was slightly more virulent (Godwin *et al.*, 1991). Among the screened disarmed *A. tumefaciens* strains average inoculation capacities of stem segments (blue inclusions in GUS assay) were, with the cointegrative Ti vector: C58C1pGV2260 – 30%, C58C1pGV3850 – 38% and, with a binary Ti vector: C58pGV3850 – 10.5%, EHA105 – 30.5% and LBA4404 – 59.5% (Kuvshinov *et al.*, 1999). EHA101 also exhibited good inoculation capacity of hypocotyl explants in two transformation protocols of turnip rape (Radke *et al.*, 1992; Takasaki *et al.*, 1997). Compared to other *Brassica* species *B. rapa* ssp. *oleifera* is more responsive to *A. tumefaciens* inoculation than the *B. oleracea* (Kuvshinov *et al.*, 1997). Acetosyringone has been tested as an inducer of virulence in *Agrobacterium* and was found having a slight positive effect, which, however, in practical terms was negligible (Godwin *et al.*, 1991; Kuvshinov, 1999). Lowering the pH of the medium to 5.2 has been found to have a positive effect for transformation of hypocotyls (Takasaki *et al.*, 1997). It is interesting that most reports describe shoot regeneration from cotyledon explants, while three of the four *A. tumefaciens* transformation protocols were developed on hypocotyl explants and the remaining one on stem segments. This can be explained by the fact that cotyledons are less responsive to *A. tumefaciens* transformation than hypocotyls (Mukhopadhyay *et al.*, 1992). Comparison of the different protocols shows that the transformation frequency of hypocotyls is lower than of segments from inflorescence carrying stems (Radke *et al.*, 1992; Kuvshinov *et al.*, 1999). To achieve substantial transformation levels, hypocotyl explants required culturing on a tobacco feeder layer, whereas the stem segments exhibited high regeneration capacities of transformed shoots without a feeder layer.

Silver nitrate application was found positive for shoot morphogenesis by all of the authors reporting on turnip rape transformation. Silver nitrate enhances both the proportion of explants producing shoots and the frequency of regenerated shoots per explant two- to five-fold. It is likely, that freshly excised explants produce more ethylene than callus or differentiated cells. This can explain the fact that silver nitrate effect decreases during cultivation of explants and becomes even negative during rooting (Radke *et al.*, 1992; Kuvshinov *et al.*, 1999). On the other hand, hypocotyl explants have to be grown on silver ion free medium the first three days during co-cultivation with *A. tumefaciens* because the silver ions have a bactericidal effect and prevent *Agrobacterium* infection. Large inflorescence carrying stem segments can be cultured on silver containing agar during co-cultivation if the explant is not embedded with its basal side into the medium (Kuvshinov *et al.*, 1999). It is possible to achieve 100% shoot regeneration on certain varieties from stem segments, whereas hypocotyls produce shoots at a frequency from 45% (Mukhopadhyay *et al.*, 1992) to 78% (Radke *et al.*, 1992). Together with silver nitrate, enhanced aeration was found very important for shoot regeneration. Sufficient aeration was achieved by taping the Petri dishes with a porous paper tape 3MM instead of Parafilm (Kuvshinov *et al.*, 1999).

Two major hormone combinations were found optimal for different types of explants. The first was continuous culturing on 2 - 5 mg/l BAP and 0,5 - 3 mg/l NAA until shoot regeneration (Mukhopadhyay *et al.*, 1992). The second was culturing explants on 0.3 - 1,5 mg/l 2.4 D during 3 - 6 days and, then 2 - 3 mg/l BAP and 0 - 0.5 mg/l NAA for 7 - 14 days until green embryogenic nodules and shoot primordia formed (Radke *et al.*, 1992; Kuvshinov *et al.*, 1999). The duration of culturing and concentration of the hormone depended on the type of explant and the variety cultured. After shoot primordia formation the explants were transferred on hormone free media. The medium with the first hormone composition as well as continuous culturing on regeneration medium containing hormones led to vitrification of shoots and decreased subsequent shoot survival and rooting. Shoots regenerated from the stem segments formed often inflorescences and attempted to flower. To avoid inflorescence development and to enhance rooting 0.05 - 0.2 mg/l NAA or 2 mg/l IBA as well as 1% sucrose supply was used. There were no differences between plant material cultured on MS (Murashige and Skoog, 1962) or B5 (Gamborg *et al.*, 1968) media.

Different antibiotics used to select for transformed tissues and to control *A. tumefaciens* had also an impact on the plant tissue responses. Of the two antibiotics commonly used for *A. tumefaciens* growth suppression, claforan (cefatoxim) suppressed shoot regeneration while carbenicillin did not (Kuvshiov *et al.*, 1999). Because claforan is light sensitive the concentrations used ranged between 300 to 500 mg/l, while the effective concentrations against *Agrobacterium tumefaciens* range from 50 to 100 mg/l. Carbenicillin is more stable in the tissue culture medium and concentrations of 100 to 200 mg/l are sufficient. Kanamycin and hygromycin were tested for selection of transformed tissues and recovered shoots. The experiments showed that both of these antibiotics can be used for selection at concentrations from 15 to 30 mg/l. The proportion of non-transgenic escapes on kanamycin selection was, however, ten times higher than on hygromycin selection (Kuvshinov *et al.*, 1999). Below are two protocols for transformation of hypocotyl and inflorescence carrying stem segments.

A. *tumefaciens* transformation protocol for hypocotyl explants of *B. rapa* ssp. *oleifera*.

(Compiled from protocols by Radke *et al.*, 1992; Mukhopadhyay *et al.*, 1992 and Takasaki *et al.*, 1997).

Surface sterilized seeds (usually 1 min in 75% ethanol, 2 - 10 min in 2% Na-hypochlorite, washing 3 times 5 min in water) were placed on agar solidified MS or B5 media without sugars and hormones. 4 - 7 days old hypocotyls were cut in 2 - 5 mm segments. The hypocotyl segments were placed onto a filter paper over a nurse culture of tobacco feeder layer cells (Radke *et al.*, 1988), which were spread onto MS medium with 3 % sucrose and 0.6 % agar. Tobacco feeder layer enhanced transformation 2-5 times (Radke *et al.*, 1992). Alternatively, hypocotyl explants can be grown with the addition of 10% coconut water instead of the tobacco feeder layer (Mukhopadhyay *et al.*, 1992). Hormone treatment can alternatively be 1 mg/l 2.4D for 6 - 10 days, then 3 mg/l BAP and 1 mg/l zeatin for three to five weeks, or continuous culturing on 2 mg/l BAP and 3 mg/l NAA. Regarding the bacterial cultures *A. tumefaciens* strains EHA101 or LBA4404 carrying the binary Ti vector were grown in LB broth supplemented with rifampicin, kanamycin or other required antibiotics. Explants were inoculated with 1×10^8 bacteria/ml suspension of *Agrobacterium* in liquid MS medium immediately after excision or after one-day pre-culturing of the explants. Co-cultivation with *A. tumefaciens* was performed for 2 to 3 days on solid MS medium or on a feeder layer in 28°C. After co-cultivation explants were washed and shifted onto solid MS or B5 media containing the same hormones, 30 - 60 μM silver nitrate, 200 mg/l cefatoxim or 500 mg carbenicillin and cultured for 3 to7 days. After that, selection was applied as 15 - 25 mg/l kanamycin. When explants started to form shoots, they were transferred onto hormone free media with 1% sucrose and the same antibiotics. Two to three weeks later shoots were shifted for rooting onto solid B5 medium with 2 mg/l IBA and 1% sucrose or on solid MS medium with 0,005 mg/l NAA and 50 mg/l casein hydrolysate. Shoots, which had not rooted, were re-cut and placed back onto the rooting medium. The media were refreshed every 10 to14 days of culturing.

A. *tumefaciens* transformation protocol for inflorescence carrying stem segments of *B. rapa* ssp. *oleifera* (updated from Kuvshinov *et al.*, 1999).

The explant source plants of *B. rapa* ssp. *oleifera* were grown in the greenhouse. The status of the source plants determines the success of the entire transformation procedure. The plants were grown for 3 weeks under intensive continuous light, high temperature (28-30/18-20°C) and low humidity (<50%). Plants should be well supplied with fertilizers and water. Stem segments from the upper one or two 1.5 - 4 cm long internodes were excised from 15 - 25 cm tall plants being at the stage before inflorescence stem elongation. The excised inflorescence-carrying stem was surface sterilized for 1 min in 70% ethanol, 10 min in Na-hypochlorite (2% active Cl) and washed 3 times for 5 min in sterilized water. The stem was cut in 4-6 mm segments. The segments were placed horizontally onto solid MS medium (0.7% agar) containing 2% sucrose, 0.5 - 1.5 mg/l 2,4D, 30 μM silver nitrate and 0 - 200 μM acetosyringone. The explants were cultured at 16/8 hour day/night light and 25/18°C day/night temperature throughout the transformation and regeneration experiments. Plates with explants were sealed with porous paper tape 3MM. Depending on the variety, hormone composition the first three to six days was 0.5 - 1.5 mg/l 2,4D and then 7 - 20 days 2 - 3 mg/l BAP and 0 - 0.05 mg/l NAA. The explants pre-cultured for 24 h were inoculated with *Agrobacterium*. *A. tumefaciens* LBA4404 strain carrying the binary Ti vector (alternatively C58C1pGV3850 with cointegrative Ti system or other bacterial strain) was grown in YEB (Lichtenstein and Draper, 1985) with 25 mg/l rifampicin and 50 mg/l kanamycin, +28°C, overnight with shaking. An aliquot of 30 μl of an overnight culture was inoculated in 3 ml of fresh YEB with the same antibiotics and 0 - 100 μM acetosyringone and grown one more night. The overnight culture was diluted 10 times with 1/2 strength liquid MS. One-day cultured explants were immersed in *Agrobacterium* suspension for a few minutes. Then the extra liquid was removed from the explants on sterilized filter paper, and explants were placed horizontally on fresh MS medium with the same composition as described above. Co-cultivation with *Agrobacterium* was performed at +28°C for two days. After that, the explants were washed from *Agrobacterium* in 300 mg/l carbenicillin water solution for 5 min. Then the explants were placed on selection medium vertically their basal side down. The correct orientation of the explants is important, because shoot regeneration occurs only on the basal side of the explant. Usually, after three days of culturing the basal side of the explant becomes wider than the upper side and is easily detectable. Orienting the explant basal side down significantly enhances shoot regeneration and selection of transformed tissues. The selection medium was supplemented with a composition of plant hormones depending on the variety, 2% sucrose, 30 μM silver nitrate, 200 mg/l carbenicillin and 15 - 20 mg/l hygromycin (Sigma). After green nodules and shoot primordia formed the explants were shifted onto hormone free MS medium with the same antibiotics and silver nitrate. Recovered shoots were cut from the callus and transferred onto rooting medium supplemented with 2% sucrose, 0.1 - 0.2 mg/l NAA and 200 mg/l carbenicillin without silver nitrate and hygromycin. Our recent experiments with cultivar Valtti have shown that the right growing conditions in the greenhouse for the explant source plants enable us to achieve 100% shoot regeneration and 25% transformation frequency.

Direct transformation of *B. rapa* ssp. *oleifera*

Data on direct transformation experiments of turnip rape are very scarce, in spite of the fact that the first transformation experiments on protoplasts of *B. rapa* ssp. *oleifera* were performed already in the eighties (Paszkovski *et al.*, 1986). Recalcitrance in regeneration and transformation allowed only one group to develop a direct transformation method using PEG mediated introduction of the DNA into protoplasts (Pauk *et al.*, 1991a; Pauk *et al.*, 1991b). The PEG induced transformation was performed parallel with two plasmids: pIDS 211 (p35S-GUS-NOS), and pIDS 411 (PRT-GUS-NOS). It was demonstrated that TR promoter produced a higher transient expression than the 35S promoter in a similar plasmid construct. After successful transient expression Pauk *et al.* (1991b) obtained transformants, which were resistant to kanamycin. The activity of the introduced gene was tested by NPT assay.

2.2. *B. rapa* ssp. *pekinensis*

2.2.1. *In vitro* culturing

The problems of recalcitrance in tissue culture of *B. rapa* (syn. *B. campestris*) ssp. *pekinensis*, the Chinese cabbage, are close to the level of B. *rapa* ssp. *oleifera*. The research group of Chi has published several reports on shoot regeneration from seedling explants (Chi *et al.*, 1989; Chi *et al.*, 1990; Chi *et al.*, 1991; Chi *et al.*, 1994). Cotyledonous explants showed better regeneration capacity than hypocotyl segments. Excised from 3 days old seedlings, cotyledonous explants produced shoots in 3 - 4 weeks on MS media supplied with 2 mg/l BAP and 1 mg/l NAA. As with turnip rape, ethylene inhibitors such as silver nitrate (20-30 μM), silver sulfate and aminoethoxyvinylglycine (5 μM), enhanced shoot regeneration of cotyledon explants of Chinese cabbage from 20-30% to 70-85%. Hypocotyl segments did not produce shoots at all without ethylene inhibitors. Amino-oxyacetic acid and 2,4-dinitrophenol did not affect shoot regeneration frequency (Chi *et al.*, 1989). Rooting was achieved on hormone free medium or in the presence of 0.1 mg/l IBA (Chi *et al.*, 1990). It was shown that aminoethoxyvinylglycine decreased ethylene synthesis, whereas silver nitrate did not have an effect on ethylene production by the cotyledon explants (Chi *et al.*, 1991). Thus, the actual mechanism of ethylene or silver nitrate action on *in vitro* shoot differentiation is not clear (Chi *et al.*, 1994).

Some varieties of Chinese cabbage have been shown to produce somatic embryos (Choi *et al.*, 1996). Zhang *et al.* (1998) reported on reproducible and effective shoot regeneration from cotyledonary explants of several cultivars of Chinese cabbage. Shoot regeneration frequency of the 123 different genotypes tested ranged from 0 to 95%. Maximum shoot regeneration was obtained in the presence of 5 mg/l BAP and 0.5 mg/l NAA. High agar concentration (1.2 - 1.6%) and addition of silver nitrate (10 - 30 μM) had also a profound positive impact on shoot regeneration (Zhang *et al.*, 1998). Cotyledonary explants excised from 8 days old seedlings of cv. 'Top Salad' produced somatic embryos at a frequency of 20% on solid MS medium supplemented with 1 - 2 mg/l 2.4 D. The cut halves of the somatic embryos produced numerous secondary embryos on 1 mg/l 2.4D in 6 weeks, whereas intact embryos produced only a few secondary embryos. Most of the embryos developed into plantlets after transfer onto basal MS medium (Choi *et al.*, 1996). Besides cotyledons and hypocotyls, other organs of *B. rapa* ssp. *pekinensis* can serve as sources for explants producing shoots *in vitro* (Cheema and Sood, 1987; Yoon and Kim, 1993).

Isolated microspores of Chinese cabbage were successfully cultured on NN medium containing 10% sucrose in darkness at 33°C for one day followed by culturing at 25°C. After 14 days of culture, microspores developed into embryos ranging from globular to cotyledonary stage. Plants were regenerated after transfer of embryos to hormone free medium containing 3% sucrose (Sato *et al.*, 1989). Addition of 0.05 mg/l BAP and 0.5 mg/l NAA hormones enhanced embryo production. Relatively abnormal plantlets regenerated from embryos on hormone free medium but recovered to normal morphology on MS medium supplemented with 3% sucrose, 1.0 mg/l BAP and 0.05 mg/l NAA. The regenerated plants were haploid, dihaploid or tetraploid or had an aneuploid chromosome composition (Hee and Seong, 1997). Microspores of tropical varieties had higher embryogenesis and regenerative ability than the domestic Japanese varieties, which had very low embryogenesis rate (Kuginuki-Yasuhisa *et al.*, 1997). Immature zygotic embryos, ovules, and anther filaments of Chinese cabbage were successfully cultured on MS medium and produced somatic embryos. Up to 80% of the zygotic embryos, 17% of the ovules, and 13% of the filaments produced somatic embryos with or without an intervening callus phase. After transfer on hormone free MS medium, most of the somatic embryos developed into plantlets (Choi *et al.*, 1998).

Yamagishi *et al.* (1988) reported successful protoplast culture of Chinese cabbage. Mesophyll protoplasts excised from *in vitro* cultured plantlets were cultured on 1/2 strength liquid MS medium with 0.5 mg /l 2.4D (alternatively 5 mg/l NAA) and 0.5 - 1 mg/l kinetin. Cell colonies developed on liquid medium were transferred onto solid MS medium supplemented with 1 - 2 mg/l 2.4D and 0.5 - 1 mg/l kinetin. Developed calli were transferred onto MS medium with 1 mg/l zeatin, where 29% of the calli differentiated into shoots. The recovered shoots were rooted on hormone free MS medium. Basing on the protoplast regeneration protocol, the same author reported successful somatic hybridization of Chinese cabbage (*B. rapa* ssp. *pekinensis*) and head cabbage (*B. oleracea* var. *capitata*). Hybrid plants exhibited both of the parental phenotypes in different proportions (Yamagishi *et al.*, 1992).

2.2.2. Genetic transformation

Protocol on *A. tumefaciens*-mediated transformation of cotyledonary explants of Chinese cabbage was reported by Jun *et al.* (1995). Five days old seedlings grown *in vitro* from surface-sterilized seeds were used as target tissue for transformation. Cotyledons with 1 - 2 mm long petioles were excised from the seedlings and co-cultured for two days with *A. tumefaciens* strain LBA4404 carrying the binary Ti vector with *nptII* selection marker gene. The explants were cultured on MS medium supplemented with 2% sucrose, 5 mg/l BAP and 0.5 mg/l NAA. A 44% shoot induction frequency was achieved in regeneration tests, but decreased to 30% after transformation and selection on 15 mg/l kanamycin. Selection of transformed tissues on kanamycin was applied immediately after washing off the *Agrobacterium* and was continued for 3 - 4 weeks. Longer selection has a negative effect on shoot growth and rooting. The final shoot transformation frequency achieved was 5%. All of the transformed plants exhibited normal phenotype and were fertile. The method was used to transfer cucumber mosaic virus (CMV) and potato virus Y (PVY) cDNA clones into the Chinese cabbage genome (Jun *et al.*, 1995).

Procedure for the production of fertile transgenic Chinese cabbage transformed with *A. rhizogenes* has been reported by Christey *et al.*

(1997). Many of the *Brassica* species are more susceptible to *A. rhizogenes* transformation than to *A. tumefaciens*. *A. rhizogenes* transformation, however, produces hairy roots phenotype and causes sterility in the transformed plants. In the report by Christey *et al.* (1997) binary Ti vectors in *A. rhizogenes* were used. Co-cultivation procedure was performed as described above. Transgenic hairy root lines were maintained by subculturing every 4 - 5 weeks on hormone free Linsmaier-Skoog medium (Linsmaier and Skoog, 1965) solidified with 0.25% Gelrite with selection on 50 - 100 mg/l of kanamycin (LS-N). Healthy root explants (1 - 2 cm) from the growing region of the hairy root cultures were placed for shoot regeneration medium. Shoot regeneration was achieved on RCC medium (Guerche *et al.*, 1987) with substitution of 0.3% Phytagel (Sigma) for agar. Hormones required for shoot regeneration were 5 mg/l NAA and 0.5 - 5 mg/l thidiazuron (TDZ). Regenerated shoots of 2-3 cm in length were transferred onto LS-N medium without hormones for shoot elongation. Regeneration frequency of transformed shoots varied from 27 to 43%. Some of the regenerated plants exhibited altered phenotype and decreased fertility (Christey *et al.*, 1997).

There is only one report on direct gene transfer to Chinese cabbage. It describes the transfer of the GUS gene carrying construct by particle bombardment (Cho *et al.*, 1994). The method requires high-level of shoot regeneration as bombardment decreases regeneration rate, especially on selective medium. Recovered plantlets often carried tandem, multiple or partially deleted insertions.

2.3. *In vitro* culturing and genetic transformation of other subspecies of *B. rapa*

The data on transformation development for other *B. rapa* subspecies are scarce, but some advances have been made during the past years. A study on microspore culture of Korean turnip was performed by Hee and Seong (1997). The regeneration procedure of microspores was close to the ones reported for Chinese cabbage, but embryo culture of Korean turnip required NLN medium supplemented with 10% sucrose, 0.05 mg/l BAP and 0.5 mg/l NAA. Plant regeneration from embryos was achieved on MS medium with 3% sucrose, 1.0 mg/l BAP and 0.05 mg/l NAA (Hee and Seong, 1997). *A. rhizogenes*-mediated transformation of leaf segments and cotyledonary petioles of turnip has been described by Christey and Sinclair (1992). Cotyledons with petioles were excised from 7 - 12 days old seedlings. Leaf segments were excised from *in vitro* grown plants. The cut explants were inoculated with A4T strain of *A.*

rhizogenes and placed on a tobacco feeder layer on LS-N medium (Linsmaer and Skoog, 1965). After 2 days of co-cultivation the explants were washed from bacterium and hairy roots were grown on LS-N medium. Shoot regeneration was achieved on RCC medium (Guerche *et al.*, 1987) from 2 - 3 mm long segments of hairy roots. After regeneration the shoots were transferred onto hormone free LS-N medium for elongation and rooting. Many of the regenerated plants exhibited altered phenotype. Plants carrying only the T-DNA of the binary vector were selected after crossing the transgenic plants with non-transgenic lines. The progeny exhibited normal phenotype (Christey and Sinclair, 1992).

There are a few reports on *in vitro* culturing of pakchoi (*B. rapa* ssp. *chinensis*) – an important leafy vegetable in China with an estimated area of approximately 600 000 ha or 20% of the total vegetable area. The pakchoi cotyledon and hypocotyl explants were excised from 3 to 4 days old aseptically germinated seedlings. The explants were grown on solid MS medium. Shoots were developed on cotyledon explants with a frequency of 63 - 67% on optimal hormone combination of 2 - 4 mg/l BAP, 1 mg/l NAA supplemented with 20 μM silver nitrate. Silver nitrate enhanced shoot differentiation by 2 – 3-fold, whereas another ethylene inhibitor aminoethoxyvinylglycine had no effect on regeneration (Chi *et al.*, 1990). Hypocotyl segments formed shoots on the same media at a lower frequency than cotyledons: 5 - 42% depending on the variety. Microspores of nine cultivars of pakchoi were cultured on NLN82 medium at 33°C for a day and then incubated at 25°C. After three weeks embryos developed and were transferred onto solid B5 medium with 2% sucrose, where plantlets developed and rooted (Cao *et al.*, 1994). Shoot regeneration capacity of *B. rapa* ssp. *parachinensis* was similar to other *B. rapa* subspecies with cotyledonous explants achieving a rate of 75% on solid MS medium supplemented with 2 mg/l BAP, 1 mg/l NAA and 30 μM silver nitrate (Chi *et al.*, 1990).

3. CONCLUSION AND FUTURE PROSPECTS

Summarizing the developed regeneration and transformation protocols, it can be concluded that reliable transformation protocols have been developed only for the two agriculturally most important subspecies of *B. rapa*: the first is the oilseed turnip rape and brown sarson (*B. rapa* ssp. *oleracea*), and the second is the vegetable crop, Chinese cabbage (*B. rapa* ssp. *pekinensis*) (Figure 1). Although, other *B. rapa* subspecies have also been widely used in agriculture, data on their *in vitro* culturing and transformation techniques are very few. Compiled experiences of

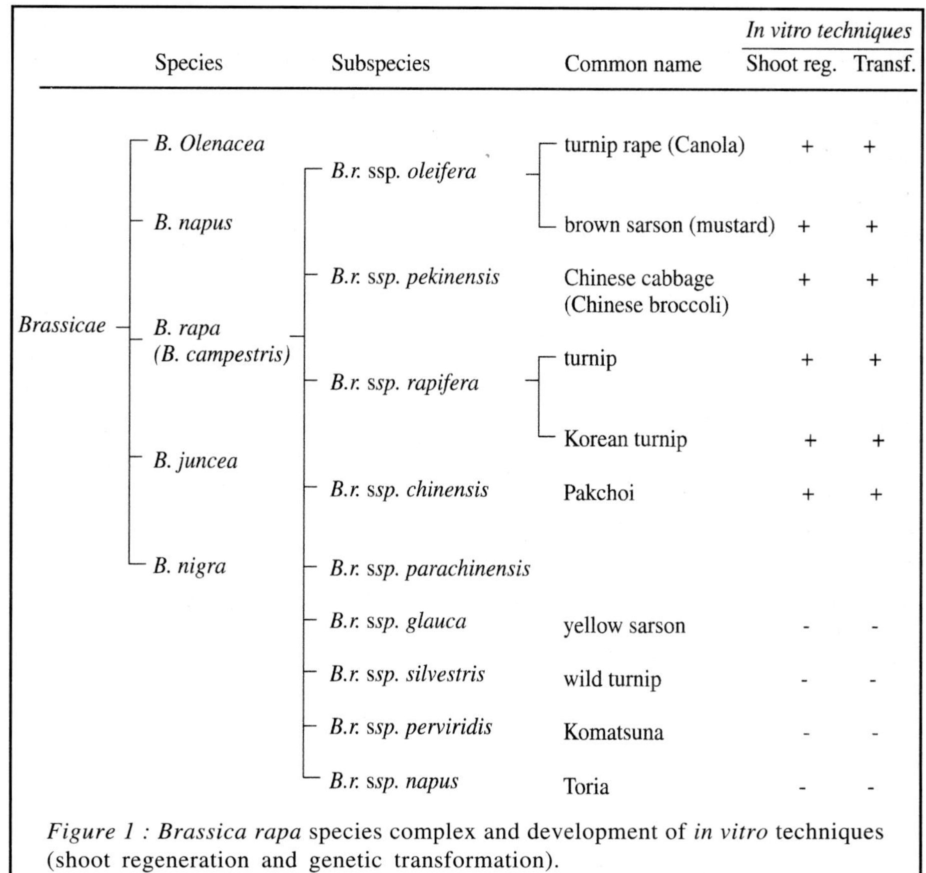

Figure 1 : Brassica rapa species complex and development of *in vitro* techniques (shoot regeneration and genetic transformation).

the reports on *in vitro* culture of *B. rapa* can probably assist in the development of transformation techniques for different subspecies of *B. rapa*. Several common features have emerged regarding *in vitro* culture of *B. rapa* subspecies. The addition of silver nitrate or other ethylene inhibitors in the growth medium significantly enhances shoot regeneration. Sufficient aeration of the culture vessel is equally important, which can be achieved by sealing culture plates with porous paper tape instead of Parafilm. Shoot regeneration often requires two phases: two to six days on 2,4-D containing medium followed by one to three weeks on BAP and NAA containing medium. Alternatively continuous culturing on BAP and NAA containing medium has proven successful. Shoot growth and elongation can be achieved on media without hormones. Difficulties in rooting are a characteristic for *B. rapa*. Lowered sucrose or nutrient content, addition of small dosage of IBA, NAA or IAA can help recovered shoots to form roots. While successfully developed microspore and protoplast cultures resulted in haploid, dihaploid and hybrid production involving *B. rapa*, they were not followed up by development of direct or *Agrobacterium*- mediated transformation of embryos or protoplasts except for two isolated reports. An often overlooked issue is the importance of the growing conditions of the explant source plants for successful regeneration and transformation efficiency. In most cases the explants with young and fast growing meristems are the best target tissues for transformation and shoot regeneration.

As the tissue culture and transformation methodologies are increasingly more available for *B. rapa* it is envisioned that reports on trait improvement will appear in the near future. In the oil seed rape the traits include improved fatty acid profiles especially increased amounts of palmitic, oleic and linoleic acid and reduction in linolenic acid. To improve the feed quality of the seed meal increased amounts of lysine, histidine and methionine are of interest. For the vegetable crops of *B. rapa* the main interests are in improving their disease and insect pest resistance.

REFERENCES

Baillie AMR, Epp DJ, Hutcheson D and Keller WA (1992) *In vitro* culture of isolated microspores and regeneration of plants in *Brassica campestris*. *Plant Cell Rep.*, **11** : 234-237.

Burnett L, Yarrow S and Huang B (1992) Embryogenesis and plant regeneration from isolated microspores of *Brassica rapa* L. ssp. *oleifera*. *Plant Cell Rep.*, **11** : 215-218.

Burnett L, Arnoldo M, Yarrow S and Huang B (1994) Enhancement of shoot regeneration from cotyledon explants of *Brassica rapa* ssp. *oleifera* through pretreatment with auxin and cytokinin and use of ethylene inhibitors. *Plant Cell Tiss. Org. Cult.*, **37** : 253-256.

Cao MQ, Li Y, Liu F and Dore C (1994) Embryogenesis and plant regeneration of pakchoi (*Brassica rapa* L. ssp. *chinensis*) via *in vitro* isolated microspore culture. *Plant Cell Rep.*, **13** : 447-450.

Cardi T and Earle ED (1997) production of new CMS *Brassica oleracea* by transfer of 'Anand' cytoplasm from *B. rapa* through protoplast fusion. *Theor. Appl. Genet.*, **94** : 204-212.

Cheema HK and Sood N (1987) *In vitro* plant regeneration of *Brassica campestris* L. on stem and leaf explants. In : *Symposium on in vitro problems related to mass propagation of horticultural plants*. Ed. Boxus P *et al.*, ISHS, Wageningen, Netherlands, p. 701-704.

Chi GL and Pua EC (1989) Ethylene inhibitors enhanced *de novo* shoot regeneration from cotyledons of *Brassica campestris* ssp. *chinensis* (Chinese cabbage) *in vitro*. *Plant Sci.*, **64** : 243-250.

Chi GL, Barfield DG, Sim GE and Pua EC (1990) Effect of $AgNO_3$ and aminoethoxyvinylglycine on *in vitro* shoot and root organogenesis from seedling explants of recalcitrant *Brassica* genotypes. *Plant Cell Rep.,* **9** : 195-198.

Chi GL, Pua EC and Goh CJ (1991) Role of ethylene on *de novo* shoot regeneration from cotyledonary explants of *Brassica campestris* ssp. *pekinensis* (Lour) Olsson *in vitro. Plant Physiol.,* **96** : 178-183.

Chi GL, Lin WS, Lee JEE and Pua EC (1994) Role of polyamines on *de novo* morphogenesis from cotyledons of *Brassica campestris* ssp. *pekinensis* (Lour) Olsson *in vitro. Plant Cell Rep.,* **13** : 323-329.

Cho HS, Lee YH, Suh SC, Kinm DH and Kim HI (1994) transformation of beta-glucuronidase (GUS) gene into Chinese cabbage (*Brassica campestris* var. *pekinensis*) by particle bombardment. *RDA J. Agri. Sci. Biothech.,* **36** : 181-186.

Choi PS, Soh WY and Liu JR (1996) Somatic embryogenesis and plant regeneration in cotyledonary explant cultures of Chinese cabbage. *Plant Cell Tiss. Org. Cult.,* **44** : 253-256.

Choi PS, Min SR, Ahn MY, Soh WY and Liu JR (1998) Somatic embryogenesis and plant regeneration in immature zygotic embryo, ovule, and anther filament cultures of Chinese cabbage. *Scientia Hort. Amsterdam,* **72** : 151-155.

Christei MC and Sinclair BK (1992) Regeneration of transgenic kale (*Brassica oleracea* var. *acephala*), rape (*B. napus*) and turnip (*B. campestris* var. *rapifera*) plants via *Agrobacterium rhizogenes* mediated transformation. *Plant Sci.,* **87** : 161-169.

Christey MC, Sinclair BK, and Braun RH and Wyke L (1997) Regeneration of transgenic vegetable brassicas (*Brassica oleracea* and *B. campestris*) via Ri-mediated transformation. *Plant Cell Rep.,* **16** : 587-593.

Dietert MF, Barron SA and Yoder OC (1982) Effect of genotype on *in vitro* culture in the genus *Brassica. Plant Sci. Lett.*, **26** : 233-240.

Dunwell JM (1981) *In vitro* regeneration from excised leaf discs of three *Brassica* species. *J. Exp.Bot.,* **32** : 789-799.

Eapen S and George L (1997) Plant regeneration from peduncle segments of oil seed *Brassica* species : Influence of silver nitrate and silver thiosulfate. *Plant Cell Tiss. Org. Cult.,* **51** : 229-232.

Gamborg OL *et al.* (1968) *Exp. Cell Res.,* **50** : 151.

Gamborg OL, Miller RA and Ojima K (1968) Nutrient requirements of suspension cultures of soybean root cells. *Exp. Cell Res.,* **50** : 151-158.

Glimelius K (1984) High growth rate and regeneration capacity of hypocotyl protoplasts in some *Brassicae. Physiol. Plant.,* **61** : 38-44.

Guerche P, Jouanin L, Tepfer D and Pelletier G (1987) Genetic transformation of oilseed rape (*Brassica napus*) by the Ri T-DNA of *Agrobacterium rhizogenes* and analysis of inheritance of the transformed phenotype. *Mol. Gen. Genet.,* **206** : 382-386.

Gudwin I, Todd G, Ford-Lloyd B and Newbury HJ (1991) The effects of acetosyringone and pH on *Agrobacterium*-mediated transformation vary according to plant species. *Plant Cell Rep.,* **9** : 671-675.

Guo YD and Pulli S (1995) *In vitro* pollen culture and the regeneration of *Brassica campestris* L. plants. *J. Agri. Sci.,* **4** : 513-518.

Guo YD and Pulli S (1996) High-frequency embryogenesis in *Brassica campestris* microspore culture. *Plant Cell Tiss. Org. Cult.,* **46** : 219-225.

Hachey JE, Sharma KK and Moloney M (1991) Efficient shoot regeneration of *Brassica campestris* using cotyledon explants cultured *in vitro*. *Plant Cell Rep.,* **9** : 549-554.

Hee KY and Seong LS (1997) Microspore culture of Chinese cabbage (*Brassica campestris* ssp. *pekinensis*) and Korean turnip (*B. campestris* ssp. *rapa*). *J. Korean Soc. Hort. Sci.,* **38** : 368-371.

Jain RK, Chowdhury JB, Sharma DR and Friedt W (1988) Genotypic and media effects on plant regeneration from cotyledon explant cultures of some *Brassica* species. *Plant Cell Tiss. Org. Cult.,* **14** : 197-206.

Javed MA and Hassan S (1992) Callusing and regeneration response of cotyledon explant of some rapeseed genotypes. *Sarhad J. Agri.,* **8** : 329-334.

Jourdan PS and Earle ED (1989) Genotypic variability in the frequency of plant regeneration from leaf protoplasts of four *Brassica* ssp. and of *Raphanus sativus*. *J. Amer. Soc. Hort. Sci.,* **114** : 343-349.

Jun SI, Kwon SY, Paek KY and Paek KH (1995) *Agrobacterium*-mediated transformation and regeneration of fertile transgenic plants of chinese cabbage (*Brassica campestris* ssp. *pekinensis* cv. 'spring flavor'). *Plant Cell Rep.,* **14** : 620-625.

Kao KN and Michailyuk MR (1975) Nutritional requirements for growth of *Vicia hajastana* cells and protoplasts at very low population density in liquid media. *Planta,* **126** : 105-110.

Keller WA and Armstrong KC (1979) Stimulation of embryogenesis and haploid production in *Brassica campestris* anther cultures by elevated temperature treatments. *Theor. Appl. Gen.,* **55** : 65-67.

Knutzon DS, Thompson GA, Radke SE, Johnson WB, Knauf VC and Kridl JC (1992) Modification of *Brassica* seed oil by antisense expression of a stearol-acyl carrier protein desaturase gene. *Proc. Natl. Acad. Sci. USA,* **89** : 2624-2628.

Kuginuki Yasuhisa, Nakamura Kohi, Hida Ken Ichi and Yosikawa Hiroaki (1997) Varietal differences in embryogenic and regenerative ability in microspore culture of Chinese cabbage (*Brassica rapa* L. ssp. *pekinensis*). *Breed. Sci.,* **47** : 341-346.

Kuvshinov V, Koivu K and Pehu E (1997) Biotechnology and genetic resources applied in oil-seed and vegetable *Brassica* improvement. In: *Plant biotechnology and plant genetic resources for sustainability and productivity.* (Eds Watanabe K and Pehu E) Academic Press, R.G. Landes Company, Austin, Texas, USA, pp. 197-207.

Kuvshinov V, Koivu K, Kanerva A and Pehu E (1999) *Agrobacterium tumefaciens*-mediated transformation of greenhouse-grown *Brassica rapa* ssp. *oleifera*. *Plant Cell Rep.,* **18** : 773-777.

Lentini Z, Mursell MA, Mutschler MA and Earle ED (1988) Ethylene generation and reversal of ethylene effects during development *in vitro* of rapid-cycling *Brassica campestris* L. *Plant Sci.,* **54** : 75-81.

Lichtenstein C and Draper J (1985) Gene engineering of plants. *In*: *DNA cloning - a practical approach.* (Ed Glover D M) Oxford IRL Press, v. **2** : 67-119.

Linsmaier EM and Skoog F (1965) Organic growth factor requirements of tobacco tissue cultures. *Physiol. Plant.,* **18** : 100-127.

Loudon PT, Nelson RS and Ingram DS (1989) Studies of protoplast culture and plant regeneration from commercial rapid-cycling *Brassica* species. *Plant Cell Tiss. Org. Cult.,* **19** : 213-224.

Menczel L, Nagy F, Kiss ZR and Maliga P (1981) Streptomycin resistant and sensitive somatic hybrids of *Nicotiana tabaccum + Nicotiana knightiana*: correlation of resistance to *N. tabaccum* plastids. *Theor. Appl. Gen.*, **59** : 191-195.

Mukhopadhyay A, Arumugam N, Nandakumar PBA, Pradhan AK, Gupta V and Pental D (1992) *Agrobacterium*-mediated genetic transformation of oilseed *Brassica campestris*: transformation frequency is strongly influenced by the mode of shoot regeneration. *Plant Cell Rep.*, **11** : 506-513.

Murashige T and Skoog F (1962) A revised medium for rapid growth and bioassays with tobacco tissue cultures. *Physiol. Plant.*, **15** : 473-497.

Murata M and Orton TJ (1987) Callus initiation and regeneration capacities in *Brassica* species. *Plant Cell Tiss. Org. Cult.*, **11** : 111-123.

Nagy J and Maliga P (1976) Callus induction and plant regeneration from mesophyll protoplast of *Nicotiana sylvestris. Z. Pflanzenphysiol.*, **78** : 453-455.

Narasimhulu SB, Prakash S and Chopra VL (1988) Comparative shoot regeneration responces of diploid brassicas and their synthetic amphidiploid products. *Plant Cell Rep.*, **7** : 525-527.

Nitsch JP and Nitsch C (1969) Haploid plants from pollen grains. *Science,* **163** : 85-87.

Palmer CE (1992) Enhanced shoot regeneration from *Brassica campestris* by silver nitrate. *Plant Cell Rep.*, **11** : 541-545.

Paszkowski J, Pisan B, Shilito RD, Hohn T, Hohn B and Ptrykus I (1986) Genetic transformation of *Brassica campestris* var. *rapa* protoplasts with an engineered cauliflower mosaic virus genome. *Plant Mol. Biol.*, **6** : 303-312.

Pauk J, Fekete S, Vilkki J and Pulli S (1991a) Protoplast culture and plant regeneration of different agronomically important *Brassica* species and varieties. *J. Agri. Sci.*, **63** : 371-378.

Pauk J, Fekete S and Stefanov I (1991b) Protoplast and direct genetic transformation experiments in rape. *Novenytermeles,* **40** : 193-202.

Radke SE, Andrews BM, Moloney MM, Crouch ML, Kridl JC and Knauf VC (1988) Transformation of *Brassica napus* L. using *Agrobacterium tumefaciens*: developmentally regulated expression of reintroduced napin gene. *Theor. Appl. Genet.*, **75** : 685-694.

Radke SE, Turner JC and Facciotti D (1992) Transformation and regeneration of *Brassica rapa* using *Agrobacterium tumefaciens. Plant Cell Rep.*, **11** : 499-505.

Sato T, Nishio T and Hirai M (1989) Plant regeneration from isolated microspore cultures of Chinese cabbage (*Brassica campestris* ssp. *pekinensis*). *Plant Cell Rep.*, **8** : 486-488.

Sato S, Katoh N, Iwai S, Hagimori M (1998) Establishment of reliable methods of *in vitro* pollen germination and pollen preservation of *Brassica rapa* (syn. *B. campestris*). *Euphytica,* **103** : 29-33.

Schenk HR and Hoffmann F (1979) Callus and root regeneration from mesophyll protoplasts of *Brassica* species: *B. campestris, B. oleracea* and *B. nigra. Z. Pflanzenzuchtg,* **82** : 354-360.

Shiba H, Hinata K, Suzuki A and Isogai A (1995) Breakdown of self-incompatibility in *Brassica* by the antisense RNA of the *SLG* gene. *Proc Japan Acad,* **71**, Ser. B : 81-83.

Songstad DD, Duncan DR and Widholm JM (1988) Effect of 1-aminocyclopropane-1-carboxylic acid, silver nitrate and norbornadiene on plant regeneration. *Plant Cell Rep.*, **7** : 262-265.

Takasaki T, Hatakeyama K, Ojima K, Watanabe M, Toriyama K and Hinata K (1997) Factors influencing *Agrobacterium*-mediated transformation of *Brassica rapa* L. *Breed. Sci.,* **47** : 127-134.

Ulrich TH, Chowdhury JB and Wildholm JM (1980) Callus and root formation from mesophyll protoplasts of *Brassica rapa. Plant Sci. Lett.,* **19** : 347-354.

United States Department of Agriculture. (1999) The PLANTS database (http://plants.Usda.gov/plants). National Plant Data Center, Baton Rouge, LA 70874-4490.

Winnie T, Prakash L, Prakash K, Jin GC and Sanjay S (1997) Direct shoot formation and plant regeneration from cotyledon explants of rapid-cycling *Brassica rapa. In Vitro Cell. Dev. Biol. Plant,* **33** : 288-292.

Yamagishi H, Nishio T and Takayanagi K (1988) Plant regeneration from mesophyll protoplasts of Chinese cabbage (*Brassica campestris* L.). *J. Japanese Soc. Hort. Sci.,* **57** : 200-205.

Yamagishi H, Hossain MM and Yonesawa K (1992) Production of somatic hybrid between southern type Chinese cabbage Keshin and cabbage Yoshin and their flowering characteristics. *J. Japanese Soc. Hort. Sci.,* **61** : 311-316.

Yoon EK and Kim HY (1993) Root and shoot regeneration from leaf discs of *Brassica campestris* ssp. *pekinensis*, *B. rapa*, and *Raphanus sativus*. *Korean J. Plant Tissue Culture,* **20** : 167-170.

Zhang FL, Takahata Y and Xu JB (1998) Medium and genotype factors influencing shoot regeneration from cotyledonary explants of Chinese cabbage (*Brassica campestris* L. ssp. *pekinensis*). *Plant Cell Rep.,* **17** : 780-786.

Zhao KN, Bittisnich DJ, Halloran GM and Whitecross MI (1995) Studies of cotyledon protoplast cultures from *B. napus, B. campestris* and *B. oleracea.* II: Callus formation and plant regeneration. *Plant Cell. Tiss. Org. Cult.,* **40** : 73-84.

Chapter 13

AGROBACTERIUM-MEDIATED GENETIC TRANSFORMATION OF *BRASSICA JUNCEA*

Arun Jagannath, Panchali Bandyopadhyay, Smriti Mehra, N Arumugam, Pradeep Kumar Burma and Deepak Pental★

Department of Genetics, University of Delhi South Campus, Benito Juarez Road, New Delhi - 110 021, INDIA

Summary

Brassica juncea is a major oilseed crop of the Indian subcontinent and is cultivated over 6 million hectares of land in rain-fed areas of Northern India. In this section, we review various regeneration and transformation studies on Brassica juncea and describe a protocol for efficient Agrobacterium-mediated transformation of this crop plant. Although regeneration in B. juncea has been reported from cotyledonary as well as hypocotyl explants, regeneration through hypocotyl explants is more amenable to Agrobacterium-mediated transformation. The current protocol is based on modifications of an earlier procedure (Bade and Damm, 1995) and has been shown to significantly enhance transformation frequencies of recalcitrant varieties (e.g. Varuna and Donskaja) which were not amenable to genetic transformation using earlier protocols. The protocol described herein has been successfully used in our laboratory for development of various transgenic lines in Brassica juncea viz., lines with herbicide resistance (using the bar gene), male sterile and restorer lines for heterosis breeding (using the barnase and barstar genes) and cre/lox lines for removal of marker genes.

Keywords : *Agrobacterium tumefaciens, Brassica juncea,* genetic transformation

★Corresponding author : E-mail : dpental@hotmail.com

1. INTRODUCTION

Amongst all major food crops, members of the genus *Brassica* have been very amenable to genetic manipulation for development of transgenic lines (James, 1999). While extensive work has been reported on *B. napus* (rapeseed), an oilseed crop of Europe and Canada, very little work has been done on genetic manipulation of *B. juncea* (Indian mustard), a major oilseed crop of the Indian subcontinent. One of the major requirements for development of a transformation system for any plant species is an efficient regeneration system involving regeneration of plantlets, preferably through callus phase, from tissue explants derived from intact plants. Several methods for *in vitro* regeneration of whole plants of *B. juncea* have been described in literature wherein different varieties have been shown to regenerate with varying frequencies on media containing different combinations of hormones. Regeneration through shoot morphogenesis has been reported from hypocotyl explants (Murata and Orton, 1987; Chi *et al.*, 1990; Pental *et al.*, 1993) as well as cotyledonary explants (George and Rao, 1980; Fazekas *et al.*, 1986; Murata and Orton, 1987; Jain *et al.*, 1988; Narasimhulu and Chopra, 1988; Jain *et al.*, 1989; Chi *et al.*, 1990; Sharma *et al.*, 1990). In subsequent studies, it was shown that inclusion of silver nitrate in the regeneration medium significantly enhanced shoot regeneration frequencies from both kinds of explants (Chi *et al.*, 1990; Barfield and Pua *et al.*, 1991; Pental *et al.*, 1993).

In addition to an efficient regeneration system, availability of suitable *Agrobacterium* strains that can induce high frequency infection of tissue explants and marker gene(s) for efficient selection of transformed tissue(s) are also important for establishing an effective transformation protocol. Earlier studies on genetic transformation of *B. juncea* plants (Mathews *et al.*, 1985) with a nopaline-type *A. tumefaciens* strain (A208) harboring a wild-type Ti-plasmid led to the production of teratomic tissues from which a few nopaline-positive shoots were obtained. However, attempts to induce roots from the same were unsuccessful. In another study by Charest *et al.* (1989), virulence of nine different *A. tumefaciens* strains on *B. juncea* var Zem was tested by stem inoculation of plants using a syringe. Tumorigenesis was very rapid and extremely efficient on *B. juncea* with all strains except LBA4404. Tumorigenesis induced by nopaline-type strains at the site of infection was more extensive as compared to that induced by octopine-type strains. Mathews *et al.* (1990) reported *Agrobacterium*-mediated transformation of *B. juncea* cotyledonary explants using a chimeric nos-*nptII* gene. However, the frequency of plant transformation was low and putative transformed shoots could not be rooted on selective media.

Barfield and Pua (1991) developed an efficient protocol for the production of normal, fertile transgenic plants in *B. juncea* using *Agrobacterium*-mediated transformation of hypocotyl explants. Regeneration frequency of shoots registered a substantial increase (90-100%) in this study. Of the four disarmed *A. tumefaciens* strains tested (GV3111-SE, A208-SE, EHA101 and A348), A208-SE was found to be most suitable for genetic transformation. Transgenic shoots could be generated at a frequency of 9-10% and these were further confirmed by Southern analysis. In a subsequent study, Pental *et al.* (1993) tested twelve different agronomically important varieties of *B. juncea* from India, Eastern Europe and Canada for their regeneration response from hypocotyl explants using media with and without silver nitrate. Presence of silver nitrate in the medium was found to be crucial for high frequency regeneration from all the tested varieties. The regeneration protocol developed in the above study was also used for *Agrobacterium*-mediated transformation of *B. juncea*. Transformation efficiency was determined by *gus* staining of infected hypocotyl ends. For *B. juncea* var RLM198, transformation frequency ranged from 16.7-60.5%, while for Pusa Bold and Skorospieka II, it was around 15.2-34.9% and 16.9% respectively. Transgenic plants were confirmed by Southern analysis and transmission of marker genes to F_1 progeny was also established.

Although regeneration has been reported from cotyledonary as well as hypocotyl explants of *B. juncea*, regeneration through cotyledonary explants is less amenable to *Agrobacterium*-mediated transformation (Mathews *et al.*, 1990). In *B. campestris*, cotyledonary explants are recalcitrant to *Agrobacterium*-mediated transformation (Mukhopadhyay *et al.*, 1992). In contrast, successful transformation of hypocotyl explants has been reported in several *Brassica* species *viz.*, *B. campestris* (Mukhopadhyay *et al.*, 1992), *B. napus*, *B. oleracea* (De Block *et al.*, 1989) and *B. juncea* (Barfield and Pua, 1991; Pental *et al.*, 1993) making it the explant of choice for *Agrobacterium*-mediated transformation. The protocol described in the following sections is based on modifications of an earlier protocol (Bade and Damm, 1995) and describes an improved method developed in our laboratory for *Agrobacterium*-mediated transformation of hypocotyl explants of *Brassica juncea* (Mehra *et al.*, 2000). Using the modified protocol, transformation frequency for the *B. juncea* vars Varuna and Donskaja [which were not amenable to the protocol developed earlier (Pental *et al.*, 1993)] could also be significantly enhanced to 74 % and 86 % respectively (as determined by *gus* staining of hypocotyl ends).

2. VARIABLES TESTED FOR EFFICIENT *AGROBACTERIUM*-MEDIATED TRANSFORMATION OF *BRASSICA JUNCEA*

Important parameters that were taken into consideration during the development of the current protocol and analyzed for their effects on regeneration and genetic transformation of *B. juncea* using hypocotyl segments as explants are outlined below.

1. ***Agrobacterium strains*** **:** The *A. tumefaciens* nopaline-type strain GV3101 (Zambryski *et al.*, 1983) which comprises of a nopaline-type helper plasmid in a C58 bacterial background and the disarmed plasmid pMO90 was found to give higher transformation frequencies in *B. juncea* than strains PGV2260, LBA4404 and A281. Infection with strains LBA4404 and A281 led to a hypersensitive reaction on cut ends of hypocotyl explants. The binary vector pGSFR-*gus*-intron (Pental *et al.*, unpublished) was used in the present study for determination of transformation frequencies and analysis of transgene expression in transformed tissues. The above vector consists of an intron-containing β-*glucuronidase* gene under transcriptional control of the CaMV35S promoter (Vancanneyt *et al.*, 1990) cloned at the HindIII site of pGSFR780A (De Block *et al.*, 1989). The vector also contains the *bar* and *nptII* selectable marker genes driven by the CaMV35S and nos promoters respectively.

2. ***Bacteriostatic agents :*** Of various bacteriostatic agents tested for their effects on regeneration of *B. juncea*, augmentin was found to have no inhibitory effect. Regenerating shoots formed on media containing augmentin were well elongated and non-vitrified. The bacteriostatic agents cefotaxime and carbenicillin inhibited regeneration in var Varuna. Regeneration was also inhibited in var Donskaja on using cefotaxime.

3. ***Selectable marker genes*** **:** Various marker genes such as *nptII* (*neomycin phosphotransferase*, conferring resistance to kanamycin), *hpt* (*hygromycin phosphotransferase*, conferring resistance to hygromycin) and *bar* [*bialaphos resistance*, conferring resistance to phosphinothricin (PPT)] could be effectively used for development of transgenics in *B. juncea* using this protocol. While kanamycin and hygromycin are primarily used as *in vitro* selectable markers, the herbicide Basta (containing PPT as the active component) can be used for both *in vitro* as well as field-level selection of transgenic plants. Kanamycin at a concentration of 50 mg/L and hygromycin at 20 mg/L were found to be suitable for selection of transformed tissues. The concentration of PPT required for efficient selection was found to be different for different varieties. A concentration of 20 mg/L PPT was found to be effective for var Donskaja while for vars Varuna and RLM 198, 10 mg/L PPT was sufficient. Reducing the concentration of PPT led to proliferation of untransformed tissues in some varieties. Regeneration frequency of transgenic plants on selection media containing PPT was 11.4 % for var Varuna while the plant conversion frequency was 87.5 %. In case of var Donskaja, frequency of generation of PPT-resistant plants was 13 % and the plant conversion frequency was 10 %.

4. ***Carbon source*** **:** In experiments using Basta as the selective agent, glucose was used as the carbon source. Use of sucrose as the carbon source in such experiments leads to death of explants due to ammonia toxicity (De Block *et al.*, 1995).

3. MEDIA AND REAGENTS

1. The compositions of various media along with supplements required are listed in Table 1 below. All reagents/media/stocks are prepared in double-distilled water unless stated otherwise.

Table 1. Media used for genetic transformation of *B. juncea* hypocotyl explants

Medium	Composition
PM	Liquid N1B1, 2 % glucose or 3 % sucrose
WM	Liquid N1B1, 2 % glucose or 3 % sucrose, 200 mg/L augmentin
SIM	Solid N1B1, 2 % glucose or 3 % sucrose, 200 mg/L augmentin, 10 mg/L PPT or 50 mg/L kanamycin or 20 mg/L hygromycin, 20 μM $AgNO_3$
RM	MS medium, 2 % glucose or 2 % sucrose, 200 mg/L augmentin, 10 mg/L PPT or 50 mg/L kanamycin or 20 mg/L hygromycin, 2 mg/L IBA

PM: preculture medium; WM: washing medium; SIM: shoot induction medium; RM: rooting medium; N1B1: MS medium (Murashige and Skoog, 1962) supplemented with 1 mg/L NAA and 1 mg/L BAP.

2. Tissue culture media used in the protocol consist of MS salts (Murashige and Skoog, 1962) and MS vitamins with either 2% glucose or 3 % sucrose (depending on the selective agent used: glucose is used for selection on PPT and sucrose for selection on kanamycin and hygromycin). pH of the medium is adjusted to 6.1 with NaOH before autoclaving. Media can be stored in glass bottles for not more than one month from the date of autoclaving. Solid media [containing 0.8 % agar (RM301, HiMedia)] are melted in a microwave oven when required and plates are poured around two days prior to use. Glucose, bacteriostatic agents, selective agents and silver nitrate (*see* Note 1) are added to the autoclaved/melted medium after cooling. All plant growth regulators used in the protocol are added prior to autoclaving.
3. **YEB medium (per litre) :** 1g yeast extract, 5g beef extract, 5g peptone, 0.5g $MgSO_4.7H_2O$, 5g sucrose, pH 7.2.
4. **Antibiotics/Selective agents :** A 100 mg/L stock of kanamycin (Sigma) is prepared in distilled water, filter sterilized and stored frozen. Phosphinothricin is used in the form of Basta (Hoechst; active ingredient PPT at 200 g/L) and a working stock of 20 mg/ml is prepared in distilled water, filter sterilized and stored at 4^0C. Hygromycin is available as a 50mg/ml stock from Roche Biochemicals. Augmentin (Time Cap Pharma Labs Pvt. Ltd) to a final concentration of 200 mg/L is directly added (in powder form) to autoclaved, cooled media. A 25 mg/ml stock of Rifampicin (Ranbaxy) is prepared in dimethylsulfoxide (DMSO) and stored at 4^0C.
5. **Sodium hypochlorite solution :** (1 %, w/v) is prepared in water from a commercially available 4 % w/v stock solution (Merck). A 0.05 % mercuric chloride (BDH) solution is prepared in distilled water and autoclaved prior to use.
6. **Growth regulators :** 1 mg/ml stocks of 6-benzylaminopurine (BAP, Sigma) and α-naphthalene acetic acid (NAA, Sigma) and a 2 mg/ml stock of Indole-3-butyric acid (IBA, Sigma) are prepared in distilled water and stored frozen.

4. A PROTOCOL FOR *AGROBACTERIUM*-MEDIATED GENETIC TRANSFORMATION OF *BRASSICA JUNCEA*

The following is a generalized protocol for tissue culture and genetic transformation of *Brassica juncea*. All tissue culture work is performed under sterile conditions in a laminar flow hood. Cultures are grown at

23 ± 1^0C, 2000 lux (Phillips cool white fluorescent lamps) with a photoperiod of 16h light/ 8h dark cycle, except when stated otherwise.

1. Surface sterilize seeds by treating in diluted Teepol (surfactant) solution for 15 min followed by washing under running water for ~30 min. Treat seeds with 70 % ethanol for 2 min under sterile condition and rinse twice with sterile distilled water. Further, treat seeds with 0.05 % mercuric chloride for 10 min followed by a treatment with 1 % w/v sodium hypochlorite for 9 min. After each treatment, rinse seeds thoroughly with sterile distilled water (*see* Note 2).
2. Surface-sterilized seeds are germinated on hormone-free, full-strength MS medium. Seeds are germinated in glass tubes (four seeds per tube) covered with a cotton plug. Seeds are initially placed in dark for two days and subsequently grown in light conditions.
3. For transformation experiments, hypocotyl segments obtained from 6 day-old seedlings are used (*see* Note 3). Place seedlings from about 6-7 tubes horizontally in a 150 mm petridish containing PM (Table 1) and remove seedling apices and roots using a sharp surgical blade. Hold several hypocotyls together with a pair of forceps and cut simultaneously into ~0.5 cm segments using a surgical blade. Care should be taken to give a single sharp cut to the hypocotyls. For each experiment ~400 explants are prepared (*see* Note 4). Transfer the explants with a sterile spatula/spoon to a 250 ml Erlenmeyer flask containing ~80 ml of PM. Take care to remove agar pieces that come along with the roots. Wash with liquid PM in case of agar contamination. Place the flask on a rotatory shaker (100 rpm) and pre-culture for 24 h (diffused light, temperature 23 ± 1^0C).
4. Inoculate *Agrobacterium* strain GV3101 containing the appropriate binary plasmid in 5 ml of liquid YEB medium (with appropriate antibiotics) in a 100 ml flask and grow overnight on a rotatory shaker (250 rpm, 28^0C, in dark). Inoculate one ml of the overnight culture (OD_{600} ~1.2) in 30 ml of liquid YEB medium in a 250 ml Erlenmeyer flask and grow for 4-5 h at 28^0C on a rotatory shaker (250 rpm, 28^0C, in dark). Grow the culture till it reaches an OD_{600} of 0.5 - 0.6.
5. Transfer the *Agrobacterium* culture to a 50 ml autoclaved centrifuge tube and centrifuge at 5000 rpm for 15 min at room temperature. Remove supernatant and suspend the pellet in 20

ml of PM by gentle shaking. Measure the concentration and dilute the suspension to an OD_{600} value of ~0.3 using PM.

6. Stop pre-culture of explants by removing PM with a pipette. Check whether the medium is almost completely clear. Excess turbidity could indicate fungal or bacterial contamination.
7. Add around 30ml of the *Agrobacterium* suspension (OD_{600} ~0.3) to the explants, mix gently, and incubate for 30 min (stationary culture).
8. Remove the *Agrobacterium* suspension after 30 min. Care should be taken to remove the liquid completely as otherwise overgrowth of *Agrobacterium* occurs during subsequent co-cultivation. Add about 80 ml of liquid PM to the flask and place on a rotatory shaker for ~18hours for co-cultivation (100 rpm, 23 ± 1^0C, in diffused light).
9. Stop co-cultivation by removing the liquid PM when the medium becomes slightly turbid due to growth of *Agrobacterium* (~18hr). Rinse explants twice for 1min with 25 ml of WM (Table 1) and twice for 30min with 80-100 ml of WM on a rotatory shaker (100 rpm). Pour WM medium from the flask along with all the explants in a 90 mm petridish and remove the medium with a pipette.
10. Prepare fresh petridishes (90 mm) with 30 ml of semi-solid SIM (Table 1) containing the appropriate selective agent. Place 30-35 explants horizontally on each SIM plate after blotting the explants dry using sterile Whatman paper (No. 1). Seal the petridishes with Urgopore tape and place them under light (*see* Note 5). It is extremely important to prevent fogging in petridishes for proper regeneration to occur from hypocotyl explants (*see* Note 6).
11. Transfer explants with developing calli to fresh petridishes after 20-25 days (*see* Note 7). The first shoots start regenerating at the end of 20 days on media containing PPT while they are seen 5 days earlier on media containing kanamycin or hygromycin.
12. Excise the regenerating shoots and the shoot cluster by cutting carefully through the underlying callus. Transfer to glass tubes containing RM (Table 1) with the appropriate selective agent and 200 mg/L of augmentin. Take only one shoot from each cut end of the hypocotyl. A cluster of shoots may represent the same transformation event. Transfer the elongated shoots to

RM. Rooting takes about 15-20 days in the presence of selective agents. Maintain plantlets by repeated sub-culturing on RM medium in the presence of the selective agent as well as the bacteriostatic agent. Several copies of each transformed event can be generated by nodal cultures.

5. NOTES

1. Silver nitrate stock solution should not be turbid. Stocks should be prepared in double distilled water.
2. The yellow-seeded varieties of *B. juncea* (for example, TM18 and Donskaja) should be given milder treatment than the brown-seeded varieties during seed sterilization. The sequence of steps for seed sterilization of the above varieties is identical to those for brown seeded varieties till alcohol treatment. However, subsequent treatment with mercuric chloride is reduced to 2 min followed by a 7 min treatment with 0.1 % sodium hypochlorite for yellow-seeded varieties.
3. Typically, 6-day-old seedlings attain a length of ~6 - 7 cm and are appropriate for preparation of hypocotyl explants for transformation experiments. Hypocotyl segments from seedlings older than 6 - 7 days show lower transformation efficiencies than younger seedlings. The light intensity should be maintained at 2000 lux for proper seedling maturation and regeneration.
4. The lowermost segments of *B. juncea* seedlings always regenerate and therefore should be retained while preparing explants for transformation.
5. Nescofilm or Urgopore tape should be used to seal petridishes. Parafilm should not be used as it inhibits gaseous exchange.
6. Fogging in petridishes should be strictly avoided. Fogging leads to vitrification of the explants and a consequent reduction in regeneration frequency. To prevent fogging, petridishes should be placed on a perforated aluminium sheet which is elevated to about one and a half inches from the shelf of the culture trolley. This arrangement allows sufficient airflow between the petridishes and the shelf of the culture trolley thereby blocking direct heat transfer to petridishes from lights on the lower shelf.
7. Regenerating explants should not be left too long on SIM plates as *Agrobacterium* growth may recur after degradation of the bacteriostatic agent. Maintenance of adequate concentrations of the selective agent is also essential to prevent the emergence

of chimeras and non-transformed escapees.

6. PROSPECTS AND PERSPECTIVES

Transgenics in *B. juncea* harboring genes *viz., bar* for herbicide resistance (Mehra *et al.*, 2000), *barnase/barstar* for development of male sterile lines (Jagannath *et al.*, 2001) and restorer lines for heterosis breeding and the *cre/lox* system for marker removal (unpublished results) have been successfully generated in our laboratory using the genetic transformation system described above. For development of herbicide-tolerant transgenic plants in *B. juncea* (Mehra *et al.*, 2000), the *bar* gene was deployed under transcriptional control of either the CaMV35S promoter with a duplicated enhancer (35Sde*bar*) or a single enhancer CaMV35S promoter with and without a leader sequence from the Alfalfa Mosaic Virus (AMVL; 35SAMVL*bar* and 35S*bar*). While transformants could be recovered on PPT-containing media at reasonable frequencies with the 35Sde*bar* (23%) and 35SAMVL*bar* (16%) constructs, transformation frequencies with the 35S*bar* construct were low (Mehra *et al.*, 2000). Transgenic plants developed using the 35Sde*bar* and 35SAMVL*bar* constructs were subjected to Southern analysis to identify plants containing a single copy of the T-DNA insert. Of 46 35Sde*bar* plants analyzed, seven were found to be single copy events while from fifty 35SAMVL*bar* plants analyzed, 10 single copy plants were identified. Single copy transgenic plants thus obtained were selfed to develop homozygous lines. Such lines could be deployed for low-till or no-till agriculture which would aid in conservation of moisture and soil fertility in rain-fed areas of the country.

Male sterile lines for heterosis breeding in *B. juncea* have been developed in our laboratory by using the TA29 or A9 promoter for tapetum-specific expression of the *barnase* gene (Jagannath *et al.*, 2001). The *bar* gene has been used as a selectable marker in *barnase* constructs to enable *in vitro* as well as field-level selection of transgenic plants. Transformation frequencies with *barnase* constructs based on use of the CaMV35Sde or CaMV35SAMVL promoter to express the *bar* gene were substantially lower (0-2%) than those observed with corresponding constructs containing the *bar* gene alone (16-23%). Moreover, all male sterile plants obtained using the above constructs were characterized by abnormalities in other agronomically important traits such as vegetative morphology, female fertility, seed germination frequencies and segregation ratios among backcrossed progeny. This reduction in transformation frequencies and failure to obtain agronomically viable male sterile plants using the above constructs was attributed to deregulated expression of the *barnase* gene under enhancing effects of the CaMV35S promoter. Use of a Spacer DNA fragment between the *barnase* and *bar* gene cassettes was found to be an effective method to prevent leaky expression of the *barnase* gene resulting in high-frequency generation of agronomically viable male sterile lines in *B. juncea* (Jagannath *et al.*, 2001). Restorer lines based on tapetum-specific expression of the *barstar* gene (an intracellular inhibitor of barnase) have also been generated in *B. juncea* (unpublished results). Following identification of suitable male sterile/restorer combinations, these would be diversified into appropriate combiners for heterosis breeding in this crop.

Other thrust areas wherein transgenic technology could be used for crop improvement in *B. juncea* include transgenics for improved oil quality (using antisense technology) and removal of marker genes (using the *cre/lox* system). With the identification of other novel genes from the sequenced *Arabidopsis* genome, genes conferring agronomically desirable traits *viz.,* disease-resistance could be deployed in *B. juncea* using the protocol outlined in this study to develop agronomically superior transgenic varieties.

REFERENCES

Bade JB and Damm B (1995). *Agrobacterium*-mediated transformatiom of rapeseed (*Brassica napus*). *In Gene Transfer to Plants.* (Eds. Potrykus I and Spangenberg G) Springer Lab. Manual.

Barfield DG and Pua EC (1991). Gene transfer in plants of *Brassica juncea* using *Agrobacterium tumefaciens*-mediated transformation. *Plant Cell Rep.*, **10** : 308-314.

Charest PJ, Iyer VN and Miki BL (1989). Virulence of *Agrobacterium tumefaciens* strains with *Brassica napus* and *Brassica juncea*. *Plant Cell Rep.*, **8** : 303-306

Chi GL, Barfield DG, Sein GE and Pua EC (1990). Effect of $AgNO_3$ and aminoethoxyvinylglycine on *in vitro* shoot and root organogenesis from seedling explants of recalcitrant *Brassica* genotypes. *Plant Cell Rep.*, **9** : 195-198.

De Block M, De Brouwer D and Tenning P (1989). Transformation of *Brassica napus* and *Brassica oleracea* using *Agrobacterium tumefaciens* and the expression of the *bar* and *neo* genes in the transgenic plants. *Plant Physiol.,* **91** : 694-701.

De Block M, De Sonville AP and Debrouwer D (1995). The selection mechanism of phosphinothricin is influenced by the metabolic status of the tissue. *Planta,* **197** : 619-628.

Fazekas GA, Sedmach PA and Palmer MV (1986). Genetic and environmental effects on *in vitro* shoot regeneration from cotyledon explants of *Brassica juncea*. *Plant Cell Tiss. Org. Cult.*, **6** : 183-187.

George L and Rao PS (1980). *In vitro* regeneration of mustard plants (*Brassica juncea* var. Rai-5) on cotyledon explants from non-irradiated, irradiated and mutagen treated seeds. *Ann. Bot.*, **46** : 107-112.

Jagannath A, Bandyopadhyay P, Arumugam N, Gupta V, Burma PK and Pental D (2001). The use of a spacer DNA fragment insulates the tissue-specific expression of a cytotoxic gene (*barnase*) and allows high frequency generation of transgenic male sterile lines in *Brassica juncea* L. *Mol. Breed.*, **8** : 11-23.

Jain RK, Chowdhary JB, Sharma DR and Friedt W (1988). Genotypic and media effects on plant regeneration from cotyledon explant cultures of some *Brassica* species. *Plant Cell Tiss. Org. Cult.*, **14** : 197-206.

Jain RK, Sharma DR and Chowdhary JB (1989). High frequency regeneration and heritable somaclonal variation in *Brassica juncea*. *Euphytica*, **40** : 75-81

James C (1999). Global review of commercialized transgenic crops. ISAAA Briefs.

Mathews H, Bharathan N, Litz RE, Narayanan KR, Rao PS and Bhatia CS (1990). Transgenic plants of mustard *Brassica juncea* (L.). Czern and Coss. *Plant Sci.*, **72** : 245-252.

Mathews H, Bhatia CR, Mitra R, Krishna TG and Rao PS (1985). Regeneration of shoots from *Brassica juncea* (Linn.) Czern and Coss. cells transformed by *Agrobacterium tumefaciens* and expression of nopaline dehydrogenase genes. *Plant Sci.*, **39** : 49-54.

Mehra S, Pareek A, Bandyopadhyay P, Sharma P, Burma PK and Pental D (2000). Development of transgenics in Indian oilseed mustard (*Brassica juncea*) resistant to herbicide phosphinothricin. *Curr. Sci.*, **78** : 1358-1364.

Mukhopadhyay A, Arumugam N, Nandakumar PBA, Pradhan AK, Gupta V and Pental D (1992). *Agrobacterium*-mediated genetic transformation of oilseed *Brassica campestris*: transformation frequency is strongly influenced by the mode of shoot regeneration. *Plant Cell Rep.*, **11** : 506-513.

Murashige T and Skoog F (1962). A revised medium for rapid growth and bioassays with tobacco tissue cultures. *Physiol. Plant.*, **15** : 473-497.

Murata M and Orton TJ (1987). Callus initiation and regeneration capacities in *Brassica* species. *Plant Cell Tiss. Org. Cult.*, **11** : 111-123.

Narasimhulu SB and Chopra VL (1988). Species specific shoot regeneration response of cotyledonary explants of Brassicas. *Plant Cell Rep.*, **7** : 104-106.

Pental D, Pradhan AK, Sodhi YS and Mukhopadhyay A (1993). Variations amongst *Brassica juncea* cultivars for regeneration from hypocotyl explants and optimization of conditions for *Agrobacterium*-mediated genetic transformation. *Plant Cell Rep.*, **12** : 462-467.

Sharma KK, Bhojwani SS and Thrope TA (1990). Factors affecting high frequency differentiation of shoots and roots from cotyledon explants of *Brassica juncea* (L.) Czern. *Plant Sci.*, **66** : 247-253.

Vancanneyt G, Schmidt R, O'connor-Sanchez A, Willmitzer L and Rocha-Sosa M (1990). Construction of an intron-containing marker gene: Splicing of the intron in transgenic plants and its use in monitoring early events in *Agrobacterium* mediated plant transformation. *Mol. Gen. Genet.*, **220** : 245-250.

Zambryski P, Joos H, Gentello C, Leemans J, Van Montagu M and Schell J (1983). Ti plasmid vector for the introduction of DNA into plant cells without alteration of their normal regenration capacity. *EMBO J.*, **2** : 2143-2150.

Chapter 14

TRANSGENIC PLANTS OF CASSAVA OBTAINED THROUGH *AGROBACTERIUM*-MEDIATED TRANSFORMATION

Paul Chavarriaga Aguirre and WM Roca*

Biotechnology Research Unit, International Center for Tropical Agriculture CIAT, AA 6713, Cali, Colombia

Summary

Cassava is the cheapest available food source with high calorie yield per hectare of all stable crops. Its starchy roots provide food for over 500 million people worldwide. Cassava yield are severely reduced by various pests (mealy bug and green mites) and diseases (African cassava mosaic virus, Cassava bacterial blight). The presence of toxic cyanogenic glycosides and the low protein content of the roots are major nutritional deficiencies of cassava as food. Traditional breeding for pest and disease resistance is difficult due to the low fertility, highly heterozygous nature and limited gene availability in the germplasm. The development of transformation systems via Agrobacterium and particle bombardment allowing the regeneration of transgenic plants makes it possible to use biotechnology to complement traditional breeding programmes in cassava improvement.

Keywords : *Agrobacterium* transformation, cassava, transgenic plants, target for transformation

---

*Present address : International Potato Center - *CIP*, Apartado 1558, Lima 12, Perú
E-mail : w.roca@cgiar.org; p.chavarriaga@cgiar.org

1. INTRODUCTION

Genetic transformation of Cassava (*Manihot esculenta* Crantz), using *Agrobacterium tumefaciens* or particle bombardment as gene-delivery systems, is finally a reality after more than 10 years of continous efforts of several research centers around the world. With both systems, it has been possible to obtain transgenic plants of cassava expressing both marker and selectable genes, as well as genes for agronomic traits. However, in this chapter, we will discuss the most relevant publications on *Agrobacterium*-mediated transformation (abbreviated from here on as Agro-trans) of cassava, in which whole plants have been proved to be transgenic. The reason for this being that Agro-trans is a technology more easily reachable by National Agricultural Research Programs in developing countries, where, ultimately, transgenic cassava with resistance or tolerance to biotic and abiotic stresses is most needed. Agro-trans does not require sophisticated and expensive equipment, like transformation using particle bombardment. Besides, Agro-trans usually produces fewer insertions, of the genes contained in the T-DNA, than particle bombardment does, which is becoming a bio-safety requisite needed for the safe release and commercialization of transgenic plants. The reader is encouraged to read the following publications if interested in transformation of cassava using micro-particle bombardment: Schopke *et al.* (1996); Raemakers *et al.* (1996); Zhang *et al.* (2000); Zhang and Puonti-Kaerlas (2000).

2. FROM THE FIRST TRANSGENIC CALLI TO THE FIRST TRANSGENIC PLANTS

The pioneering experiments that culminated with the production of the first transgenic cassava calli, expressing selectable and useful genes, were developed towards the end of the 80's by Calderon-Urrea (Calderón-Urrea, 1988). In these experiments, Calderón-Urrea used somatic embryos from the cultivar Mcol1505 to transform them with the *Agrobacterium tumefaciens* strains C58C1rif^R(pGV2260), A281, and EHA101, carrying the plasmid pGV1040. Selection of transgenic tissues was applied using the herbicide Phosphinotricin (PPT, commercially known as Basta®) at a concentration of 2 mg/l as selective agent. This selection was possible due to the presence of the *bar* gene, which confers resistance to PPT. In addition to the bar gene the *uid*A gene (for β-glucuronidase expression, also known as *gus*), was also introduced. The presence of the introduced genes was demonstrated by selection in the presence of PPT, histochemical determination of *gus* gene expression, and by Southern blot analysis. This was the first demonstration, at the phenotypic and molecular level, of the expression of bacterial genes, controlled by bacterial promoters, in cassava cells and tissues.

Around 1990, another series of transformation experiments were initiated using a wild type Agro-strain (CIAT 1182), isolated from cassava plants, to introduce the same pGV1040 plasmid into somatic-embryo-derived cotyledons (SECs) of the cultivar Mper183 (Sarria *et al.*, 1993, 1995, 1997 and 2000). After transformation and selection of SECs on media with 8 to 32 mg/l PPT, somatic embryos started to appear, which were then converted to plants. Several putatively transgenic lines (based on *gus* expression and PCR-based tests) were rescued, although only in one of them (line 53-5.2) the insertion of the T-DNA in at least three different sites was proved at the molecular level (hybridization by Southern). Basta spraying of plants in the greenhouse showed that the line 53-5.2 was highly resistant to the herbicide, probably not chimeric, while other lines had chimeric resistance patterns (patchy yellowing of leaves; Figure 1). It seemed like PCR-based selection allowed for some escapes (false positives), which is also fairly common in transformation of cassava using antibiotic as selective agents (see *i.e.,* Zhang *et al.*, 2000). The first transgenic plants of cassava were so established, expressing a gene of potential commercial use. Most of these lines are still kept *in vitro* and in the greenhouse at CIAT after many cycles of propagation, which may have eliminated the chimerism, producing fully transgenic or non-transgenic plants. It would be interesting to analyze these lines to establish if they still express the *bar* gene, if they "lost" it by segregation of chimeric cells, or if the gene is still present but silent. Such information may be relevant to know more about the long-term expression of transgenes in cassava.

The successful use of SECs as explants for transformation was confirmed by Li *et al.* (1996), although several reports had previously confirmed their transformability and potential to regenerate plants (Arias-Garzón and Sarria, 1995; Chavarriaga-Aguirre *et al.*, 1993). In the experiments developed by Li *et al.* (1996), SECs from cultivar MCol22 were transformed with the following Agro-strains: LBA4404-pTOK233 (Hiei *et al.*, 1994), a super-virulent strain with multiple copies of the *vir*G, *vir*B and *vir*C genes that can transform even rice; C58C1-pIG121Hm (Hiei *et al.*, 1994); LBA4404 and EHA105 (Sciaky *et al.*, 1978), both with plasmid pBin9GusInt (Holtorf *et al.*, 1995). All gene constructs carried a GUS-gene interrupted by an intron (*gus*-intron) to avoid expression in bacteria and thus eliminate false positives. Selection of transgenic tissues was achieved by using hygromycin and geneticin, both from the kanamycin family of antibiotics, at 15 and 20 mg/l respectively. Transgenic plants were regenerated using media with the cytokinin BAP. The authors claimed that regeneration of these plants was direct, through organogenesis, although no histological evidence was provided. Regardless of the route of regeneration, they obtained nine plants expressing the gus-intron and resistant to hygromycin, from the expression of the *hpt* gene, under the control of the 35S and 35SΩ (enhanced 35S) promoters. Five out the nine plants were analyzed molecularly. In fact, all contained three or more copies of the inserted T-DNA, probably due to the supervirulence of the strains. In some cases T-DNA insert rearrangements were observed. The differential intensity of bands in the Southern blot, and the expression of the *gus*-intron gene, indicated a possible chimerism in some of the transgenic lines.

3. TRANSFORMATION EFFICIENCY USING SECS AS INITIAL EXPLANTS

The number of transgenic plants obtained in the experiments described above indicated that the efficiency of transformation was relatively low. In the case of Sarria *et al.* (2000), it was 1% (one transgenic plant, proved at the molecular level, out of 100 initial explants). For Li *et al.*, the frequency was even lower: 0.3 or 0.5% (6 plants out of 1735 explants, or 3 plants out of 563 explants). This low transformation frequency is understandable if we assume that somatic embryogenesis in cassava is mostly of multicellular origin (Stamp, 1984), which makes then necessary to transform a large number of cells simultaneously, in the same explant, to obtain fully transgenic embryos. This is probably why chimerism is frequent in this transformation system. Chimeras in cassava are not desirable since transgenic plants should not go through the sexual cycle (to clean the chimera by producing true seeds). It would probably result in the loss of the clone (cultivar or genotype) since cassava is a highly heterozygous species.

Massive, genetic transformation systems in crops like maize and soybean usually require the production of between 50 to more than 200 independent, transgenic lines. It ensures the selection of fewer lines, may be just one, which comply with all the characteristics needed for commercial use and/or for transfer of the transgenes to other cultivars. This fact certainly could represent a bottleneck in cassava transformation, although some solutions are sighted in the horizon. One of them may be the use of the Friable Embryogenic Callus (FEC) as explant for transformation (Taylor *et al.*, 1996). This system seems to be much more efficient for the regeneration of plants in several cultivars and is discussed later.

4. TRANSFORMATION OF FEC USING *AGROBACTERIUM*

The first reported Agro-trans experiments using FEC were those of González *et al.* (1998). However, several reports have appeared describing the biolistic-mediated transformation of FEC, and the regeneration of whole transgenic plants (Schopke *et al.*, 1996; Raemakers *et al.*, 1996; Zhang *et al.*, 2000; Zhang and Puonti-Kaerlas, 2000). Gonzáles *et al.*, described a protocol to transform FEC from cultivar TMS60444 (MNig11), with the strain ABI-pILTAB188. It is important to mention at this point that the embryo-to-plant frequency of conversion in TMS60444 is fairly high (ranges between 18 and 27%) when compared with other highly embryogenic cultivars like MCol2215, where embryos convert to plants at frequencies between 8 and 20% (López, 2000). Transforming FEC via *Agrobacterium* may take up to 30 to 34 weeks (8 to 9 months), which seems quite long, using selection with Paramomycin at 25 μM (another antibiotic of the Kanamycin family). One of the factors that strongly delays the recovery of transgenic plants seems to be the sensitivity of young, globular embryos to infection with *Agrobacterium* per se (Schopke *et al.*, 1993; Li *et al.*, 1996; Gonzáles *et al.*, 1998); growth rate decreases and thus the rescue of transgenic plants. Sensitivity to infection with *Agrobacterium* seems to be also cultivar dependent, with some cultivars like MCol2215 being more sensitive than TMS60444 (Chavarriaga and Ladino, unpublished results).

Although González *et al.* argue that the limiting step in the FEC transformation system is the low frequency of calli that produce embryos, and the low conversion rate of transformed embryos to plants, the latter seems to be the highest reported so far for Agro-trans and biolistic-mediated transformation. The frequency of transgenic plants oscillated between 1 to 10% (2 plants out of 180 or 20 embryogenic units selected on Paramomycin). Chimerism may be also reduced due to the likely unicellular origin of globular embryos in FEC. According to the authors, hundreds of transgenic lines (FECs) may be produced relatively easy from 0.5 ml of SCV★. In other aspects the FEC system behaves more

★One unit (in ml) of SCV (Settled Cell Volume) is determined by leaving undisturbed the FEC (which is in suspension) for 30 min in a graduated 15 ml tube.

or less similar to the SEC for Agro-trans: the number of insertions varies (1 to 4 in the case of Gonzáles *et al.*), antibiotic selection is also used, and the same gene contsructs, including promoters, work well.

The FEC system has been developed for 13 cassava cultivars, from which plants have been regenerated in seven of them (Taylor *et al.*, 1998). López (2000) developed FEC for MCol2215 (also known as Venezolana), a locally grown variety in Colombia, of economic importance mostly for small farmers. Plants were also regenerated from MCol2215. Transient gene expression of the *gus*-intron gene proved the feasibility of transforming FEC from MCol2215 using the strains LBA4404 and C58C1 (Chavarriaga *et al.*, 2000; Figure 1). Researchers in Holland have also used *Agrobacterium* to transform FEC, probably from the TMS60444 cultivar, using antibiotics as selective agents (Raemakers *et al.*, 1998). In general, the Agro-trans system using FEC could be applied to a range of cassava cultivars of regional importance, especially in developing countries where the biolistic approach is still expensive and unaffordable. Most probably the key to obtain a large number of transgenic plants with these cultivars relays on optimizing the selection and regeneration steps. In the case of TMS60444, it is now possible to recover up 10 and 50 embryogenic lines per each infection with *Agrobacterium* (Taylor *et al.*, 1998). This, coupled to the higher capacity of TMS60444 to regenerate plants (about 100 per gram of FEC, Taylor *et al.*, 1998), represents a significant advance in transformation of cassava. Similar approaches should be sought for other cultivars.

The number of cultivars adapted to farmers' needs, and ecosystem requirements, is probably over 1000. The task of using all these varieties for Agro-trans will not be an easy one, but it has to start soon. Using model cultivars, like TMS60444, to transform and then transfer, through classical breeding, the transgene to the desired clone, is not the most adequate technology for cassava. The time required to recuperate the genotype of the commercial cultivar may vary between 6 to 10 years.

The time required to develop FEC for cassava may vary depending upon the genotype. In some cases it could take up to one year. Since long exposure of cells to *in vitro* conditions may increase the chances of somaclonal variation in plants regenerated from them, it is desirable to develop methods to preserve cell lines in time to guarantee genetic stability. One such method could be cryo-preservation of FEC. Preservation of cassava shoot tips has been accomplished in liquid nitrogen, and regeneration of plants from cryo-preserved meristematic

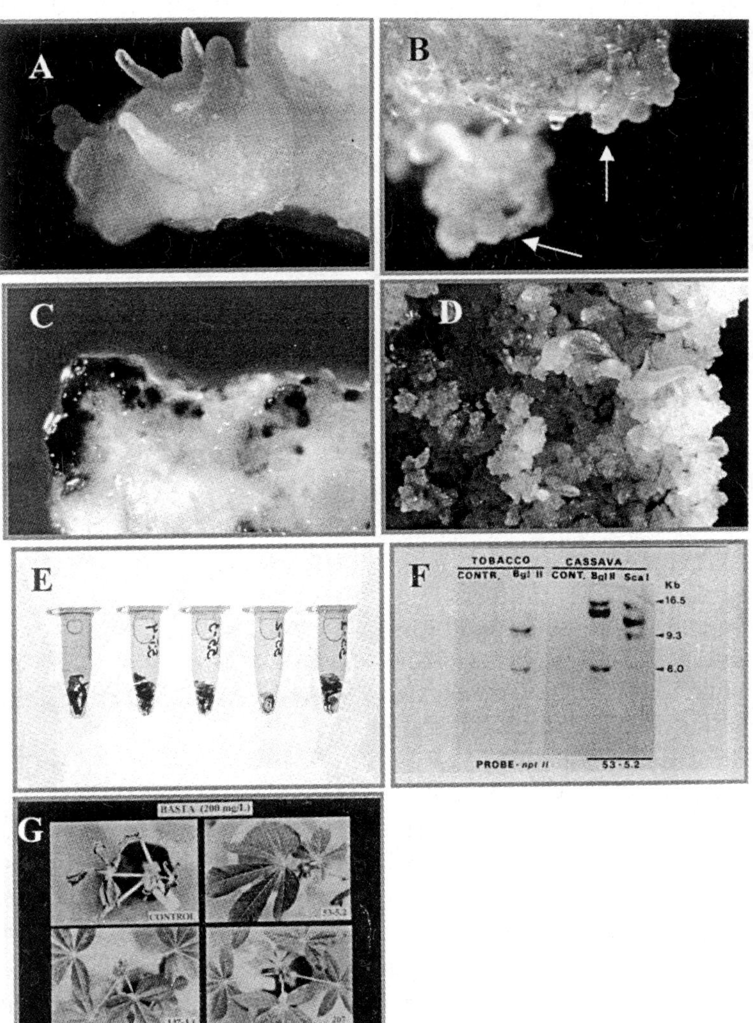

Figure 1 : Steps towards *Agrobacterium*-mediated transformation of cassava. (A) Induction of somatic embryos. (B) Induction and proliferation of Friable Embryogenic Callus (arrows). (C) Transformation with *Agrobacterium* strains carrying marker genes (i.e., *gus*-intron and *npt*II) and genes of interest (*i.e., cry*1Ab gene for insect resistance). (D) Selection and proliferation of transformed embryos. (E) Transgenic embryos expressing scorable genes (i.e., *gus*-intron). (F) Molecular confirmation of stable transformation (Southern blot for transgenic cassava expressing herbicide resistance; see Sarria *et al.,* 2000). (G) Phenotypic evaluation of the trait of interest (plant lines resistant to Basta).

tissues has been successful (Escobar *et al.*, 1996). We are presently working on applying and/or modifying current cryo-preservation protocols to use them with FEC from different cassava genotypes.

5. TECHNICAL NEEDS TO IMPROVE TRANSFORMATION IN CASSAVA

Scaling-up the system for transformation of cassava requires careful consideration of the following issues:

1) Availability of genes and promoters of interest (see below),
2) Improvement of tissue culture technologies to regenerate plants,
3) Efficient transformation and selection systems,
4) Capacity for greenhouse and field handling,
5) Public perception of transgenic cassava, and
6) Availability of qualified personnel.

Among all the issues numbered above, it is of great importance to count on selection systems widely accepted by the public (the consumer, the producer, etc.). One of those systems is that of positive selection with sugars like Mannose (Joersbo *et al.*, 1998). It has been successfully tested in cassava (Zhang and Puonti-Kaerlas, 2000), using particle bombardment, although it is not yet available for most laboratories that work on transformation of this crop. Even though selectable markers are necessary to differentiate between transgenic and non-transgenic tissues, the trend today is to produce plants containing only the gene of interest (clean transformants, precision genetic engineering), preferable with a single copy of it, and with no genes for selection. Therefore, it would be necessary to use site-specific recombination techniques to eliminate selectable genes during or after selection of transgenic plants. Some examples of the usefulness of these technical approaches to eliminate specific genes are documented (Qin *et al.*, 1994; Bayley *et al.*, 1992; Odell *et al.*, 1990).

Another anticipated need to improve cassava transformation would be tissue-specific and inducible promoters. There are already some tissue-specific genes cloned in cassava, like the β-glycosidase from roots (Liddle *et al.*, 1997; see also Marello *et al.*, 1997, for root-specific clones). Using specific and inducible promoters would allow us to concentrate, and activate, the expression of the transgene of interest at will. For instance, an ideal situation would be to limit the expression of a starch-modifying enzyme to the roots, or to switch on insecticidal genes (like *cry* genes for insect resistance) only when the insect attacks (activating the gene by wounding). There are promoters inducible with ions like copper, or with molecules like steroids (Mett *et al.*, 1993; Schena *et al.*, 1991) that have been tested in plants. Searching for inducible promoters, and tissue-specific genes in cassava, may be a priority if we want transgenesis to stay updated, acceptable for consumers and usable as a tool for breeding.

Searching for cassava's own genes, promoters and constructs, will help researchers in cassava transformation to deal with Intellectual Property Right-related issues. Not having access to genes/promoters of interest is often a major bottleneck for cassava transformation. Being limited to use genes and promoters only for research purposes, but not for commercial applications will be also a major constraint for an efficient delivery of transgenic cassava to farmers and consumers, especially if products derived from transgenic cassava, like flour or dry chips, reach international markets where patents apply.

6. MAIN TRAITS AS TARGETS FOR TRANSGENESIS AND FUTURE PROSPECTS

There are five main areas in cassava research where transgenesis could play a role: viral diseases, insect pests, cyanide toxicity, starch quality and post-harvest root deterioration (Roca *et al.*, 1992). Genes modifying these traits are susceptible of modification through Agro-trans or particle bombardment. For instance, resistance to the African Cassava Mosaic Disease, and the Common Cassava Mosaic Virus, may be achievable by the insertion of viral genes (Taylor *et al.*, 1998; and personal communication). Similarly, resistance to insects like stem borers (*i.e., Chilomima clarkey*) may be also possible through transformation with *cry* genes (from *Bacillus thuriengiensis*; Chavarriaga *et al.*, 2000). Other insects, like hornworms, mealybugs, and whiteflies are pests of regional or world importance, which might be controlled using insecticidal genes. Cassava Bacterial Blight (CBB) is another pest of concern among cassava growers. Genes responsible for resistance to CBB have been mapped (Jorge *et al.,* 2000), and, eventually, they may be cloned and transferred to susceptible varieties via Agro-trans or bombardment.

In the category of traits that have a direct impact on the starch industry or on the consumers, and which are also targets for genetic transformation, we could find: modifying starch quality (Munyikwa *et al.*, 1998), reducing cyanogenic potential increasing root and leaf protein content and extending post-harvest shelf life among others. Depending upon the trait, the time required to released transgenic plants to the fields may vary from 3 to more than 7 years. Among cassava growing countries in Latin America, which posses Biosafety regulations to use or test transgenic plants in the fields, are Colombia, Brazil, Argentina, Mexico, Cuba and Costa Rica. Thus, it seems evident that the first transgenic cassava plants might be tested in the fields of these countries. Research Institutes like CIAT, in Cali, Colombia, already have governmental authorization to conduct field trials for transgenic rice lines (with resistance to the Hoja Blanca virus). It is hoped that, within a year or less, permits will be granted for field-testing of transgenic cassava plants.

ACKNOWLEDGMENTS

The authors thank the Department for International Development -D$_f$ ID (UK); the Ministry of Foreign Affairs, Directorate General, International Cooperation -DGIS (The Netherlands); the Centro de Estudios Ganaderos y Agrícolas -CEGA (Colombia) and the Ministerio de Agricultura y Desarrollo Rural -MADR (Colombia) for funding research on cassava genetic transformation at CIAT.

REFERENCES

Arias DI and Sarria R (1995) New *Agrobacterium tumefaciens* plasmids for cassava transformation. In: *The Cassava Biotechnology Network*: Proceedings of the Second International Scientific Meeting, Bogor, Indonesia, August 1994, p 245

Bayley CC, Morgan M, Dale EC and Ow DW (1992) Exchange of gene activity in transgenic plants catalyzed by the Cre-*lox* site-specific recombination system. *Plant Mol. Biol.*, **18** : 353-361

Calderon-Urrea A (1988) Transformation of *Manihot esculenta* (cassava) using *Agrobacterium tumefaciens* and expression of the introduced foreign genes in transformed cell lines. MSc Thesis, Vrije Universiteit Brussel, Belgium, 37 pp.

Chavarriaga-Aguirre P, Schopke C, Sangare A, Fauquet C and Beachy RN (1993) Transformation of cassava (*Manihot esculenta* Crantz) embryogenic tissues using *Agrobacterium tumefaciens*. In: *Proceedings of the first international scientific meeting of the Cassava Biotechnology Network,* Cartagena, Colombia, p 222

Chavarriaga P, Mancilla L, Ladino JJ, Ramirez C, López D, Herrera CJ and Roca WM (2000) Towards genetic transformation of cassava for insect resistance. Poster at the symposium "From germplasm bank to farmer fields, role of CIAT biotechnology in research and training in Latin America", December 4 of 2000, CIAT, Cali, Colombia

Escobar R., *et al.* (1997) A methodology for recovering cassava plants from shoot tips maintained in liquid nitrogen. *Plant Cell Rep.,* **16** : 474-478

González AE, Schopke C, Taylor NJ, Beachy RN and Fauquet CM (1998) Regeneration of transgenic cassava plants (*Manihot esculenta* Crantz) through *Agrobacterium*-mediated transformation of embryogenic suspension cultures. *Plant Cell Rep.,* **17** : 827-831

Hiei Y, Ohta S, Komari T and Kumashiro T (1994) Efficient transformation of rice (*Oryza sativa* I.) mediated by *Agrobacterium* and sequence analysis of the boundaries of the T-DNA. *Plant J.,* **6** : 271-282

Holtorf S, Apel K and Bohlmann H (1995) Comparison of different constitutive and inducible promoters for the overexpression of transgenes in *Arabidopsis thaliana. Plant Mol Biol.,* **29** : 637-646

Joersbo M, Donaldson I, Kreiberg K, Guldager Petersen S, Brunstedt J and Okkels FT (1998) Analysis of mannose selection used for transformation of sugar beet. *Mol. Breed.,* **4** : 111-117

Jorge V, Fregene MA, Duque MC, Bonierbale MW, Tohme J and Verdier V (2000) Genetic mapping of resistance to bacterial blight disease in cassava (*Manihot esculenta* Crantz). *Theor. Appl. Genet.,* **101** : 865-872

Li HQ, Sautter C, Potrykus I and Puonti-Kaerlas J (1996) Genetic transformation of cassava (*Manihot esculenta* Crantz). *Nature Biotechnol.,* **14** : 736-740

Liddle S, Hughes J and Hughes MA (1997) Analysis of a cassava root β-glycosidase promoter. *African J Root Tuber Crops,* **2** : 158-162

López D (2000) Inducción de callo embriogénico friable "CEF" y regeneracion de plantas de la variedad de yuca (*Manihot esculenta* Crantz) MCol 2215. Thesis, Universidad Nacional de Colombia, Palmira, Colombia, 85 pp

Mett VL, Lochhead LP and Reynalds PH (1993) Copper-controllable gene expression system for whole plants. *Proc. Natl. Acad. Sci. USA,* **90** : 4567-4571

Munyikwa TRI, Raemakers CJJM, Schreuder M, Kok R, Schippers M, Jacobsen E and Visser RGF (1998) Pinpointing towards improved transformation and regeneration of cassava (*Manihot esculenta* Crantz). *Plant Sci.,* **135** : 87-101

Odell JJ, Caimi PG, Sauer B and Russell SH (1990) Site-directed recombination in the genome of transgenic tobacco. *Mol. Gen. Genet.,* **223** : 369-378

Quin M, Bayley C, Stockton T and Ow DW (1994) Cre recombinase-mediated site-specific recombination between plant chromosomes. *Proc. Natl. Acad. Sci. USA,* **91** : 1706-1710

Raemakers CJJM, Sofiari E, Taylor NJ, Henshaw GG, Jacobsen E and Visser RGF (1996) Production of transgenic cassava (*Manihot esculenta* Crantz) plants by particle bombardment using luciferase activity as selection marker. *Mol. Breed.,* **2** : 339-349

Raemakers CJJM, Schreuder M, Munyikwa TRI, M, Jacobsen E and Visser RGF (1998) Towards a routine transformation procedure for cassava. *Proceedings fourth international meeting of the Cassava Biotechnology Network,* Salvador, Brazil (in press)

Roca WM, Henry G, Angel F and Sarria R (1992) Biotechnology research applied to cassava improvement at the International Center for Tropical Agroculture (CIAT). *AgBiotech News and Information,* **4** : 303N-308N

Sarria R, Gomez A, Catano ML, Herrera PV, Calderon A, Mayer JE and Roca WM (1993) Towards the development of *Agrobacterium tumefaciens* and particle bombardment-mediated cassava transformation. In: *Proceedings of the first international scientific meeting of the Cassava Biotechnology Network,* Cartagena, Colombia, August 1992, p 216

Sarria R, Torres E, Balcazar N, Destefano L and Roca WM (1995) Progress in *Agrobacterium*-mediated transformation of cassava (*Manihot esculenta* Crantz). In: *Proceedings of the second international scientific meeting of the Cassava Biotechnology Network*, Bogor, Indonesia, p 241

Sarria R, Torres E, Angel F and Roca WM (1997) Transgenic cassava. In: *Abstracts of the third international scientific meeting of the Cassava Biotechnology Network,* Kampala, Uganda, p 61

Sarria R, Torres E, Angel F, Chavarriaga P and Roca WM (2000) Transgenic plants of cassava (*Manihot esculenta* Crantz) with resistance to Basta obtained by *Agrobacterium*-mediated transformation. *Plant Cell Rep.,* **19** : 339-344

Schena M, Lloyd AM and Davis RW (1991) A steroid-inducible gene expression system for plant cells. *Proc. Natl. Acad. Sci. USA,* **88** : 10421-10425

Schopke C, Franche C, Bogusz D, Chavarriaga P, Fauquet CM and Beachy RN (1993) Transformation in cassava (*Manihot esculenta* Crantz). In: *Biotechnology in Agriculture and Forestry,* Vol. 23 Plant Protoplasts and Genetic Engineering, (Ed. Bajaj YPS), Springer-Verlag Berlin Heidelberg, p 273-289

Schopke C, Taylor NJ, Carcamo R, Konan NK, Marmey P, Henshaw GG, Beachy RN and Fauquet CM (1996) Regeneration of cassava plants (*Manihot esculenta* Crantz) from microbombarded embryogenic suspension cultures. *Nat. Biotechnol.,* **14** : 731-735

Sciaky D, Montoya AL and Chilton MD (1978) Fingerprints of *Agrobacterium* Ti plasmids. *Plasmid,* **1** : 238-253

Stamp JA (1984) *In vitro* plant regeneration studies with cassava (*Manihot esculenta* Crantz) PhD Thesis, University of Birmingham, UK, 340 p.

Taylor NJ, Edwards M, Kiernan RJ, Davey CDM, Blakesley D and Henshaw GG (1996) Development of friable embryogenic callus and embryogenic suspension culture systems in cassava (*Manihot esculenta* Crantz). *Nat. Biotechnol.,* **14** : 726-730

Taylor NJ, Schopke C, Masona MV, Carcamo R, Ho T, Beachy RN and Fauquet CM (1998) Production of transgenic cassava plants with marker genes and genes of interest by microparticle bombardment and *Agrobacterium*-mediated gene transfer. *Brazilian Cassava Journal,* Supplement, Vol 17, pp 34

Zhang P, Legris G, Coulin P and Puonti-Kaerlas J (2000) Production of stable transformed cassava plants via particle bombardment. *Plant Cell Rep.,* **19** : 939-945

Zhang P and Puonti-Kaerlas J (2000) PIG-mediated cassava transformation using positive and negative selection. *Plant Cell Rep.,* **19** : 1041-1048

AUTHOR INDEX

SUBJECT INDEX